# Chipless RFID Reader Architecture

# Chipless RFID Reader Architecture


Nemai Chandra Karmakar
Randika V. Koswatta
Prasanna Kalansuriya
Rubayet E-Azim

**ARTECH HOUSE**

BOSTON | LONDON
artechhouse.com

Library of Congress Cataloging-in-Publication Data
A catalog record for this book is available from the U.S. Library of Congress.

British Library Cataloguing in Publication Data
A catalog record for this book is available from the British Library.

ISBN-13: 978-1-60807-561-4

Cover design by Vicki Kane

*To my beloved wife, Shipra, and daughters,*
*Antara and Ananya.*
—N. C. K.


*To my parents, Ranjani and Bandula, my sister, Harshani,*
*and my loving wife, Ruwani.*
—R. V. K.


*To my wife, Sarangi, and my parents.*
—P. K.


*To my husband, Rajib, my mom, dad, and brother,*
*who are the strength of my life.*
—R. E-A.

# Contents


Preface                                                                                    *xiii*

Acknowledgments                                                                            *xxi*

Foreword                                                                                   *xxiii*

**CHAPTER 1**

Introduction                                                                                 1
1.1   Chipless RFID                                                                           1
1.2   Chipless RFID Tag Reader                                                                6
1.3   Executive Summaries                                                                     7
    1.3.1   Operating Principle of Chipless RFID Systems                   7
    1.3.2   Reader Architecture for Chipless RFID Systems                   8
    1.3.3   Physical Layer Development of Chipless RFID Tag Readers          9
1.4   Conclusion                                                                             11
    Questions                                                             12
    Qualitative and Descriptive Questions                                 12
    Numerical Questions                                                   13
    References                                                            13

**CHAPTER 2**

Chipless RFID System Operating Principles                                                   15
2.1   Chipless RFID Tags                                                                     15
    2.1.1   Time Domain (TD) Based Chipless RFID Tags                       16
    2.1.2   Frequency Domain Based Tags                                     20
    2.1.3   Image Based Tags                                                28
    2.1.4   Hybrid Domain Chipless RFID Tags                                29
    2.1.5   Summary of the Review of Chipless RFID Tags                     30
2.2   Multiresonator Based Chipless RFID Tag                                                 31
    2.2.1   Operating Principle for Reading of Multiresonator Based
    Chipless RFID Tags                                                     33
2.3   Methods for Reading RFID Tags                                                          35
    2.3.1   Reading Time Domain Based Chipless RFID Tags                    36
    2.3.2   Reading Frequency Domain Based Chipless RFID Tags               36
    2.3.3   Reading Hybrid Domain Based Chipless RFID Tags                  37
    2.3.4   SAR Based Reading Process                                       37

|  |  |  |
|---|---|---|
| 2.4 | Conclusion | 38 |
|  | Questions | 39 |
|  | References | 39 |

**CHAPTER 3**

**Chipless RFID Readers**                                                                    **43**

| 3.1 | Introduction to Chipless RFID Readers | 43 |
| 3.2 | Chipless RFID Reader System Architectures | 44 |
|  | 3.2.1 General Overview of Chipless RFID Reader Architecture | 44 |
| 3.3 | Chipless RFID Readers and Tag Reading Techniques | 47 |
|  | 3.3.1 Time Domain Readers and Tag Reading Techniques | 48 |
|  | 3.3.2 Frequency Domain Readers and Tag Reading Techniques | 50 |
|  | 3.3.3 Hybrid Domain Readers and Tag Reading Techniques | 56 |
|  | 3.3.4 SAR Based Readers and Tag Reading Techniques | 57 |
| 3.4 | Limitations and Issues with Current Chipless RFID Readers | 60 |
|  | 3.4.1 Cost of Readers | 60 |
|  | 3.4.2 Read Range | 61 |
|  | 3.4.3 Tag Reading Speed | 61 |
|  | 3.4.4 Anticollision, Error Correction, and Data Integrity | 62 |
|  | 3.4.5 Orientation of the Tag | 63 |
| 3.5 | Conclusion | 63 |
|  | Questions | 63 |
|  | References | 64 |

**CHAPTER 4**

**Frequency Domain Based RFID Reader Development**                                           **67**

| 4.1 | Introduction | 67 |
|  | 4.1.1 Organization of This Chapter | 67 |
| 4.2 | Operation of Frequency Domain Based Chipless RFID Readers | 69 |
|  | 4.2.1 Detecting the Features of the Frequency Signatures of Chipless Tags: Type-1 Readers | 71 |
|  | 4.2.2 Recovering the Frequency Signature of Chipless Tags: Type-2 Readers | 76 |
| 4.3 | Design of Frequency Domain Based Chipless RFID Readers | 81 |
|  | 4.3.1 Design of a Reader That Detects the Features of Frequency Signatures of Chipless RFID Tags | 81 |
|  | 4.3.2 Design of a Reader That Recovers the Frequency Signatures of Chipless RFID Tags | 82 |
| 4.4 | Results | 86 |
|  | 4.4.1 Results: Detecting the Features of Frequency Signatures | 86 |
|  | 4.4.2 Results: Recovering the Frequency Signature of Chipless RFID Tags | 88 |
| 4.5 | Conclusion | 88 |
|  | Questions | 90 |
|  | References | 91 |

# Contents

ix



## CHAPTER 5

**Time Domain Based Chipless RFID Reader**　　93

| | | |
|---|---|---|
| 5.1 | Introduction | 93 |
| 5.2 | Theory of Operation of a Time Domain Based Chipless RFID Reader | 96 |
| | 5.2.1　Generation of UWB Short Pulses | 96 |
| | 5.2.2　UWB Short Pulse Based Interrogation | 100 |
| 5.3 | Design of UWB TD Based Chipless RFID Reader | 105 |
| | 5.3.1　Simulation Setup in Agilent ADS2009 | 106 |
| 5.4 | Results | 110 |
| | 5.4.1　Discussion of Results | 111 |
| 5.5 | Conclusion | 115 |
| | Questions | 117 |
| | References | 118 |

## CHAPTER 6

**Hybrid Chipless RFID Reader**　　121

| | | |
|---|---|---|
| 6.1 | Introduction | 121 |
| 6.2 | Frequency-Phase Based System | 125 |
| | 6.2.1　Working Principle of the Encoding Element | 125 |
| | 6.2.2　Hybrid Frequency-Phase Chipless RFID Based on C-Sections | 127 |
| | 6.2.3　Chipless RFID Reader for Hybrid Frequency-Phase Chipless RFID | 128 |
| 6.3 | Frequency-Polarization Based System | 131 |
| | 6.3.1　Resonant and Polarization Characteristics of the Split-Ring Resonator | 131 |
| | 6.3.2　Encoding Information in a SRR Based Frequency-Polarization Tag | 132 |
| | 6.3.3　Chipless RFID Reader for Hybrid Frequency-Polarization Based Chipless RFID | 134 |
| 6.4 | Frequency-Time Based System | 136 |
| | 6.4.1　Group Delay Based Frequency-Time Hybrid Chipless RFID | 136 |
| | 6.4.2　Non-Group-Delay Based Approaches for Hybrid Frequency-Time Chipless RFID | 141 |
| | 6.4.3　Chipless RFID Reader for Hybrid Frequency-Time Based Chipless RFID | 142 |
| 6.5 | Conclusion | 143 |
| | Questions | 144 |
| | Reference | 145 |

## CHAPTER 7

**Antennas for Chipless RFID Readers**　　147

| | | |
|---|---|---|
| 7.1 | Introduction | 147 |
| 7.2 | Antenna Parameters | 151 |
| 7.3 | Review of UWB Antennas | 155 |
| 7.4 | Practical Design Procedure for UWB Antennas | 157 |
| 7.5 | UWB Antenna Development | 159 |
| | 7.5.1　UWB Disk-Loaded Monopole Antennas | 159 |

       7.5.2   UWB Dipole Array Antennas   165
       7.5.3   Log-Periodic Dipole Antennas   176
       7.5.4   Log-Periodic Dipole Array Antennas   180
       7.5.5   UWB Horn Antennas   191
  7.6   Applications   200
       7.6.1   Disk-Loaded Monopoles as Reader Antennas   200
       7.6.2   Elliptical Leaf Dipole Arrays as Reader Antennas   200
       7.6.3   LPDAAs as Reader Antennas   201
       7.6.4   Horns as Reader Antennas   201
  7.7   Conclusion   201
       Questions   202
       References   204


**CHAPTER 8**

## Microwave and Millimeter-Wave Active and Passive Components   207


  8.1   Introduction   207
  8.2   Passive Components   208
       8.2.1   Antennas   208
       8.2.2   Filters   209
       8.2.3   Couplers   215
       8.2.4   Circulators   217
       8.2.5   Power Divider/Combiner   218
       8.2.6   Resonators   221
       8.2.7   Mixers   224
  8.3   Active Components   227
       8.3.1   Voltage-Controlled Oscillators   227
       8.3.2   Low-Noise Amplifiers   233
  8.4   Discussion and Conclusions   234
       Questions   236
       References   237


**CHAPTER 9**

## Digital Module for Chipless RFID Readers   239


  9.1   Introduction   239
  9.2   Digital Control Board   242
       9.2.1   Power Supply Unit   243
       9.2.2   IR Tag Sensor   243
       9.2.3   RF Power Supply Control Switch   246
       9.2.4   VCO Control Voltage Generator   246
  9.3   Digital Signal Processing Board   247
  9.4   Middleware   249
  9.5   Performance of Digital Control Section   251
  9.6   Conclusion   252
       Questions   253
       References   253

**CHAPTER 10**

**RFID Reader System Integration and Applications** 257
10.1 Introduction 257
10.2 The Integrated Chipless RFID Reader 259
10.3 Specifications of a Generic Chipless RFID Reader 259
    10.3.1 Chipless RFID Tag 260
    10.3.2 Reading Techniques 260
    10.3.3 Reader Device 264
    10.3.4 Application Software 267
10.4 Applications 268
    10.4.1 Conveyer Belt Application 269
    10.4.2 Vehicle Tracking 270
10.5 Conclusion 273
    Questions 273
    References 274

List of Acronyms 277

Glossary 281

About the Authors 287

Index 289

# Preface

## Introduction to Radio Frequency Identification (RFID)

RFID is a wireless modulation and demodulation technique for the automatic identification of objects, the tracking of goods, smart logistics, and access control. RFID is a contactless, usually short distance transmission and reception technique for unique ID data transfer from a tagged object to an interrogator (reader). The generic configuration of an RFID system comprises: (i) an ID data-carrying tag; (ii) a reader; (iii) middleware; and (iv) an enterprise application. The reader interrogates the tag with the RF signal, and the tag, in return, responds with an ID signal. Middleware controls the reader and processes the signal, and finally feeds into the enterprise application software in the IT layer for further processing. The RFID technology has the potential to replace barcodes due to its large information-carrying capacity, flexibility in operation, and versatility of application [1]. Due to its unique identification, tracing, and tracking capabilities, RFID also gives value-added services incorporating various sensors for real-time monitoring of assets and public installations in many fields. However, the penetration of RFID technology is hindered due to its high price tag, and many ambitious projects have stalled due to the cost of the chips in the tags. Chipless RFID tags mitigate the cost issues and have the potential to penetrate mass markets for low-cost item tagging [2]. Due to its cost advantages and unique features, chipless RFID will dominate 60% of the total RFID market with a market value of \$4 billion by 2019 [3]. Since the removal of the microchip causes a chipless tag to have no intelligence-processing capability, the signal processing is done only in the reader. Consequently, a fully new set of requirements and challenges need to be incorporated and addressed, respectively, in a chipless RFID tag reader. This book addresses the new reader architecture and integration techniques for reading various chipless RFID tags.

## Recent Development of Chipless RFID Tags

IDTechEx (2009) [3] predicts that 60% of the total tag market will be occupied by the chipless tag if the tag can be made at a cost of less than one cent. However, the removal of an application-specific integrated circuit (ASIC) from the tag is not a trivial task, as it performs many RF signal and information-processing tasks. Intensive investigation and investment are required for the design of low-cost but robust passive microwave circuits and antennas using low-conductivity ink on low-cost and lossy substrates. Some types of chipless RFID tags are made of microwave resonant structures using conductive ink. Obtaining high fidelity (high quality factor) responses from microwave passive circuits made of low-conductivity ink on low-cost and lossy materials is very difficult [4]. Great design flexibility is required to meet the benchmark of reliable and high fidelity performance from these low-grade laminates and poor conductivity ink. This exercise has opened up a new discipline in microwave printing electronics in low grade laminates [5].

The low-cost chipless tag will place new demands on the reader as new fields of applications open up. IDTechEx [3] predicts that, while optical barcodes are printed in only a few billion per year, close to one trillion (>700 billion) chipless RFIDs will be printed each year. The chipless RFID has unique features and much wider ranges of applications compared to optical barcodes. However, very little progress has been achieved in the development of the chipless RFID tag reader and its control software. This is because conventional methods of reading RFID tags are not implementable in chipless RFID tags. For example, hand-shaking protocol cannot be implemented in chipless RFID tags. Dedicated chipless RFID tag readers and middleware [6] need to be developed to read these tags reliably.

The development of chipless RFID has reached its second generation with more data capacity, reliability, and compliance to some existing standards. For example, RF-SAW tags have new standards, can be made smaller with higher data capacity, and currently are being sold in millions [7]. Approximately thirty companies have been developing TFTC tags. TFTC tags target the HF (13.65 MHz) frequency band (60% of existing RFID market) and have read-write capability [7]. However, neither RF-SAW nor TFTC is printable and cannot reach sub-cent level prices. In generation-1 of conductive ink-based fully printable chipless RFID tag development, few chipless RFID tags—which are in the inception stage—have been reported in the open literature. They include: a capacitive gap coupled dipole array [8]; a reactively-loaded transmission line [9]; a ladder network [10]; and finally, a piano and a Hilbert curve fractal resonators [11]. These tags are in prototype stage and no further development to commercial grade has been reported to date. A comprehensive review of chipless RFID can be found in Author's most recent books [12].

Following twenty years of RF and microwave research experiences, we have pioneered multi-bit chipless RFID research [13–14]. For the last eight and a half years at Monash University, our research activities include numerous chipless tag and reader developments as follows.

At Monash University, our research group has developed a number of printable, multi-bit chipless tags featuring high data capacity. These tags can be categorized into two types: *retransmission-based* and *backscattered-based*. The *retransmission-based* tag, originally presented by Preradovic et al. [13], uses two orthogonally-polarized wideband monopole antennas and a series of spiral resonators. The RFID reader sends a UWB signal to the tag, and the receiving antenna of the tag receives it and passes through the microstrip transmission line. The gap-coupled spiral resonators-based stop band filters create attenuations and phase jumps in designated frequency bands. This magnitude and phase encoded signal is then retransmitted back to the reader by the tag's transmitting antenna.

The attenuation in the received signal due to the resonator encodes the data bits. In two Australian Research Council (ARC) Grants (DP0665523: *Chipless RFID for Barcode Replacement*, 2006–2008 and LP0989652: *Printable Multi-Bit Radio Frequency Identification for Banknotes*, 2009–2011), we developed up to 128 data bits of chipless RFID with four slot-loaded monopole antennas and wideband feed networks [15]. This chipless tag is fully printable on polymer substrate.

## Backscatter-Based Chipless Tag

Balbin et al. [13] have presented a multi-antenna backscattered chipless tag. Here, only the resonator structure is present on the tag, and as no transmitter-receiver tag antenna exists, it is more compact than retransmission type tags. The interrogation signal from a reader is backscattered by the tag. By analyzing this backscattered signal's attenuation at particular frequencies, the tag ID is decoded.

## Monash University Chipless RFID Systems

Under various research grant schemes, the authors have developed various chipless RFID tag reader architectures and associated signal processing schemes. To date four different varieties of chipless RFID tag readers have been developed for the 2.45, 4–8, and 22–26.5 GHz frequency bands. Feasibility studies of advanced level detection [13] and error correction algorithm have been developed.

As stated above, our group has developed five different varieties of chipless RFID tag readers in various frequency bands at 2.45, 4–8, and 22–26.5 GHz frequency bands. The readers comprised an RF transceiver section, a digital control section, and middleware (control and processing). The reader transmitter comprises a swept frequency, voltage-controlled oscillator (VCO) [6, 16]. The VCO is controlled by a tuning voltage that is generated by a digital-to-analog converter (DAC). Each frequency over the ultra-wide band (UWB) from 4–8 GHz is generated with a single tuning voltage from the ADC. In addition, the VCO has a finite settling time to generate a CW signal against its tuning voltage. Combining all these operational requirements and linearity of the devices, the UWB signal generation is a slow process (taking a few seconds to read a tag). To alleviate this problem and improve the reading speed, some corrective measures can be taken. They are: (i) high speed ADC and (ii) low settling time VCO. These two devices will be extremely expensive if available in the market. The reader cost will be very high to cater for the requirement specifications of the chipless RFID reader. In this regard, signal processing for advanced detection techniques alleviate the reading process in greater details. Also, the sensitivity of the reader architecture using dual antenna in bi-static radar configuration and I/Q modulation techniques can be greatly enhanced. Highly sensitive receiver design is imperative in detecting very weak backscattering signal from the chipless tag. With the presence of interferers and movement and the variable trajectory of the moving tags, this situation is worsened. In this regard, a highly sensitive UWB reader-receiver needs to be designed. Designing such a receiver is not a trivial task where the power transmission is limited by UWB regulations. I\Q modulation in the receiver will improve the sensitivity to a greater magnitude.

In addition to this high-sensitivity receiver design, a high-end digital board with a powerful algorithm will alleviate the reading process. The digital board serves as the centerpiece of the reader, where data would be processed and numerous control signals to the RF section of the reader would be generated. The digital board has a Joint Test Action Group (JTAG) port where a host PC can be connected to monitor, control, and reprogram the reader if necessary. In addition, it is also the host to the power supply circuit that is used to generate the necessary supply voltages for most components of the reader. The digital board consists of: (i) an FPGA board with ADC; (ii) a power supply circuit; and (iii) a digital-to-analogue converter (DAC). High sampling rate A\D and D\A converters and a FPGA controller will augment the powerful capturing and processing of backscattering signals. The digital electronics and interface with a PC will accommodate custom-made powerful algorithm, such as singular value decomposition for improved detection [17], time frequency analysis [18] for localization [19], or anticollision [20] of chipless RFID tags. All these control algorithms and signal processing software will be innovations in the field. This book has addressed these advanced level analog and digital designs of the chipless RFID reader.

Antennas form a significant part of the chipless RFID tag readers [21]. Polarization mismatch and low gain of antenna are the two pressing problems in chipless RFID tag reading. Dual-polarized and circularly-polarized tags and reader antennas will alleviate the mismatch in the polarization. Mismatch of the signal by the presence of interferers will create major challenges for reading the tag. Similar to 3-D scanning optical barcode readers, multistate reader antennas with polarization diversity and beam scanning may alleviate the problem to some degree. Array antennas have solved low-gain issues. This will need a significant innovation in the development of reconfigurable antennas that will generate both polarization and beam reconfigurabilities. In the physical layer development, a smart antenna with beam and/or polarization-agile reconfigurability is required to track and read the mobile tag. In the UWB chipless RFID reader systems, two varieties of smart antennas, such as switched beam (SB) and phased array with polarization diversity, can be employed. The SB antenna comprises broadband antenna elements, a switching network, and an RF input port. The phased array has a more complex feed network with phase shifter modules and summing circuits. Our research group has developed a multilayered shared aperture dual-polarized patch antenna for mm-wave UWB band [22]. The phase shifter and switching electronics modules have also been developed in other projects led by us. A full chapter is dedicated to address the various challenges of broadband high gain fixed beam and smart antennas for the chipless RFID reader.

In a conventional chipped RFID system, established protocols are readily available for tag detection and collision avoidance. Reading hundreds of proximity tags with the flick of an eye is commonplace. However, as stated above, reading multiple chipless RFID tags in close proximity has not been demonstrated as yet. RFID middleware is an IT layer to process the captured data from a tag by a reader. Middleware applies filtering, formatting, or logic to tag data captured by a reader so that the data can be processed by a software application. For chipped RFID, there are established protocols for these tasks. However, in chipless RFID, tasks such as hand-shaking are not possible. Therefore, a completely new set of IT layers needs to be developed. Raw data obtained from a chipless tag needs to be processed and denoised, and new techniques need to be developed. A companion book will address these issues of chipless RFID tag detection, data integrity, and anticollision.

This book aims to provide the reader with comprehensive information with the recent development of chipless RFID tags, associated chipless RFID tag readers, and their relevant detection and signal processing both in the physical layer development and the software development algorithm and protocols. To serve this goal, this book features ten chapters in three sections. They offer in-depth descriptions of terminologies and concepts relevant to chipless RFID tags, and RF transceivers, antennas, digital design, system integration, detection algorithm and anticollision protocols related to the chipless RFID reader system.

The chapters of the book are organized into three distinct topics:

- Operating Principle of Chipless RFID System;
- Reader Architecture of Chipless RFID;
- Physical Layer Development of Chipless RFID Tag Reader.

In first Chapters 1 and 2, chipless RFID fundamentals with a comprehensive overview are discussed. The overview and the recent development of chipless RFID reader systems are discussed in Chapter 3. These chapters form the foundation for the rest chapters of the book. Usually, researchers tend to ignore the physical layer development of reader architecture with the perception that RFID reader architecture is an established discipline. However, a physical layer implementation of the chipless RFID reader is a fully new domain of research. This is the first book in a new discipline. To fulfill the goal of this book, the comprehensive physical layer development of the reader components such as antennas, RF components, digital design, and integration of these blocks into a complete physical chipless RFID reader are discussed in Chapters 7–10. Finally, a state-of-the-art chipless RFID tag reader that incorporates all the physical and IT layer developments stated above are discussed in the last few sections of Chapter 10. Chapter 10 also demonstrates how the reader can mitigate tag readings in challenging environments such as retails, libraries, and warehouses.

In this book, utmost care has been paid to keep the sequential flow of information related to the chipless RFID reader architecture and signal processing. We hope that the book will serve as a good reference of chipless RFID systems and will pave the ways for further motivation and research in the field.

# References


[1]    K. Finkenzeller, *RFID Handbook: Fundamentals and Applications in Contactless Smart Cards, Radio Frequency Identification and Near-field Communication, 3rd Revised edition,* NJ: Wiley, 2010

[2]    S. Preradovic and N. C. Karmakar, "Chipless RFID: Bar Code of the Future," *IEEE Microwave Magazine,* Vol. 11, pp. 87–97, Dec 2010

[3]    IDTechEx (2009), *Printed and Chipless RFID Forecasts, Technologies & Players 2009–2029*.

[4]    R. E. Azim, N. C. Karmakar, S. M. Roy, R. Yerramilli, and G. Swiegers, "Printed Chipless RFID Tags for Flexible Low-cost Substrates," in *Chipless and Conventional Radio Frequency Identification: Systems for Ubiquitous Tagging*, Hoboken, NJ: IGI Global, 2012

[5]    R. Yerramilli, G. Power, S. M. Roy, and N.C. Karmakar, "Gravure Printing and its Application to RFID Tag Development," *Proc. MRS Fall Meeting*, Boston, MA, November 28–December 2, 2011.

[6]    S. Preradovic and N. C. Karmakar, "Multiresonator Based Chipless RFID Tag and Dedicated RFID Reader," *Digest 2010 IMS*, Anaheim, CA, 23–28 May 2010.

[7]    IDTechEx, *RFID Forecasts, Players and Opportunities 2009-2019, Executive Summary*, 2009.

[8]    I. Jalaly and I. D. Robertson, "Capacitively-tuned split microstrip resonators for RFID barcodes," in *EuMC*, 2005 Paris, France, 2005, p. 4.

[9]    L. Zhan, H. Tenhunen, and L. R. Zheng, "An Innovative Fully Printable RFID Technology Based on High Speed Time-Domain Reflections," *HDP'06*, Stockholm, Sweden, 2006.

[10]   S. Mukherjee, "Chipless Radio Frequency Identification by Remote Measurement of Complex Impedance," in *European Conference on Wireless Technologies*, 2007 Munich, Germany, 2007, pp. 249–252.

[11]   J. McVay, et al., "Space-filling curve RFID tags," *IEEE Radio and Wireless Symposium*, 2006, San Diego, CA, 2006, pp. 199–202.

[12]   S. Preradovic, and N.C. Karmakar, *Multiresonator-Based Chipless RFID: Barcode of Future*. NY: Springer, 2012.

[13]   S. Preradovic, I. Balbin, N. C. Karmakar, and G. F. Swiegers, "Multiresonator-based chipless RFID system for low-cost item tracking," *IEEE Trans. Microwave Theory & Tech.*, Vol. 57, Issue 5, Part 2, May 2009, pp. 1411–1419.

[14]   I. Balbin and N. C. Karmakar, "Multi-Antenna Backscattered Chipless RFID Design," in *Handbook of Smart Antennas for RFID Systems*, Wiley Microwave and Optical Engineering Series, pp. 415–444, 2010.

[15]   I. Balbin, *Chipless RFID Transponder Design*, PhD Dissertation, Monash University, 2010.

[16]   R. V. Koswatta and N. C. Karmakar, "A novel reader architecture based on UWB chirp signal interrogation for multiresonator-based chipless RFID tag reading," IEEE Trans. MTT, Vol. 60, No. 9, pp. 2925–2933, Sep. 2012.

[17]   A. T. Blischak and M. Manteghi, "Embedded Singularity Chipless RFID Tags," *IEEE Transactions on Antennas and Propagation*, Vol. 59, pp. 3961–3968, 2011.

[18]   B. Boashash (ed), *Time Frequency Signal Analysis and Processing: A Comprehensive Reference*, Oxford, UK: Elsevier, 2003.

[19]   Y. Zhang, M. G. Amin, and S. Kaushik, "Localization and Tracking of Passive RFID Tags Based on Direction Estimation," *Proceedings of SPIE*, Vol. 6248 (1) SPIE, May 2006.

[20]   R. Azim, and N. Karmakar, "A Collision Avoidance Methodology for Chipless RFID Tags," *Proc. 2011 Asia Pacific Microwave Conference*, Melbourne, Australia, 5–8 December 2011, pp. 1514–1517.

[21]   N. C. Karmakar (ed), *Handbook of Smart Antennas for RFID Systems*, NY: Wiley, NY, 2010.

[22]   M. A. Islam, N. C. Karmakar, and A. K. M. Azad, "Aperture Coupled UWB Microstrip Patch Antenna Array for mm-Wave Chipless RFID Tag Reader," *IEEE RFID-TA*, Nice, France, November 5–7, 2012.

[23]   P. Kalansuriya, N.C. Karmakar and E. Viterbo, "'Signal space representation of chipless RFID tag frequency signatures," *Proc. IEEE GLOBECOM*, Houston, TX, 2011.

[24] P. Kalansuriya, N. C. Karmakar, and E. Viterbo, "On the detection of frequency-spectra based chipless RFID using UWB impulsed interrogation," *IEEE Trans. Microwave Theory Tech.*, Vol. 60, No. 12, Dec. 2012, pp. 4187–4197.

[25] M. Manteghi, "A novel approach to improve noise reduction in the Matrix Pencil Algorithm for chipless RFID tag detection," *Digest IEEE APSURSI2010*, pp. 1–4.

[26] A. Lazaro, A. Ramos, D. Girbau, and R. Villarino, "Chipless UWB RFID Tag Detection Using Continuous Wavelet Transform," *IEEE AWPL*, Vol. 10, pp. 520–523, 2011.

# Acknowledgments

I would like to thank Mr. Mark Walsh, senior editor, Artech House Publishers, Inc., for his invitation to write a book on chipless RFID reader architecture and signal processing. Thank you also to the reviewers who reviewed the book proposal and chapters. I must acknowledge Ms. Samantha Ronan, developmental editor at Artech House, for her continuous support and patience throughout the preparation, submission, and reviewing processes of the manuscript. Finally, thanks also go out to Ms. Lindsay Moore, production editor at Artech House for her patience in proof settling and corrections of this manuscript.

Dr. Shivali Bansal's and Dr. A. K. M. Azad's generous support in coordinating the book project and collating information were instrumental to the book. Without their support at the end of the book project, the preparation of the final manuscript would have been delayed significantly. Dr. Bansal systematically collected all materials and organized the whole manuscript in an appropriate sequence to meet the publisher's manuscript preparation requirements. It was a huge undertaking for Dr. Bansal in addition to her commitments to work and family. Dr. Azad contributed to Chapter 9 by collating information on digital design of chipless RFID reader and writing the initial draft of the chapter. Generous support from the authors' group members and their timely responses when reviewing chapters and preparing question sets for each chapter are gratefully acknowledged. These contributors include Abdur Rahim, A. K. M. Azad, Aminul Islam, Pouya Man, Chamath Divarathne, Emran Amin, Mohammad Zomorrodi, Nahina Islam, Nathan Hu, Shakil Bhyuan, Shivali Bansal, Uditha Bandara, Wan Wan Mohd Zamri, and Yixian Yap. Special thanks go to my current and former students Sushim Roy, Isaac Balbin, Uditha Bandara, Shakil Bhuyan, Aminul Islam, Hamza Mshiek, Stevan Preradovic, Parisa Zakavi, and Mohammad Zomorrodi for their contributions to the material on antennas in Chapter 7. Finally, the research funding support from the Australian Research Council's Discovery Project Grants and Linkage Project Grants, Victoria Government's Smart SMEs MVP, and Monash University's internal research grants are very gratefully acknowledged.

This book is my vision and I guided my Ph.D. student coauthors to fulfill the vision throughout the project. Therefore, the book is a culmination of the hard work of three dedicated student authors: Mr. Randika Koswatta, Mr. Prasanna Kalansuriya, and Ms. Rubayet E-Azim. Without their continuous motivation, dedication, and perseverance, the book would not have become a high-quality piece of scientific work.

*Nemai Chandra Karmakar*
*Department of Electrical and Computer System Engineering*
*Monash University*
*July 2013*

# Foreword

Radio-frequency identification (RFID) technology is a powerful identification technology that has many useful applications. However, high implementation cost limits its applicability in many cases. The most expensive item in an RFID tag is the application-specific integrated circuit (ASIC) microchip. Considerable effort has been made to remove the ASIC chip from RFID tags to make them cheap and fully printable. However, printing microwave circuits on a low-cost material is not a trivial task. A considerable amount of design flexibility and high-precision printing with high-conductivity ink are needed to fabricate microwave circuits with these types of materials. Furthermore, reading of the tag is highly challenging due to its low-fidelity response. No conventional protocols can be invoked in the chipless RFID tag because they have no on-board signal processing capability. To make the technology operational, a new type of RFID reader, a chipless reader, is needed. The proposed book deals with chipless RFID reader architecture and integration.

The book is organized into three sections covering various aspects of chipless RFID tag design and reader architecture. Sections 1 and 2 present a general overview and then move on to in-depth coverage of the operating principles of frequency, time, and hybrid domain chipless RFID tag technologies and their dedicated reader architectures. The third section covers the component-level designs, including ultra-wideband antenna design and development of the physical layer and information technology (IT) layer. It also considers the passive and active components that are required to develop fully integrated transceivers at the RF, microwave and millimeter-wave frequency bands. The design of the digital components is also included in this section. The section ends with a discussion of the integration of all components into a complete physical layer and IT layer technology for chipless RFID readers. New concepts are proposed for a chipless RFID reader with frequency domain swept frequency interrogation signals and ultra-wideband impulse radio signals. The final chapter of the book presents the integration of the chipless RFID reader and a few specific applications.

I believe this well-written book is the first published book on the chipless RFID reader architecture. I sincerely hope that this book with its comprehensive coverage will be a good resource for the RFID research community.

*Dr. Arun Bhattacharyya (IEEE Fellow)*
*Northrop Grumman Space Technology, California*
*July 2012*

# Introduction

## 1.1 Chipless RFID

In the current era of information communication technology, radio-frequency identification (RFID) has been going through tremendous development. RFID technology has the potential to replace barcodes due to its large information carrying capacity, flexibility in operations, and versatility in applications. With the advent of low-powered ICs, energy scavenging techniques, and low-cost chipless tags, RFID technology has achieved significant advancements [1]. Due to its unique features of identification and its tracing and tracking capabilities, RFID also provides value-added services incorporating various sensors for real-time monitoring of assets, public installations, and people from various backgrounds. However, the penetration of RFID technology is hindered due to its high price tag, and many ambitious projects have been stalled due to the cost of chipped tags. Chipless RFID tags, however, mitigate the cost issues and have the potential to penetrate mass markets for low-cost item tagging [2]. Since the removal of the microchip causes a chipless tag to have no intelligence processing capabilities, signal processing is done only in the reader. Consequently, a fully new set of requirements needs to be incorporated and addressed, respectively, in a chipless RFID tag reader. This book addresses the new reader architecture and signal processing techniques of the novel chipless RFID tag readers in various chapters and should be a very useful resource in this new field.

RFID is an emerging wireless technology for automatic identification, access control, asset tracking, security and surveillance, database management, inventory control, and logistics. A generic RFID system has two main components: a tag and a reader [3]. As shown in Figure 1.1, the reader sends an interrogating radio-frequency (RF) signal to the tag. The interrogation signal comprises a clock signal, data, and energy. In return the tag responds with a unique identification code (data) to the reader. The reader processes the returned signal from the tag into a meaningful identification code. Some tags coupled with sensors can also provide data on the surrounding environment such as temperature, pressure, moisture content, acceleration, and location. The tags are classified into active, semiactive, and passive tags based on their on-board power supplies. An active tag contains an on-board battery to energize the processing chip and to amplify signals. A semiactive tag also contains a battery, but the battery is used only to energize the chip; hence it yields better longevity compared to an active tag. A passive tag does not have a battery. It scavenges power for its processing chip from the interrogating signal emitted by a reader; hence it lasts forever. However, the processing power and reading distance are limited by the transmitted power (energy) of the reader. The middleware does the back-end processing, command and control, and interfacing with enterprise application as shown in Figure 1.1.

As mentioned earlier, the main constraint of mass deployment of RFID tags for low-cost item tagging is the cost of the tag. The main cost comes from the application-specific integrated circuit (ASIC) or the microchip of the tag. According to the respected RFID research organization IDTechEx [4], in 2019 an ASIC-based chipped tag will cost approximately 4 cents compared to 0.4 cent for a chipless tag. The costs for materials, the curing procedure, and lamination are of similar magnitudes for both types of tags. Therefore, the ASIC makes up 90% of the cost of an RFID tag. In addition, a chipless tag does not have any chip placement costs and soldering is not required. This will reduce the cost of a chipless tags even further.

If the chip can be removed without losing functionality of the tag, then chipless tags can have the potential to replace optical barcodes. The optical barcode has several operating limitations: (1) each barcode is individually read, (2) their use requires human intervention, (3) they have less data handling capability, (4) soiled barcodes cannot be read, and (5) barcodes require line-of-sight operation. Despite these limitations, the low cost benefit of the optical barcode makes it very attractive because it is printed almost without any extra cost. Therefore, there is a pressing need to remove the ASIC from RFID tags to make them competitive in mass deployment situations. If the chipless tag can be produced with costs in the sub-cent range, it will have the potential to provide low-cost item-level tagging, thus coexisting with or replacing the trillions of optical barcodes printed each year. The solution is to make the RFID tag chipless. Similar to optical barcodes, the tag should be fully printable on low-cost substrates such as papers and plastics [5]. A reliable prediction by IDTechEx [2] advocates that 60% of the total tag market will be occupied by the chipless tag if the tag can be made for less than 1 cent. However, removal of an ASIC from the tag is not a trivial task because it performs many RF signal and information processing tasks. Tremendous investigation and investment will be needed to design low-cost but robust passive microwave circuits and antennas using conductive ink on low-cost substrates. Additionally, obtaining

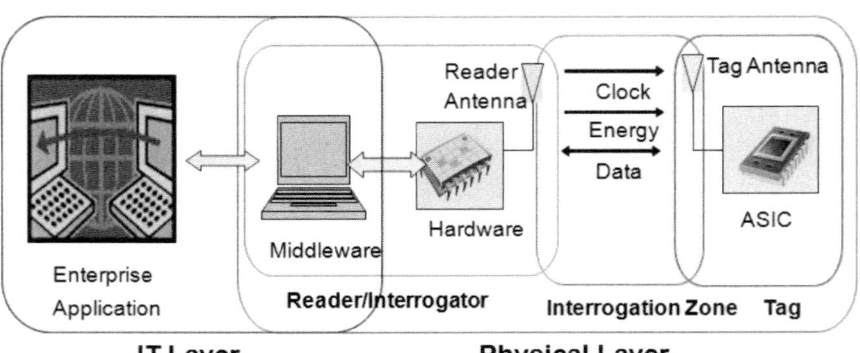


**Figure 1.1**  Architecture of conventional RF identification system.

a high-fidelity response from low-cost lossy materials is very difficult [5]. In the interrogation and decoding aspects of the RFID system, we must take into account the development of the RFID reader, which must be capable of reading the chipless RFID tag. Little progress has been achieved in the development of a chipless RFID tag reader. This is due to the fact that conventional methods of reading RFID tags cannot be implemented with chipless RFID tags. Dedicated chipless RFID tag readers [6] need to be developed to read those tags reliably. This book is the first of its kind to report on the most recent developments in reader architectures, physical layer developments, detection methods, and signal processing of the various types of chipless RFID tag readers.

The quest to develop low-cost RFID tags in the very low cost sub-cent range has opened up a new discipline. A very comprehensive account of the most recent developments in chipless RFID tags in the commercial domain can be obtained from [2]. Due to price advantages and some unique features that are not obtainable in silicon tags, the chipless RFID tag market is estimated to grow exponentially. As forecast by IDTechEx, the market volume for chipless RFID technology was less than $5 million in 2009. However, this market will grow to approximately $4 billion in 2019 [4, 7]. In contrast to the 4 to 5 billion optical barcodes that are printed yearly, approximately 700 billion chipless tags will be sold in 2019. However, the price of the chipless RFID system (readers and middleware) will remain the same. A significant proportion of the chipless tag types are thin-film transistor circuit (TFTC) tags, surface acoustic wave (SAW) tags, microwave passive circuits (radar arrays), and ink strips. This book presents reader architectures and signal processing designs that target the latter two chipless RFID tag types. These two tag types will offer the lowest price of all chipless tag types and hence development of the interrogators of these tags will occupy significant space in the research and development sectors.

The development of chipless RFID technology has reached its second generation with more data capacity, reliability, and compliance with some existing standards. For example, RF SAW tags have new standards, can be made smaller with a higher data capacity, and currently are being sold in the millions. Approximately 30 companies have been developing TFTC tags. TFTC targets the HF (13.65-MHz) band (60% of existing RFID market) and has read-write capability [7].

In the first generation, only a few chipless RFID tags, which were at the inception stage, were reported in the open literature. They include a capacitive gap-coupled dipole array [8], a reactively loaded transmission line [9], a ladder network [10], and a Peano and a Hilbert curve fractal resonator [11]. These tags were in the prototype stage and no further developments for commercial use have been reported so far.

Realizing that chipless RFID technology is the only path forward for competing with optical barcodes, and foreseeing the lucrative market potentials in chipless RFID tags, big industry players such as IBM, Xerox, Toshiba, Microsoft, HP, and new players such as Kavio and Inksure have been investing tremendously in the development of low-cost chipped and chipless RFID. Figure 1.2 shows the motivational factors for developing chipless RFID tags and reader systems. The data shown in the figure have been approximated from two sources [4, 6]. Currently, conventional chipped tags cost more than 10¢ each if purchased in large quantities. This high tag price hinders mass deployment of RFID in low-cost, item-level

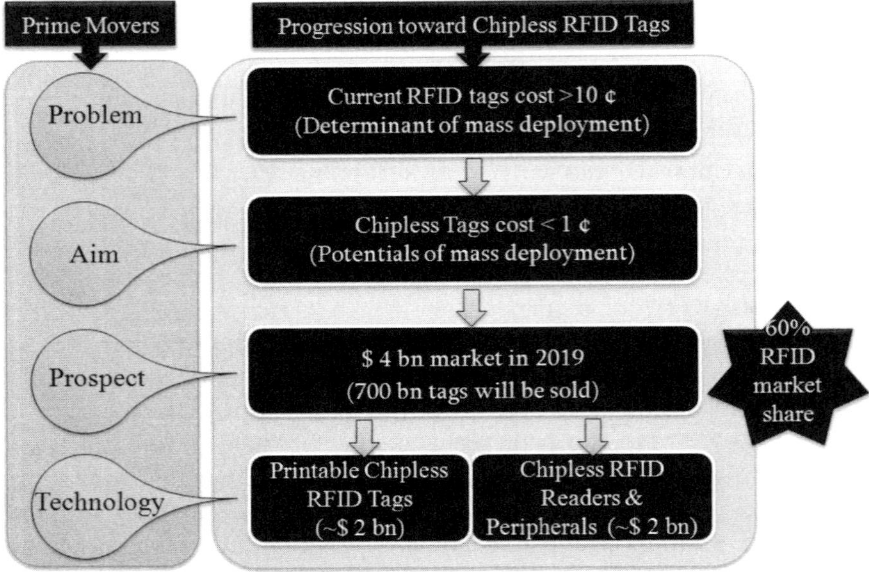


**Figure 1.2**  Prospect of chipless RFID technology.

tagging. The goal is to develop sub-cent chipless tags that will augment the low-cost, item-level tagging. The technological advancements in both chipless tags and their readers and peripherals will create an approximately $4 billion market by 2019 [4, 7].

According to [7], the revenue generated in the global chipless RFID market is expected to reach $3,925 million in 2016 from $1,087 million in 2011, at an estimated combined annual growth rate of 29.3% from 2011 to 2016. The targeted market sectors for chipless RFID technology include retail stores, supply chain management, access cards, airline luggage tagging, aged care and general healthcare, public transit, and library database management systems.

Significant strides have been achieved in tagging not only polymer banknotes but also many low-cost items such as books, postage stamps, secured documents, bus tickets, and hanging cloth tags. The technology relies on encoding spectral signatures and decoding the amplitude and phase of the spectral signature [12]. The other methods are phase encoding of backscattered spectral signals [13] and time domain delay lines [14]. So far more than 20 varieties of chipless RFID tags and five generations of readers have been designed by this team. The proof-of-concept technology is being transferred to the banknote polymer and paper for low-cost item tagging. These tags have the potential to coexist or replace the trillions of optical barcodes printed each year. To this end it is imperative to invest in low-loss conducting ink, high-resolution printing processes, and characterization of the laminates on which the tag will be printed. The design of a spectral signature-based tag needs to push in higher-frequency bands to accommodate the increased number of data bits in the chipless tag to compete with the optical barcode. In this space, the reader design needs to accommodate large reading distance and high-speed reading, multiple tag reading in proximity, error correction coding, and anticollision protocols. Also wide acceptance of RFID technology by consumers and businesses requires robust privacy and security protection [15]. The goal of this book is to address all of the issues mentioned above to make the chipless RFID system a viable commercial product for mass deployment.

Figure 1.3 shows the salient features of a chipless RFID tag and Figure 1.4 shows the burdens a chipless RFID tag reader faces to meet market demands. While industry (market) focuses on compatibility of the new tags and readers with the existing standards and operational cultures, available technologies, and on-the-shelf products, and finally, huge profit margin, it is not a trivial task for a dedicated chipless RFID tag reader to meet all requirements. Figure 1.4 shows the chipless RFID system, which needs to address a whole spectrum of technical and regulatory requirements such as the number of data bits to be read and processed, operating frequencies, radiated (transmitted) power levels and hence reading distance, mode for reading (time, frequency, or hybrid domain), compatibility with existing technological framework, simultaneous multiple tag reading, and resulting anticollision

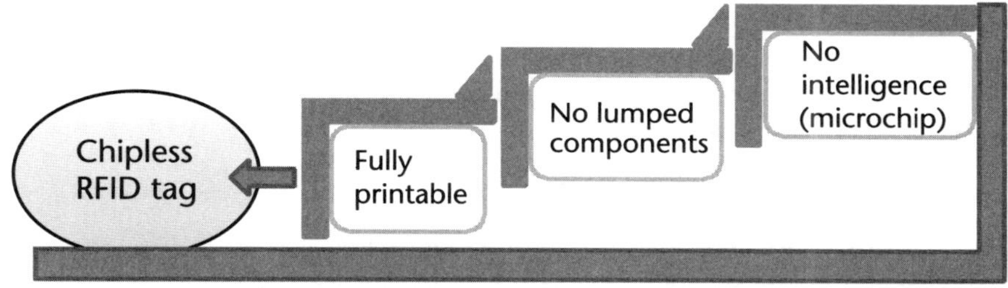


**Figure 1.3**   Salient features of chipless RFID tags.

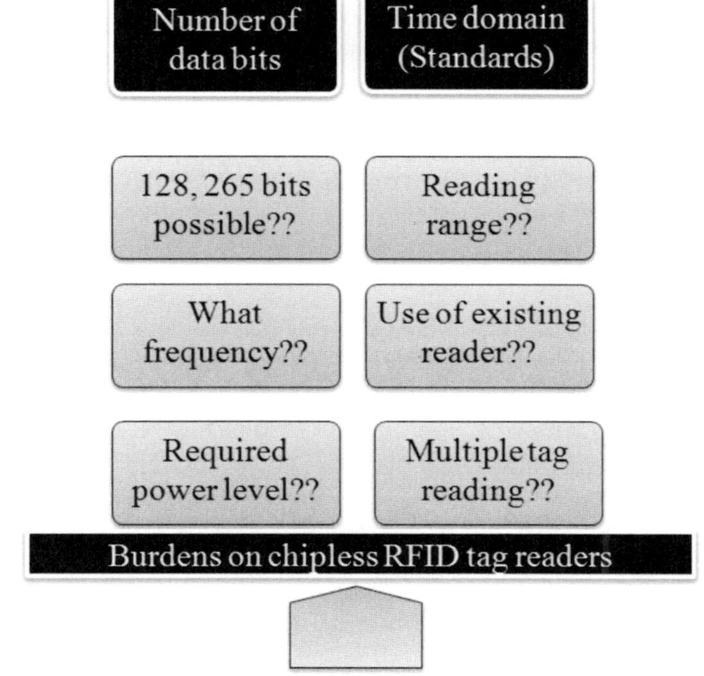


**Figure 1.4**   Design considerations for a chipless RFID tag reader.

and security issues. All of these considerations will impact the development and commercialization of the new technology. IDTechEx [4] reports on chipless RFID tag development by commercial entities and highlights the synergies required to address these issues and make the chipless RFID a commercially viable and competitive technology.

## 1.2  Chipless RFID Tag Reader

As advocated by IDTechEx [4], the chipless RFID tag readers and data processing software will have costs similar to those for their chipped counterparts. The market segment of hardware and related middleware is more than the cost of tags used. Therefore, there is a huge commercial potential to invest in readers and related software development. However, very few resources on chipless RFID reader systems are available in the open literature. This is the first initiative by an author to introduce the potential field in a combined and comprehensive body of literature. The book covers the following topics in the field:

1.  Introduction to chipless RFID;
2.  Chipless RFID reader architecture;
3.  Physical layer development of chipless RFID reader.

As mentioned above the chipless RFID will impact low-cost item tagging in this millennium. We provide readers with comprehensive coverage of chipless RFID tags and reader systems as shown in Figure 1.5. The book reports on the most recent developments and challenges ahead in the chipless RFID field. With the popularity of RFID applications, the demand for chipless RFID tags has been growing exponentially [4, 6]. The demand will further accelerate research and development on chipless RFID technology. These endeavors will lead to successful mass commercialization of chipless RFID tags. This book features:

1.  A single comprehensive and most recent source of the chipless RFID technologies applied to low-cost, item-level tagging for large-scale deployment to replace or coexist with the trillions of optical barcodes printed each year

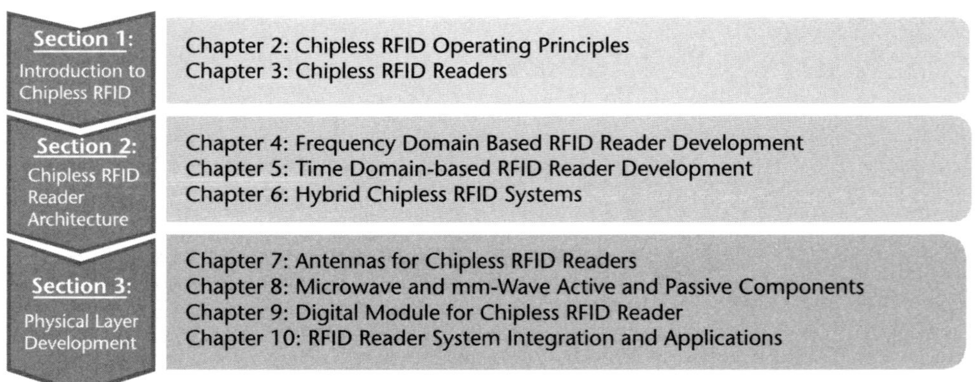


**Figure 1.5**  Organization of sections and chapters on chipless RFID systems.

2.  Broad and integrated coverage of subject matter not usually covered in a single reference in the RFID open literature
3.  The latest achievements in the design and application of chipless RFID tags and RF reader developments
4.  The presentations of basic concepts, terms, protocols, systems, architectures, and case studies for chipless RFID tags and readers
5.  Identification of fundamental problems, key challenges, future directions in designing chipless RFID tag and reader systems
6.  Coverage of a wide spectrum of topics, including signal processing algorithms, hardware architectures, test-bed evaluations, and practical applications of chipless RFID tag and reader systems.

## 1.3  Executive Summaries

Executive summaries of the chapters are given.

### 1.3.1  Operating Principle of Chipless RFID Systems

These chapters include two very high quality chapters on the review of chipless RFID tags and reader systems. The most recent developments of the various types of chipless RFID tags and readers and their operating principles are presented in Chapter 2. Chapter 3 presents the most recent developments for various chipless RFID tag reader systems that can read these chipless RFID tags. Specifics are given next.

### Chapter 2: Chipless RFID System Operating Principles

This chapter presents an overview of the operating principles of chipless RFID systems. The contents are focused on the different techniques used to encode data on a tag that does not have a microchip. A comprehensive review of chipless RFID tags is presented followed by a brief description of different reading methods used in chipless RFID tag readers. The objective of this chapter is to provide the basic operating principles of chipless RFID systems so the reader can understand the rest of the chapters of the book.

### Chapter 3: Chipless RFID Readers

Chapter 3 provides a comprehensive review of various chipless RFID readers and their operating principles. Because the readers used for reading conventional tags cannot read chipless tags for various reasons, totally new and dedicated readers need to be developed for the interrogation of chipless tags. The chapter mainly provides the fundamental operating principles of chipless RFID readers developed for different kinds of chipless tags as discussed in Chapter 2. Many theories that are the bases for the new tag reading processes are highlighted. This chapter provides the foundation information for planning and integration of the components developed in the following chapters.

### 1.3.2 Reader Architecture for Chipless RFID Systems

In these chapters, the reader architectures of the fundamental reading principles are discussed: frequency domain (or also called frequency signature) based chipless RFID readers. time domain (TD) based chipless RFID readers. and, finally, the hybrid or combined time and frequency domain based chipless RFID tag readers. All three categories of chipless RFID tag readers are under development or in their inception phases. To the best of the authors' knowledge, besides the authors' research group, no new information is publicly available in the field. Therefore, this section presents some niche ideas and principles to interrogate and decode the information data bits for the various types of chipless RFID tags. The chapters are as follows:

### Chapter 4: Frequency Domain Based RFID Reader

Multiresonator based chipless RFID tags have the potential to achieving very low cost tags. A frequency signature is used to encode data in multiresonator based chipless tags. Therefore, the reading process requires characterization of the tag in the frequency domain. This chapter provides detailed information for two chipless tag readers. The two readers operate based on two different concepts. First, a detailed description of the theoretical framework used in the reading process is explained, then the design details of the two readers are presented.

Frequency signature based chipless tags are required to operate over a very wide frequency band for multibit data capacity. The frequency domain based readers presented in this chapter use a linear frequency sweep (linear chirp) to interrogate the chipless tags. A microcontroller delivers the tuning voltage, which in turn generates the swept frequency in a voltage-controlled oscillator (VCO). The VCO needs some settle-in time to generate linear sweep. Therefore, this technique requires a considerable amount of time (a few hundred milliseconds to a few seconds) to read multibit data capacity chipless tags. The reading time of this reader depends on the frequency band of operation, the number of data bits, and the performance of the VCO to interrogate and decode the data bits. High-performance VCOs have low settle-in times, which results in high-speed reading. This problem is overcome by the time domain based chipless RFID tag reader as described in the next chapter.

### Chapter 5: Time Domain Based Chipless RFID Reader

Frequency signature based chipless tags need to operate in very wide frequency bands, as discussed in Chapters 2 and 3. The frequency domain based readers presented in Chapter 4 use a linear frequency sweep (linear chirp) to interrogate the chipless tags. To reduce the reading time, the concepts used in time domain ultrawideband (UWB) communication technology can be applied in the fast reading process of chipless RFID tags. The UWB impulse radio based reader illuminates the frequency signature based chipless tags with its RF impulse energy. The data decoding process is accomplished by applying the Fourier transform for the retransmitted signal from the tag to extract the encoded data in frequency domain. This chapter provides the fundamental theory for a time domain based chipless RFID reader design using the UWB impulse radio principle followed by the reader architecture and simulated results of the reader design.

## Chapter 6: Hybrid Chipless RFID Reader

Both frequency and time domain based tags have data encoding capacity limitations. To enhance the capacity, a multiplexing technique with a combination of time and frequency domains is used to read the tag in the frequency domain in multiple time slots similar to the time domain multiplexing technique (TDMA). This chapter presents some recent progresses in chipless RFID tag reading techniques. Both time and frequency domain encoding improves the data bit capacity, detection, and accurate estimation of information.

### 1.3.3  Physical Layer Development of Chipless RFID Tag Readers

For any enabling technology, the physical layer is the heart of the system. Software protocols, middleware, and digital interfaces all augment the hardware development and improve the efficacy of the hardware's performance. For chipless RFID systems there is no exception. Efficiently designed hardware improves the signal quality, speed, and accuracy of operation. This also reduces the load on other parts of the systems. Figure1.1 earlier in this chapter shows the physical layer of a generic RFID system. As can be seen the physical layer occupies the most significant part of the total system. For chipless RFID the exception is in the removal of the ASIC chip from the tag. Otherwise the rest of the system contains generic components and subsystems such as tag and reader antennas, hardware, and middleware for the reader.

This book elegantly covers the physical layer development with comprehensive design details of system components, integration of the components in operational subsystems, and evaluation of these subsystems into integrated reader architecture. The salient components of a reader architecture are reader antennas, microwave and millimeter-wave passive designs such as passive lowpass, bandstop and bandpass filters, power dividers, couplers, resonators, and microwave and millimeter-wave active devices such as VCOs, power amplifiers (PAs), low-noise amplifiers (LNAs), impulse generators, and switches. These passive designs and active devices cover the RF transceiver development of the chipless RFID reader.

The second part of the physical layer design is the digital section and their interfaces with the RF transceiver and middleware. The efficient design of all components and systematic evaluations of all sections can ensure the working prototype of a mixed signal system engineered chipless RFID tag reader. Three very high quality and technically comprehensive chapters cover every detail of the design and development of the components and subsystems of the various types of the chipless RFID tag readers as stated in the preceding section.

## Chapter 7: Antennas for Chipless RFID Readers

The three main types of chipless RFID systems—frequency, time, and hybrid domains—cover the UWB spectrum in both microwave frequency bands from 3 to 11 GHz and millimeter-wave frequency bands from 20 to 30 GHz, and, finally, around 60 GHz. To address the diverse requirements of such large operational bandwidth and operating frequency ranges, many types of reader antennas have been designed, fabricated, and evaluated. The antenna types vary from a simple disk-loaded ground reflector assisted printed monopole in both microstrip line and coplanar waveguide topologies to a broadband printed dipole array antenna and a broadband microstrip line fed aperture coupled patch array antenna. These comprehensive designs of antennas generated a huge depository of knowledge in both microwave and millimeter-wave frequency bands. These antennas are not only UWB but also of high gain. Therefore, the developments are significant in the field of antenna design. This chapter presents the design and fabrication of microwave and millimeter-wave band UWB antennas and their evaluation as used in their intended RFID tag reader systems.

## Chapter 8: Microwave and Millimeter-Wave Active and Passive Components

While antennas are the eyes of the system that communicate with the outer world wirelessly, the microwave and millimeter-wave passive and active components are the hearts of chipless RFID tag reader systems. This chapter formulates the specifications for various passive and active components for the reader systems followed by the fundamental theories and design details of these components. The passive designs such as microwave and millimeter-wave power dividers, resonators, and filters are designed and developed from scratch and then fabricated and measured with high-precision microwave and millimeter-wave measuring equipment. The pieces of equipment were the Agilent Performance Vector Network Analyzer PNA E8361A, Agilent Performance Spectrum Analyzer E4408A, and Tektronix high sampling rate (up to 20 GHz) digital storage oscilloscopes TDS6604.

The active devices were incorporated into design circuits to evaluate their specific functionalities and then integrated in a system. Moreover, these components are used for both chipless tags and RFID readers. Passive components are used for chipless tags and the reader's RF front-end design; the active components are used in the receiver and transmitter sections of the reader's RF front-end. Therefore, the whole chapter presents a comprehensive body of knowledge in the field and offers excellent guidelines for the designers of active and passive microwave and millimeter-wave components and systems.

## Chapter 9: Digital Module for Chipless RFID Readers

Because RFID systems use a mixed signal enabling technology, they deal with signals of various types such as microwave and millimeter-wave RF carrier signals, down-converted information carrying baseband signals and digital signals. The readers also process the signals for modulations and demodulations and encoding and decoding purposes. Therefore, all RFID readers consist of three major sections: antennas, RF transceiver, and digital control sections. The first two were covered in the two preceding chapters. Each of the three different types of chipless RFID readers (frequency, time, and hybrid domains) has a different RF and digital section for controlling its operations. This chapter provides information on the requirements of digital interfaces for chipless RFID readers. This chapter also provides two digital interfaces that are based on microcontroller and field programmable gate arrays (FPGAs).

Chapter 10: RFID Reader System Integration and Applications

This chapter presents the integration of the UWB antennas, RF reader front-ends, and digital sections in the formation of a complete chipless RFID system. Finally, the chapter includes field trials and case studies of the various integrated reader systems. The implementation challenges of these developed readers in commercial domains and essential tasks for efficient detection, multiple access, and signal integrity to improve the efficacy of the developed readers will also be addressed. These topics are covered in a companion book, Chipless RFID Reader Signal Processing.

## 1.4   Conclusion

RFID is an emerging technology for automatic identification, tracking, and tracing of goods, animals, and personnel. In recent decades, the exponential growth of the RFID market has signified its potential in numerous applications. The advantageous features and operational flexibility of RFID have attracted many innovative application areas. Therefore, there is a need for tremendous development and open literature on RFID to report new results. However, the main constraints to mass deployment of RFID technology are the cost of the tag and the reading techniques and processes of the chipless tag. This is the first book in the field that covers comprehensively many significant aspects of chipless RFID reader architecture and signal processing.

This book project is an initiative to publish the most recent results of research and development on the chipless RFID tag reader system. It aims to serve the needs for a broad spectrum of readers. It starts with a comprehensive review of various chipless RFID tag technologies developed so far, and various reading methods in three main domains: time, frequency, and hybrid. A full section is dedicated to the component-level design and physical layer development of chipless RFID tag readers, their system-level integration, and field trials. It also covers the system aspects of the chipless tag readers, system architectures, detection techniques, multiaccess protocols, and signal integrity. The book has proposed a few novel detection techniques dedicated to radar array based chipless RFID tags [12, 13]. First, signal space representation based chipless RFID tag detection is used for the frequency signature based chipless tags. Second, a UWB impulse radio detection technique is used to interrogate a frequency signature based chipless tag. Third, a singularity expansion method is used to separate poles and residues of the tag responses, and, finally, various filtering techniques such as wavelet transformation and the prolate spheroidal wave function are used for noise filtering. All of these detection, denoising, and filtering techniques improve the efficiency of the reader.

The book has also proposed a few state-of-the-art multiaccess and signal integrity protocols to improve the efficacy of the system in multiple tag reading scenarios. Comprehensive studies of anticollision protocols for chipless RFID systems and various revolutionary techniques to improve their signal integrity have also been presented. It has also identified future challenges in the sphere.

Finally, an industry approach to integration of the various systems of the chipless RFID reader technology is given: integration of physical layers, middleware, and enterprise software are the main features of the book. Overall, the book has become a one-stop-shop for a broad spectrum of readers who have interests in the area of emerging chipless RFID and sensor technologies.

## Questions

### Qualitative and Descriptive Questions

1. Define a chipped and a chipless RFID. What is the main constraint of mass deployment of RFID technology?
2. What are the different components of an RFID system? What components and\or blocks constitute the physical layer and IT layer of an RFID system?
3. What are the advantageous features of an RFID tag over an optical barcode? Add as many points as you can imagine.
4. Removing the ASIC from an RFID tag makes the tag chipless. Explain the challenging issues for developing low-cost functional chipless RFID tags.
5. What is the predicted cost of a chipless RFID tag compared to a chipped tag?
6. Name a few chipless RFID tag technologies in commercial use and in development phases.
7. Chipless tags need to be read error free. Explain the burdens on a chipless RFID tag reader to be competitive with respect to a conventional RFID reader.
8. Explain why a chipless RFID tag reader might not be made cheaper compared to a conventional reader.
9. Why does a frequency signature based chipless RFID tag operate over the ultra-wideband (UWB) frequency spectra? How can the reading time of a chipless tag be reduced in a time domain based chipless tag reader?
10. How are information data bits in terms of frequency signatures extracted from an impulse radio based chipless RFID tag reader?
11. Explain the operating principle of a hybrid time-frequency domain based chipless tag and reader system.
12. What components and\or building blocks constitute a chipless RFID reader physical layer?
13. Why is the systematic evaluation of the individual building blocks of the radio and digital sections of a chipless RFID reader required?
14. Why is an efficient UWB antenna needed in a chipless RFID reader? Name various UWB frequencies used in chipless RFID reader antenna design. Name a few developed UWB reader antennas.
15. Name a few active and passive RF\microwave components used for chipless RFID readers. Which of the components can be used for both tag and reader designs?
16. Name some RF\microwave test equipment that is used to test active and passive components and building blocks of a chipless RFID reader system. Visit the Internet to get more details about these pieces of equipment.
17. Which parts of a chipless RFID reader use microcontrollers and FPGAs? Where are they located in the reader design?

### Numerical Questions

18. Worldwide civil aviation industries lose \$4 billion/yr for misreading of optical barcode enabled luggage tags and subsequent mishandling of luggage. In the existing optical barcode reading system, an average 20% reading error occurs. Using conventional RFID tags, this error in the reading process can be reduced to less than 1%. Estimate the cost savings by implementation the RFID technology assuming the costs of the RFID reader and tags are similar in magnitude.

19. According to IDTechEx's prediction, a chipless tag will cost less than 0.4¢. Current luggage RFID tags cost 75¢. In 2003 the Las Vegas airport handled 70,000 bags each day through their various terminals. Estimate the cost savings of per day if chipless RFID tags were implemented. Assume the costs of the readers for both types of tags are similar in magnitude.

20. Monash University's libraries have approximately 3 million items, which are tagged with optical barcodes. Each barcode cost 10¢. Estimate the cost of barcodes. Estimate the replacement cost of barcodes with chipped RFID tags that cost 50¢ per tag. A chipless RFID tag cost 0.1¢. Estimate the saving in tags purchased compared to the chipped tags.

# References


[1] IDTechEx, *Piezoelectric Energy Harvesting 2012–2022: Forecasts, Technologies, Players*, 2012.

[2] IDTechEx, *RFID Forecasts, Players and Opportunities 2009–2019*, Executive Summary, 2009.

[3] K. Finkenzeller, *RFID Handbook: Fundamentals and Applications in Contactless Smart Cards, Radio Frequency Identification and Near-Field Communication*, 2nd ed., New York: Wiley, 2003.

[4] IDTechEx. *Printed and Chipless RFID Forecasts, Technologies & Players 2009–2029*.

[5] R. E. Azim et al., "Printed Chipless RFID Tags for Flexible Low-Cost Substrates," in *Chipless and Conventional Radio Frequency Identification: Systems for Ubiquitous Tagging*, Hoboken, NJ: IGI Global, May 2012.

[6] S. Preradovic and N. C. Karmakar, "RFID Readers—Review and Design," in *Handbook of Smart Antennas for RFID Systems*, Wiley Microwave and Optical Engineering Series, New York: Wiley, 2010, pp. 85–122, 2010.

[7] *Chipless RFID Market (2011–2016)—Global Forecasts by Applications (Retail, Supply Chain, Aviation, Healthcare, Smart Card, Public Transit & Others)*, http://www.marketsandmarkets.com/Market-Reports/chipless-rfid-market-501.html (accessed June 9, 2012).

[8] I. Jalaly and I. D. Robertson, "Capacitively Tuned Microstrip Resonators for RFID Barcodes," *Proc. 35th EUMC*, 2005, pp. 1161–1164.

[9] L. Zhan et al., "An Innovative Fully Printable RFID Technology Based on High Speed Time-Domain Reflections," *Proc.HDP'06*, 2006.

[10] S. Mukherjee, "Chipless Radio Frequency Identification by Remote Measurement of Complex Impedance," *Proc. Wireless Technologies, European Conference*, 2007, pp. 249–252,.

[11] J. McVay, A. Hoorfar, and N. Engheta, "Theory and Experiments on Peano and Hilbert Curve RFID Tags," *Proc. SPIE*, Vol. 6248, No. 1, pp. 624808-1-10, 2006.

[12] S. Preradovic et al., "Multiresonator-Based Chipless RFID System for Low-Cost Item Tracking," *IEEE Trans. Microwave Theory & Tech.*, Vol. 57, No. 5, part 2, pp. 1411–1419, May 2009.

[13]   I. Balbin and N. C. Karmakar, "Multi-Antenna Backscattered Chipless RFID Design," in *Handbook of Smart Antennas for RFID Systems*, Wiley Microwave and Optical Engineering Series, New York: Wiley, pp. 415–444, 2010.

[14]   R. Woolf, *Development of A 2.45 GHz Chipless Transponder for RFID Application*, Final Year Project Thesis, ECSE, Monash University, 2009.

[15]   J. Kamruzzaman et al., "Security and Privacy in RFID System," in *Advanced RFID Systems, Security, and Applications*, Hoboken, NJ: IGI Global, Hoboken, 2012.

## CHAPTER 2

# Chipless RFID System Operating Principles

## 2.1  Chipless RFID Tags

Much effort has been expended by researches all over the world to reduce the cost of RFID tags. Because the tag antenna and application-specific integrated circuit (ASIC) assembly result in a costly manufacturing process, it is still not possible to achieve RFID tags at a cost of less than 1 cent (or sub-cent) each. To address the challenge of reducing the cost of an RFID tag, numerous efforts have been made to encode data on a tag without an ASIC. Those tags have been named chipless RFID tags. The generic term chipless RFID covers all types of passive transponders that have no on-board ASIC or "chip."

Most chipless RFID tags use the electromagnetic properties of materials and various circuit layouts to achieve a specific electromagnetic behavior (or property). The identification or extraction of data from chipless tags is done by analyzing the reflected signal from the tag. This technique is similar to the operation of some types of chipped RFID tags [1]. Although some chipless RFID tag developments have been reported on in recent years, most of them are still reported as prototypes. Only a handful are considered to be commercially available or viable. The tag ID is incorporated within the structure of the tag. Such tags can be divided into different categories. The main focus of this chapter is to review the techniques used to encode data in different types of chipless tags and their reading processes. A broad classification is presented in Figure 2.1.

With the maturity of technology, thin-film transistor circuit (TFTC) tags will cover a large market segment of the chipless RFID tags in the near future. These tags use the HF (13.65-MHz) band, have a high data capacity, and are compatible with the existing time domain RFID standards. However, the cost of TFTC chipless tags will be an issue. Once fully printable passive microwave and millimeter-wave chipless tags, namely, *radar arrays*, are fully developed and can be printed in the sub-cent cost rage, the radar arrays will dominate the chipless RFID market. The author presented a comprehensive review on chipless and chipped tags in his previous book[1]. Therefore, this review focuses only on *radar array* based chipless tags.

---

1.  N. C. Karmakar (Ed.), *Handbook of Smart Antennas for RFID Systems,* New York: Wiley, 2010.

The following section presents information about different chipless RFID tags based on three categories (see also Figure 2.1):

1. Time domain TDR based chipless RFID tags
2. Frequency domain (spectral signature) based chipless RFID tags
3. Hybrid domain based chipless tags.

### 2.1.1 Time Domain (TD) Based Chipless RFID Tags

TD based tags are interrogated by sending a short pulse and observing the received trains of pulses reflected from the structure that is designed to reflect incoming pulses. Data are encoded in the time domain based on the delay in arrival of the transmitted pulse at different times. There are printable and nonprintable TD based chipless RFID tags. An example of a nonprintable chipless RFID tag is the surface acoustic wave (SAW) tag [2]. These are excited with a chirp of RF pulse centered around 2.4 GHz. SAW tags are the most common type of commercially available chipless RFID tag [3]. Most of the printable TD based tags are still in the experimental stage [1, 3]. Various types of time domain based tags are described next.

#### 2.1.1.1 SAW Chipless RFID Tags

The SAW based tag is one of the few commercially available chipless RFID tags. To date a few hundred thousand SAW tags are commercially sold each year. The sales are exponentially growing and in 2019 approximately 200 million SAW tags will be sold according to the IDTechEx predictions [4]. Their market includes electronic

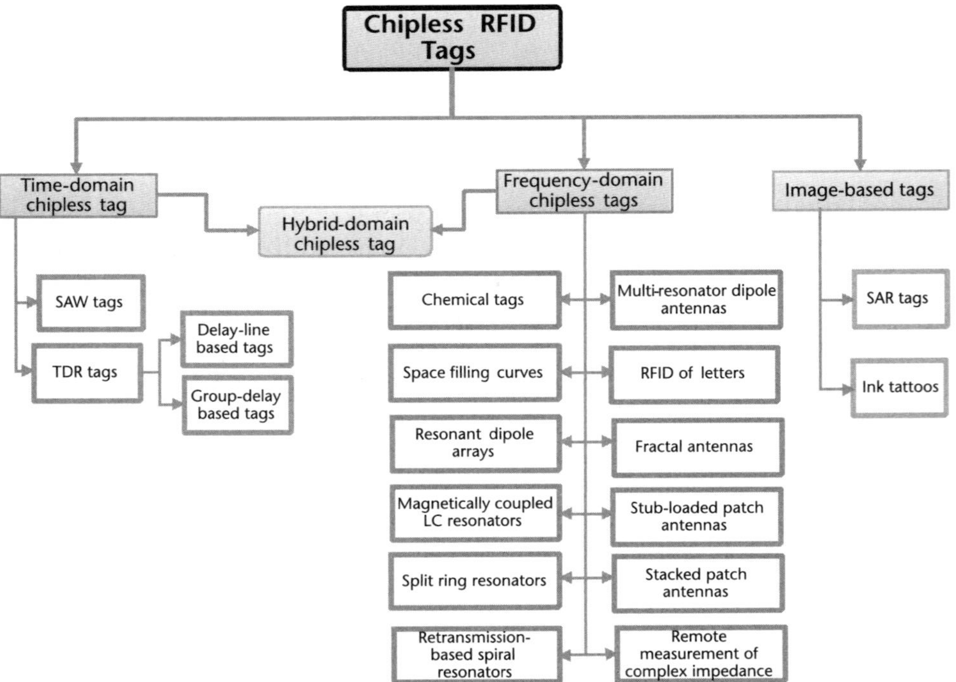


**Figure 2.1** A broad classification based on existing chipless RFID tags.

toll collection and defense and space applications. Data are encoded using time delays in the reflected pulses, which can be termed time position encoding or pulse position modulation (PPM) [4]. An interrogation pulse is sent from the reader antenna to the tag. The tag receives the pulse signal with its omnidirectional antenna and guides it to an interdigital transducer (IDT). The IDT is made of two interlaced comb-like metal structures deposited on the microacoustics piezoelectric substrate (LiNO3, etc.). It converts the guided electromagnetic wave into a surface wave (mechanical wave), which propagates slowly (100,000 times slower than the speed of light in free space [5]) over the piezoelectric surface of the tag. Then the surface wave encounters a series of bandstop filters made of aluminum strips and reflected back toward the IDT, which carries a code based on the positions of the reflectors. The reflected SAW pulses are then reconverted to electrical form again by the IDT and retransmitted to the reader by the tag antenna. Currently, however, the pulse signal is rarely used in practical reader systems for SAW tags. The principle of operation of a SAW tag is shown in Figure 2.2(a). In real reader systems a simple frequency domain reading method is used, in which the reflected signal (S11) of the tag is measured at a particular frequency and then Fourier transformed to the time domain. This technique lowers the cost of the reader. The SAW tag has some advantages over chipped RFID tags. It can be used as a temperature sensor because its piezoelectric crystal has a predictable change in its properties with change in temperature [6]. It can be detected at a distance of few meters using 2 to 3 orders

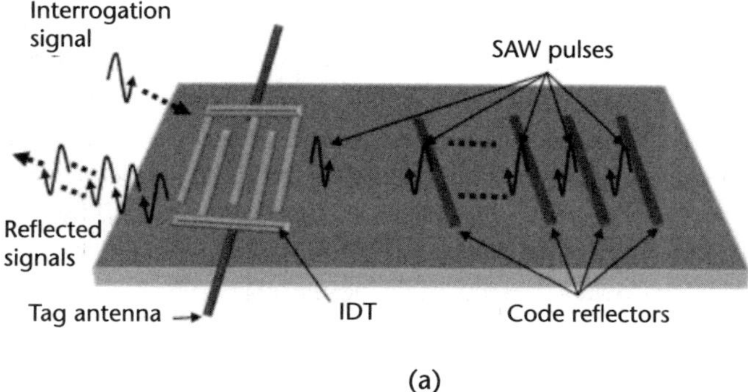


(a)

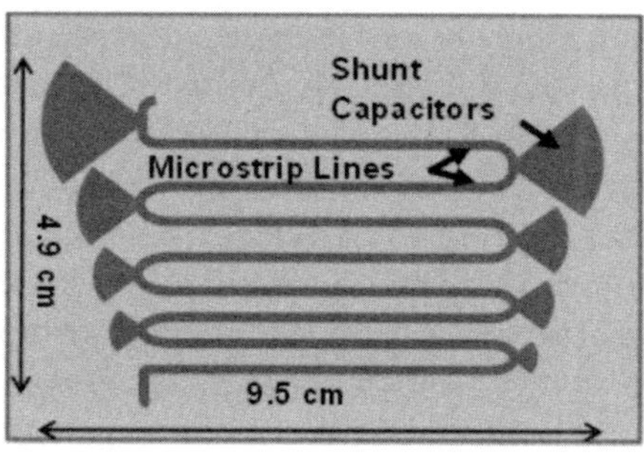


(b)

**Figure 2.2**   Schematic of (a) a SAW tag [10] and (b) a TDR tag using shunt capacitance [12].

of lower power (10 mW [5]) than the chipped RFID tags. SAW tags also work on metal surfaces and have a higher resistance to interference. This tag works more reliably in harsh environments than do chipped tags. Moreover, multiple SAW tags can also be identified, as is true of multiple chipped tags, by applying a multiple access spread spectrum signaling scheme [7, 8].

To increase the data capacity, some other techniques are now added to the IDT based SAW tags. Many of them added advantages to the current system—along with a few disadvantages too. One technique uses multiple IDT [9] to increase the number of bits. But it generates a spurious signal, which causes code confusion. Another one is reflector based X-Cyte tags [10], which use 16 reflectors distributed in four acoustic tracks. This reduces the tag size but increase the losses inside the tag where the tag cannot be tested before the final stage. An Epcos tag [4] uses two tracks on the same side of the IDT. Here bits are changed by removing the tracks completely. But it makes the reflected signal dependent on the encoded bits. Siemens developed a 31-bit tag operating in the 2.45-GHz ISM band [10]. It uses 33 equidistantly placed reflectors. But it suffered from relatively high insertion loss (attenuation) of the symbols due to the many reflections required for its amplitude shift keying (ASK) coding scheme. The BaumerIdent tag developed by TAGIX uses split-finger IDT with five open circuit $\lambda/2$ reflectors [6]. But the data capacity of this tag is very low.

Hartman et al. [10] proposed a global SAW tag using a time position coding scheme. Here the phase increment in a given group of slots of the total slot array is used as a code. A 128-bit measurement result was also shown. Orthogonal frequency coding [11] and SAW resonators with slightly different frequencies are also proposed to generate IDs for the SAW tags. At first glance, it seems that the SAW tags are fully functional and could replace the chipped tag easily. However, the main drawbacks of this SAW based tag are its costly piezoelectric substrate and its IDT and reflectors, which need to be deposited slowly by submicron photolithographic tools. These lithographic tools make the initial fixed cost for SAW tags very high. Another drawback is the incompatibility in terms of printing the tag directly on substrates such as banknotes, postage stamps, security documents, and plastic packets.

### 2.1.1.2    Time Domain Reflectometry (TDR) Based Tag

The second type of TD based tag is the TDR chipless tag. Here the transmitted signal through a transmission line is interrupted in a cascade of short circuit discontinuities along the line. The returned echoes are time sampled and the delays are recorded as stream of data bits as 1's and 0's in time. A fully printable tag has been presented in [12, 13] where the time domain reflectometry (TDR) method is used for encoding data. Here the reader transmits a train of short time pulses modulated at high frequencies. These pulses are received by the tag antenna and then travel through the attached long meandering transmission line. Impedance discontinuities are put at an interval determined by the pulse width in the transmission line to encode the logic 1 and removed the encoded logic 0. The pulse is reflected back to the reader at different times according to the position of the impedance mismatches. If an impedance mismatch is removed from one point, then there will be no reflection from that point, which makes that bit zero. The total length of the transmission line is determined by the number of bits, which is equal to the number of discontinuities and the equivalent of the physical length of the pulse width of the interrogation signal. For a typical 4-bit tag with 2-ns interrogation pulse, an approximately 54-cm transmission line is required [see Figure 2.2(b)]. This long transmission line makes the tag very large and unsuitable for commercial use. A few techniques to overcome the size issue are proposed in [14] and [15].

A 2.45-GHz TDR based fully printable tag has been developed by the authors' research group (Figure 2.3). Four stubs are placed at regular intervals. A disk-loaded UWB monopole antenna with modified ground plane [Figure 2.3 (b)] is added to the delay line [Figure 2.3(a)]. The complete tag and reader system is shown in Figure 2.3(c). Here another monopole is used as a reader antenna. The time domain response is shown in Figure 2.3(d).

Stelzer et al. [15] proposed a delay line using left-handed (LH) negative refractive index (NRI) transmission line sections. The lines are described as LH, since the sum of the electric and magnetic fields associated with a Poynting vector follow a "left-hand" rule, rather than the traditional "right-hand" (RH) rule of classical EM theory. Although this line allows the delay lines to be shorter, they exhibit significant distortion, which limits the number of usable bits for the system. It is possible that a usable transponder may come from this method, but it will still take a significant amount of research to produce a viable design.

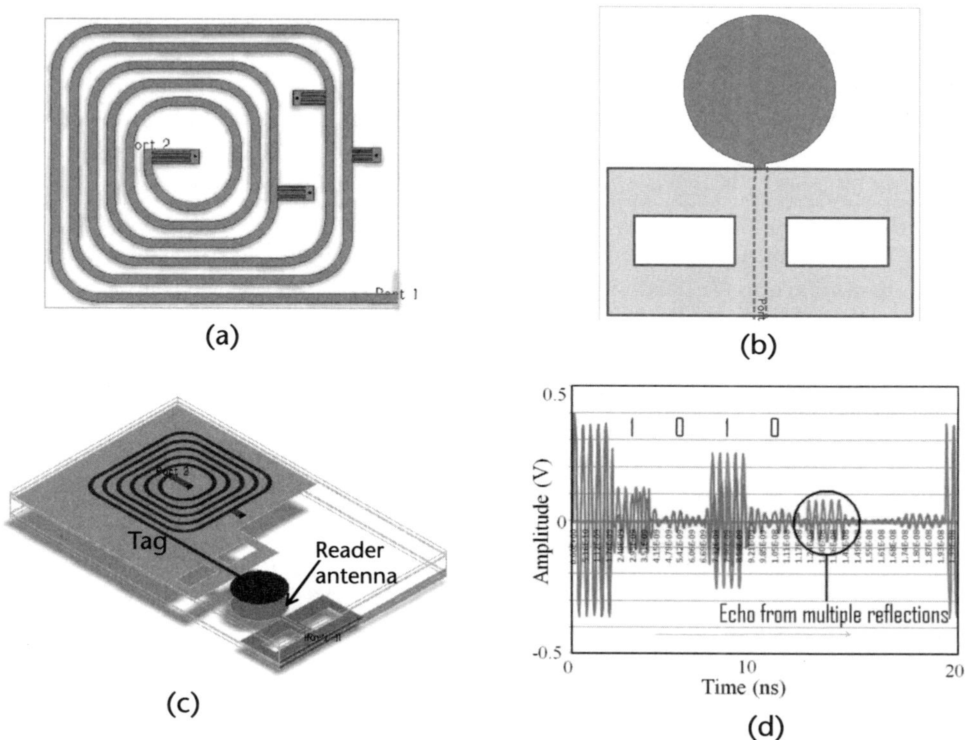


**Figure 2.3**   (a) A TDR based transmission line with 4-bit delays, (b) a UWB antenna, (c) a complete 4-bit (1010) chipless tag with a disk-loaded monopole reader antenna, and (d) the time domain response of a 4-bit(1010) tag as shown in part (c) [16].

Another printed passive RFID tag on the paper substrate with high data rate is proposed by Shao et al. [12]. They used a UWB antenna and a tapered microstrip transmission line loaded with distributed shunt capacitance as the information coding element. A tapered microstrip line is used to overcome the limitations of low ink conductivity and thin-film thickness of the printed metal lines. The silver ink used for printing the tag has conductivity of $1 \times 10^7$ S/m and a thickness of 0.5 µm, which is much less than those of copper (conductivity $5.8 \times 10^7$ S/m and thickness of 17 µm). From measurement results it is seen that the tag features a robust readability with a greater than 80-cm reading distance. But the data density is very low for this tag. Only 8 bits are encoded using eight different shunt capacitors and bits are changed from logic 1 to 0 by removing the connections between the microstrip line and shunt capacitors. The tag size will increase linearly with the increment in the number of bits and it also needs a ground plane, which will increase the cost of printing.

### 2.1.2  Frequency Domain Based Tags

Data are encoded in the frequency spectrum received from the tag. Different types are discussed next.

#### 2.1.2.1  Chemical Tags

Chemical tags are mainly made from the deposition of nanometric resonating fibers and electronic ink on a substrate [3]. These materials resonate at different frequencies and create a distinct frequency signature in the backscattered signal to the reader. These tags are cheaper than chipped ones and have the advantage of being able to be printed on plastic items. Seventy different chemicals are claimed to be used by the CrossID company to create 70 frequency signatures for a 70-bit tag [17]. This CrossID system can work with existing barcode systems. The Tapemark company proposed another tag, which consists of an antenna made of very small fibers (as small as 5 µm in diameter and 1 mm in length) [18]. The antennas are made spatially unique by changing the combinations of the fibers and read by a reader using the radar technique. These low-cost tags can work on paper and plastic packaging materials. Though this tag is printable, it requires very high accuracy, which keeps its cost at more than 5 cents each. Moreover, no further reporting about the tag development after the first reports in 2004 from these two companies infer that their claims of data capacity and reliable tag operation are still in the research stage.

#### 2.1.2.2  Space-Filling Curves

RFID tags that use space-filling curves have been proposed by McVay et al. [6]. Peano and Hilbert curves are used to generate different resonant structures to resonate in a frequency band centered around 900 MHz. The Hilbert and Peano curves are used due to their spatial efficiency over half-wave dipoles at the same frequency. A 5-bit tag is made using five Hilbert/Peano curves and in a backscattered radar cross-section (RCS) measurement result five picks are present that represent the 5 bits. The drawback of this tag is that removal of one curve is required to change that bit, which affects the radiation pattern of the other bits. Also, the size of the tag will increase with the number of bits because bits have a 1-to-1 correspondence with the number of space-filling curves.


### 2.1.2.3   Dipole Array Based Chipless Tags

Jalaly and Robertson [19] present an RF barcode that uses arrays of capacitively tuned identical microstrip dipoles. The dipoles resonate at different frequencies within the ISM bands based on the gap capacitance. When the tag is hit by a frequency swept signal from the reader, each dipole absorbs their resonant frequencies. The tagged item ID is determined from the presence or absence of the resonant frequencies. Measured results for an 11-bit prototype are presented using 11 dipole elements in the 5.8-GHz band [Figure 2.4(a)]. A further increase in the number of bits is possible using the UWB band. But to encode each extra bit, this tag needs an extra dipole. That means the size of the tag will increase linearly with the increment in the number of bits and it will be unusable where higher number of bits are required. Moreover, the resonant dipoles have a linearly polarized directional far-field radiation pattern, which also limits its practicality.

Another novel approach is proposed in [19] to achieve a RF barcode [Figure 2.4 (b)] that consists of three sets of half-wave resonant dipole–like structures that resonate at three bands (2.4, 5.25, and 5.8 GHz). The width of the dipole array of each band is then varied to increase the number of bits. The tagged item ID is determined by the presence or absence of the resonant frequencies. Measured results of a 5-bit prototype at 5.8 GHz are shown to prove the concept. The number of encoded bits is determined by the Q factor of each dipole. But the Q factor or the bandwidth mainly depends on the conduction loss and the radiation loss. Surface wave and dielectric loss are almost negligible. Q factors on the order of 100 can be achieved on Taconic TLY-5 substrate. But with this high Q factor only 4 to 5 bits can be coded in a single ISM band. Moreover, to increase the number of bits using the UWB band we need many dipoles. This will increase the size of the tag.

Deepu et al. [20] proposed a theoretical concept for a 9-bit RFID tag within a credit card size area using three diploe antennas [Figure 2.4(c)]. The single-sided tag was printed on a 0.8-mm-thick FR4 sheet with 3-mm-wide dipoles with lengths of 23.5, 16.8, and 11 mm and the dipoles were placed 9 mm apart. Three dipoles give resonances at three different frequencies according to their lengths (L1, L2, and L3). Later lengths of the dipoles are varied by a small amount ($\Delta L$) to shift their resonant frequencies to encode a larger number of bits. These shifts need to be measured reliably and accurately to decode the encoded information in the tag. The maximum number of bits that can be encoded by varying the dipoles' lengths are determined by the minimum frequency resolution of the reader. But the frequency resolution in bistatic antenna measurements will be much higher than the frequency resolution shown in the paper by metallic cavity measurements. Therefore, in the practical field we cannot achieve 23 bits within a credit card size area using this technique.

An RFID tag that uses a slot-loaded planar elliptical dipole is presented in [21]. In this type of arrangement, notches are created in the backscattered frequency signature by adding slots inside the dipole. But the tag is detected by looking at the poles and residues because poles are unique and independent of excitation and

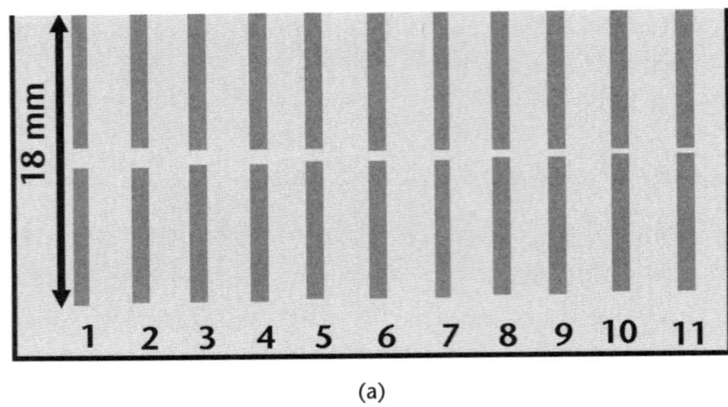

(a)

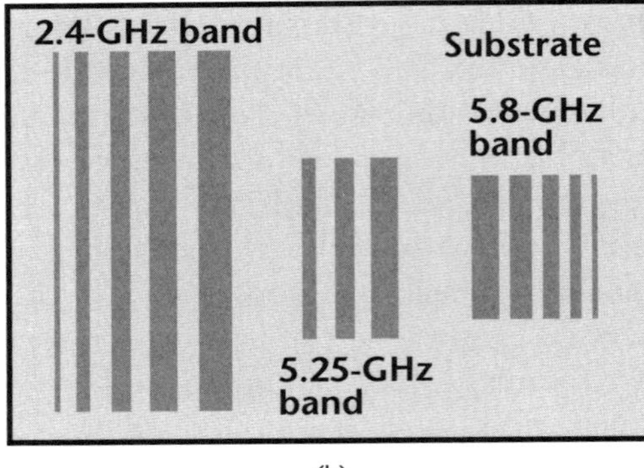

(b)

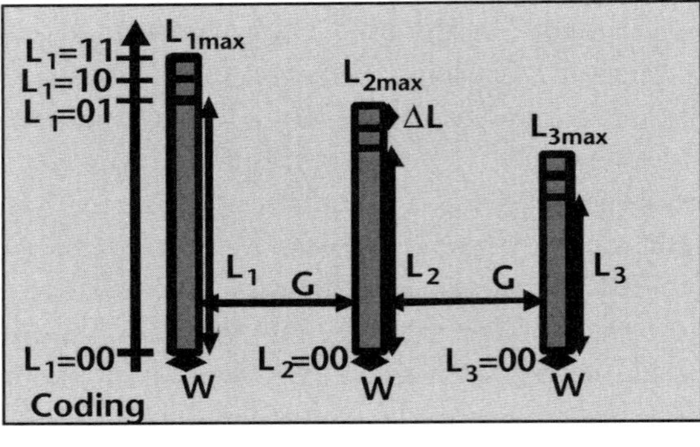

(c)

**Figure 2.4**  (a) Capacitively tuned dipoles, (b) multiband dipole based RF barcode, and (c) RF barcode for secured application [33].

observation point. But the slots do not behave independently for this tag, which is seen in the change of pole location.

## 2.1.2.4  Magnetically Coupled LC Resonators

A chipless RFID tag with a very low data capacity that uses magnetically coupled LC resonators is used for electronic article surveillance (EAS) [22]. It has a simple coil that resonates at a particular frequency. The operating principle is based on the magnetic coupling between the reader antenna and the LC resonant tag. The reader continuously transmits the resonant frequency of the tag. When the tag enters inside the sweeping range, it will resonate and absorb power at the resonant frequency, which will produce a voltage dip across the reader's antenna ports. The interrogation frequency is swept across a band of frequencies to detect the multiple coils of a multibit transponder. The advantage of this tag is that it is a very cheap and simple solution. Therefore, it is mainly used for security purposes to check for the absence or presence of a tagged item. But due to its very limited data capacity and multiple tag collision issues, it is currently losing its market day by day.

### 2.1.2.5   SRR Based Chipless Tags

Hyeong-Seok et al. [23] proposed a 4-bit printable chipless tag that uses split-ring resonators (SRRs) [Figure 2.5(a)]. The electromagnetic code is encoded using the frequency selective behavior of the SRR array. A different combination of SRR arrays generates different bit patterns. The combinations are changed by changing the resonant frequencies of SRRs by changing gap directions (Lgap) and the different parameters of the SRRs (W, H, WPatterns, DSRRs). The tag can be printed on one side of paper or plastic based items, but we need multiple sets of SRR arrays to increase the number of bits. We also need multiple waveguides to measure the multiple SRR arrays. Alignment between them should be strictly maintained.

### 2.1.2.6   Coplanar Strip Based Chipless Tags

Another novel compact (15-mm × 20-mm) printable chipless RFID tag was recently discussed by Vena et al. [24]. It uses quarter-wavelength coplanar strips [Figure 2.5(b)]. The proposed single-sided 10-bit tag has a higher data density (3.3 bits/cm2) than other techniques. It uses four strip resonators. Among them, the frequency of three resonators (1, 2, 4) can be varied independently. To encode data, the length of three of these strip resonators was varied, which shifts the resonant frequency around a reference frequency. Strip lengths are changed by adding some short circuits (L1, L2, and L4) in the slots to get better results. Both time domain and frequency domain measurements were done to demonstrate the robustness of the designed tag. However, the maximum data encoding capacity of this design totally depends on the frequency resolution of the reader and no further data capacity improvement technique is mentioned there. So it cannot be used where a data capacity of more than 10 bits is required.

### 2.1.2.7   RFID of Letters

A frequency signature based method for RF identification of letters has been presented by Singh et al. [25], where RF is used for identifying 24-mm-long letters in the Arial font. When a metallic printed letter is being hit by an electromagnetic wave, then free electrons produce a surface current. Due to their different structures, different letters have different surface current paths and different radiation characteristics are seen in the backscattered frequency signatures. But some of their resonances are very close to each other. The authors proposed to measure those letters in both horizontal and vertical polarities to differentiate them. The main limitation of this technique is that only one letter can be identified at a time and we cannot use multiple letters together as a chipless RFID tag to identify many items.

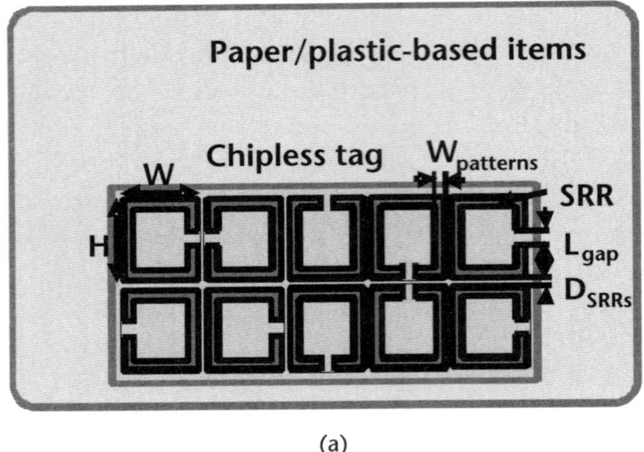


(a)

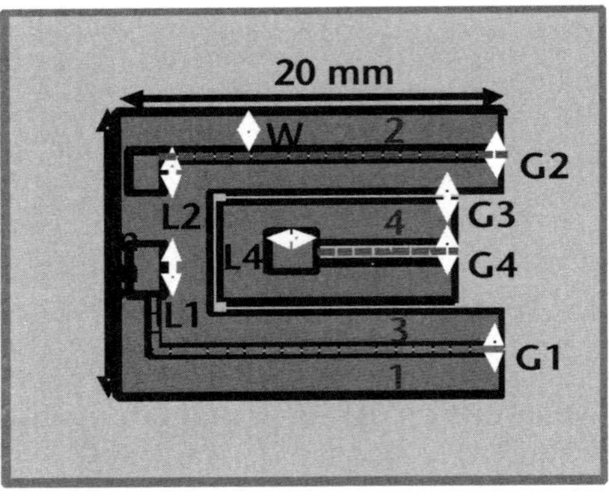


(b)

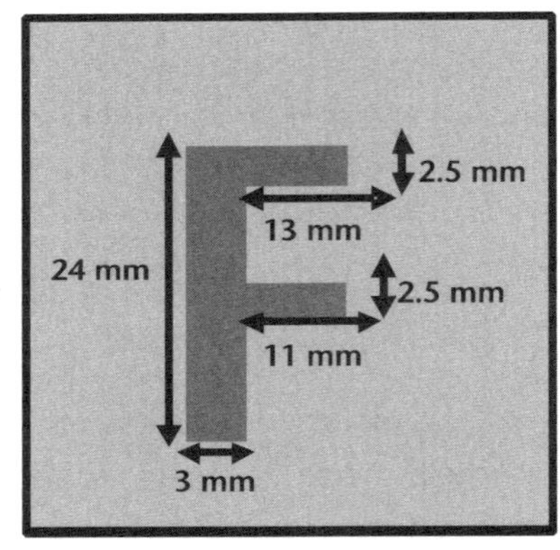


(c)

**Figure 2.5**  (a) SRR array based tag [37], (b) strip line based tag [38], and (c) letter ID based tag [39].

### 2.1.2.8  Spiral Resonator Based Chipless Tags

The authors' research group developed three types of frequency signature based chipless RFID tags and different types of RFID readers for their designed chipless tags [26–38]. The initial designed tag [Figure 2.6(a)] was made by using two

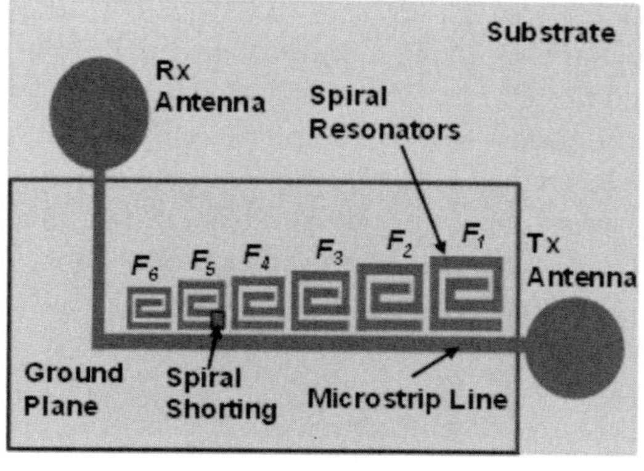



(a)

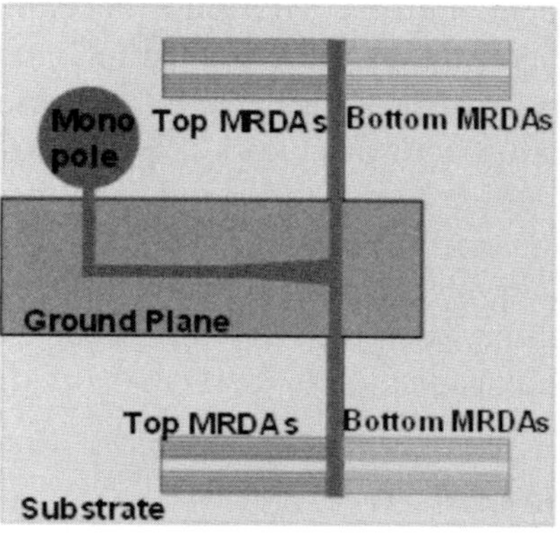



(b)

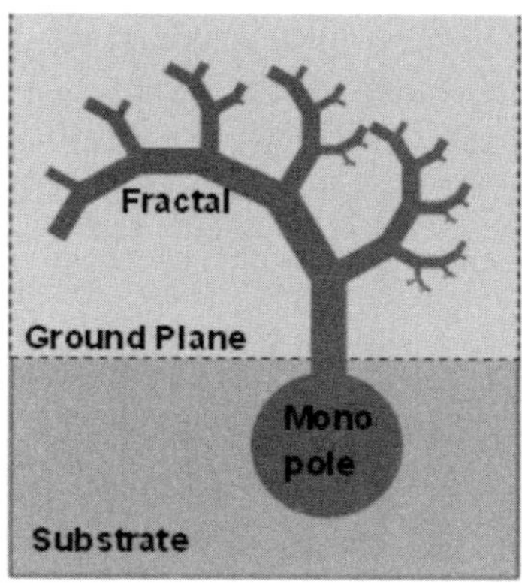



(c)

**Figure 2.6**   (a) Spiral resonator based tag [40], (b) MRDA loaded tag [52], and (c) fractal loaded tag [53].

orthogonally polarized wideband monopole antennas and a series of spiral resonators. These resonators are gap coupled to a microstrip transmission line, which connects the orthogonally polarized monopoles [27]. The reader and the tag antennas are cross-polarized to provide isolation between the transmitting and receiving signals. The RFID reader sends a UWB signal to the chipless RFID transponder [34]. The wideband receiving monopole antenna of the tag receives the signal and it passes through the microstrip transmission line. The gap-coupled spiral resonators create magnitude attenuations and phase jumps at their resonant frequencies in the UWB signal received and this magnitude and phase coded signal is retransmitted back to the reader by the tag's transmitting monopole antenna. The attenuation in the backscattered signal due to a resonator is denoted as logic 0 and the absence of attenuation at that frequency is denoted as logic 1. The attenuation was removed by simply shorting one end of that resonator, which shifts the resonant frequency outside of the desired UWB band. A 6-bit and a 35-bit prototype tag are shown in [27] and [30]. But these first-generation tags are double sided and made on Taconic TLX-0; hence, they are not fully printable. Later second-generation single-sided fully printable tags are proposed in [32]. But all of these proposed tags were not rotation independent. Later a rotation-independent tag using a single monopole antenna and multiple spiral DGSs coupled to a microstrip line was researched [33]. A temperature sensing feature can also be added to this design [31] (but it becomes double layered again). Moreover, all of the proposed tags have a 1-to-1 correspondence between each spiral resonator and each bit. So the size will be much higher for tags with a higher number of bits. Besides, the performance of the tags is also very much dependent on the printing accuracy of the complex spiral structures and small coupling gap between the CPW line and resonators. The operating principle of spiral resonator based chipless RFID tags is explained in detail in Section 2.2.

### 2.1.2.9  MRDA Based Chipless Tags

Balbin and Karmakar [39] proposed a new design using the concept presented in [29]. They used a linearly polarized UWB circular receiving monopole antenna coupled to an orthogonally polarized novel dual multiresonant dipole antenna (MRDA). The dual multiresonant loop antenna creates a frequency signature in the orthogonal polarization of the interrogation signal. They verified their proposed concept with a 6-bit double-layer prototype using two sets of MRDAs each having six dipoles [Figure 2.6(b)]. This design has advantages over the design proposed in [29] because only MRDAs are used instead of the stopband spiral resonators and the second antenna used in [29]. Here the MRDAs operate as set of parallel loop antennas. The size of the tag can be reduced because its total tag area is determined by the outer loop element. Higher frequency elements were placed inside the outer loop, which significantly improves the spatial efficiency over the tag in [29]. However, each bit is encoded here by adding an additional dipole to the circuit. Therefore, the size of the tag increases linearly with the number of bits and it also requires printing alignment on both sides, making the printing process much more sophisticated.

### 2.1.2.10  Fractal Based Chipless Tags

Another novel transponder concept using fractal multiband loading element was proposed in [39]. Data are encoded in the backscattered frequency spectrum received from the tag. When a wideband monopole antenna is loaded with the fractal structure [Figure 2.6(c)], its resonant peaks detectable in the backscattered power spectrum correspond to a set of resonant frequencies that can be directly mapped to the generating parameters of the fractal. If multiple fractal resonators are added through a power divider to a monopole antenna, then the number of resonant picks will increase, which will increase the number of bits. Although the number of combinations is large, it will be limited by the fabrication process and materials of the transponder. Moreover, the design is very large and it requires very precise double-sided fabrication and a large amount of conductive ink due to the large ground plane of the tag.

### 2.1.2.11   Slot-loaded Monopole Based Chipless Tags

A chipless RFID tag with a high data capacity that uses two orthogonally polarized slot-loaded UWB monopole antennas has been researched [40]. Data are encoded by the frequency notches produced by the open-ended slot resonators in the retransmitted frequency signature. When the tag receiver monopole antenna is hit with a UWB signal from the reader transmitting antenna, frequency notches are generated at the resonant frequencies of the slots inside it. This frequency notched signal then travels through a real frequency transfer (RFT) matching section to an orthogonally polarized transmitting antenna. This signal is again frequency notched by the slots presents at the transmitting monopole at different frequencies and retransmitted back to the reader receiver antenna. Because the slots are cut inside the monopoles, the size of the tag is more compact than the tag with a similar concept proposed in [28].

### 2.1.2.12   Stub-Loaded Patch Based Chipless Tag

A chipless RFID tag that uses an encoding technique in the backscattered phase has been discussed by Balbin and Karmakar [41, 42]. Their proposed transponder included open circuit high-impedance stub-loaded patch antennas. The patch antennas are resonant at nearby frequencies. When the patch antennas are hit by their corresponding resonant frequencies, they reradiate the signal in cross polarity with distinct phase characteristics. This cross-polarized backscattered phase at the resonant frequency of each patch can be controlled by changing the stub impedance, which can be changed by changing its length. Thus data are encoded as hexadecimal bits in the backscattered phase of the patch. The cross polarized phase measured by the phase difference between backscattered E-plane and H-plane signals in the reader antenna. Because it is a polarization-dependent technique, the misalignment between the tag and the reader may decrease the phase resolution. Moreover, data are encoded only in the phase domain at a particular frequency and the backscattering from other patches at that frequency may cause undesired non-resonant structural scattering. If the backscattered phase is varied due to multipath, then it might cause a bit encoding error as well [43].

### 2.1.2.13   Stacked Multilayer Patch Based Chipless Tags

A chipless RFID tag that uses stacked multilayer patches was proposed by Mukherjee [44]. Three stacked rectangular patches are used as a scattering element for the prototype tag. When the upper patch resonates, the middle patch acts as a ground plane. Similarly, when the middle patch resonates, the bottom patch acts as a ground plane. As a result two dips are seen in the backscattered magnitude when it is hit by a frequency swept signal. But the structural scattering tends to maintain the RCS relatively constant over the frequency band. So the dips become very small at higher distances. However, the phase (and therefore delay) undergoes significant changes at resonances and is more robust than the magnitude signal at higher distances. This time-delay (phase) measurement along with a soft computing technique [45] made the tag robust in the reading range, but its data capacity is very low. Moreover, it has multiple layers. Hence, it cannot be printed roll to roll for item-level tagging.

### 2.1.2.14    Remote Measurement of Complex Impedance

This technique was used to implement a chipless RFID by Mukherjee and Chakraborty [43]. The proposed tag does not require a semiconductor element and can be printed directly on a low-cost dielectric substrate. A narrowband chirp signal is used to illuminate multiple tags and the combined response is received by the reader antenna. The received signal is then mixed with the transmitted chirp and filtered to create the intermediate frequency (IF) for detecting a particular tag. A reference channel is used to recover the phase-frequency profile for identifying the tag. The proposed detection method requires less power and is reluctant against multipath propagation and interference. But the tag used (L-C Ladder) for verifying the detection concept cannot be used commercially because of its larger size.

### 2.1.3    Image Based Tags

#### 2.1.3.1    SAR Based Chipless Tags

Synthetic aperture radar (SAR) based chipless RFID is one of the recently proposed promising techniques. Referred to as a printed radar array by IDTechEx, three companies (InkSure, Nicanti, and Vubiq) reported up to 96-bit chipless tags. Ink-Sure [46] proposed a new design for their SAR coded tag and a reader for their tag, which can read up to 25 cm in 240 ms of time. Here the radar technique is used for generating an image of the tag made of conductive ink. The operating principle is almost similar to the barcode reading process. It uses a multiantenna array to generate a very narrow millimeter-wave beam to detect multisymbol patterns and their intersymbol diffraction effects on the tag. Then the beam orientation and sensitivity are precisely controlled to generate a map of radar cross section (RCS) from multiple observation points. A digitized image of the tag is created where the presence of conductive strips is mapped as 1's and the absence of conductive strips is mapped as 0's. To encode a higher number of bits in a small place and to get higher resolution, an ISM band of 60 GHz is used. Multiple types of inks are also used to improve the information resolution of the transponder. If slots in the metallic plane are used instead of metallic strips, then the tag will work on metallic bodies too. Though a printable tag with high data capacity was proposed long ago, it is not yet available commercially due to the very complex reader antenna and high processing power requirement for the reader for the reliable detection of the tag. Nicanti reported on a resistive ink radar array based chipless tag (NiCode™) that can be read from a few millimeters away [4]. The applications are brand protection and anticounterfeiting. Vubiq reported on a millimeter-wave chipless RFID label that can be read from a large distance [4].

### 2.1.3.2   Ink Tattoo Based Chipless Tags

An ink tattoo based chipless RFID tag can be directly printed on the surface of tagged items using electronic ink [47]. The reader interrogates the tag with a high-frequency (>10-GHz) microwave signal and receives the reflected signal [48]. This reflected signal is different for different tags due to their unique pattern of ink depositions. The company Somark Innovations claimed the development of this kind of tag and a reader that can read the tag up to 1.2 m [49]. To tag animals, permanent ink can be deposited on the dermal layer to create the tag, which can then be used for the whole life span of the animal. The tags are claimed to be readable on metal surfaces as well. Though this tag is printable, its cost is still high and there is no further reporting about the claims, which infers that the claims about tag data capacity and reading range are still in the experimental phase.

## 2.1.4   Hybrid Domain Chipless RFID Tags

In addition to the two main categories of chipless tags discussed above, it is possible to introduce another type of chipless RFID tag. A hybrid tag uses two or more features of frequency domain tags and of time domain tags to encode data on it. Therefore, these types of tags cannot be directly classified into either the frequency domain or time domain. Hence, in this book a new category is defined, the hybrid domain. The operation of a hybrid domain chipless RFID tag is described in Figure 2.7. Figures 2.7(a) and (b) show the conventional method of encoding data in the

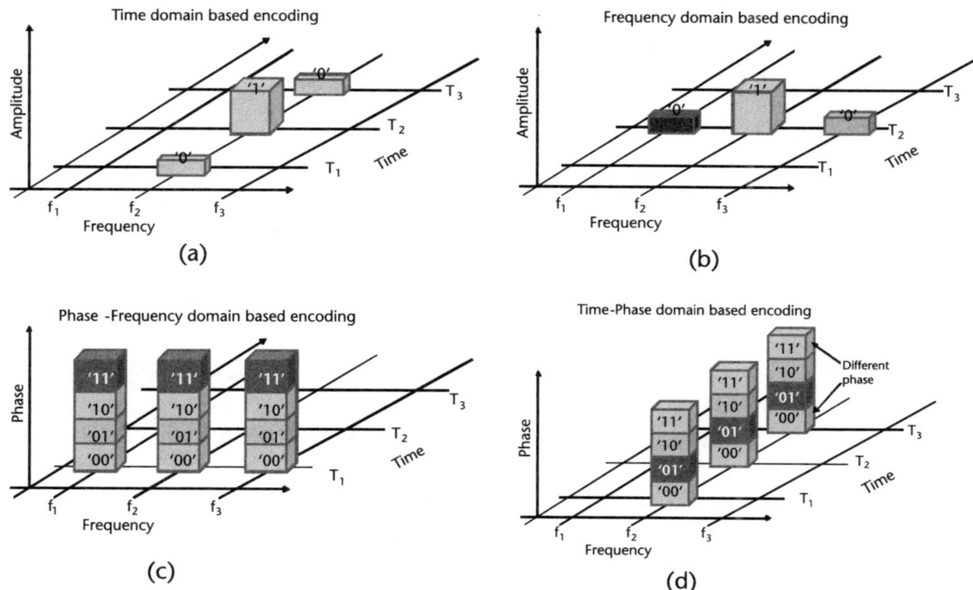


**Figure 2.7**   Illustration of data encoding methods used in the chipless tags discussed.

time domain and frequency domain, respectively. In time domain reflectometric tags, usually a short-duration RF pulse is used and echoes from the tag are captured in time. In frequency domain tags, the data are encoded using the spectral features and the data encoding does not depend on the time information. However, as illustrated in Figure 2.7(c), phase-frequency domain based encoding of data uses two features from the frequency domain—phase and frequency—to encode data. On the other hand, time-phase domain based encoding uses phase and time information from a signal to represent data. A tag that uses frequency-time domain encoding of data is reported on in [50]. The tag contains two sets of resonators that resonate at the same frequency. The length of the transmission lines coupled to the spiral resonators is different as shown in Figure 2.8. Therefore, the interrogation UWB pulse signal received from port 1 reaches the transmitting antennas connected to port 2 and 3 at different times. By analyzing the two pulses received at two different times at the reader, the data can be decoded. More detailed descriptions of hybrid domain tags and their reader are presented in Chapter 6.

### 2.1.5  Summary of the Review of Chipless RFID Tags

Based on our review of the different types of chipless RFID tags, they can be divided into two main categories: time domain and frequency domain. Time domain tags use the time information of a RF signal such as time of arrival of a reflected RF signal or position of a pulse in a stream of RF pulses. Spectral signature based chipless tags encode data using resonant structures. Based on the technique used in each tag, either the amplitude or phase of an input RF signal is modified to encode data. In some types of tags, both amplitude and phase are modified to encode data. The new term hybrid domain was introduced to categorize the tags that use both time and spectral information to encode a larger amount of compressed data compared

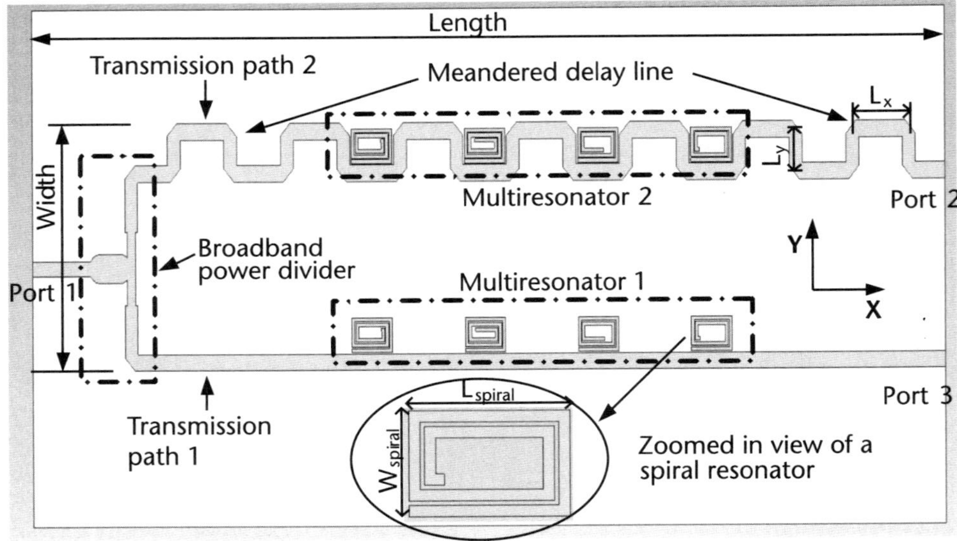



**Figure 2.8**  Layout of the ID generation circuit (length = 161 mm, width = 34 mm, Lx = 10 mm, Ly = 5.72 mm, Lspiral = 7.2 mm, Wspiral = 4.7 mm, substrate Taconic TLX-0 with $\varepsilon_r$ = 2.45, h = 0.787 mm, tan δ = 0.0019).

to that of their predecessors, time domain and frequency domain tags. Table 2.1 shows a comparison of the various chipless tags investigated in the open literature. The study reveals that chipless RFID tags may contain large data bits and small physical size as compared to conventional RFID tags. These promises with unique features will make the chipless RFID the frontier disruptive technology in the new millennium.

The next section presents a detailed description of the principle of operation of multiresonator based chipless RFID tags. Multiresonator based chipless RFID tags reported on in [29] can be implemented on low-grade flexible substrates such as paper or plastics. Further, the manufacturing of the tag can be done using conductive inks and existing printing technologies such as ink-jet and screen printing. Therefore, this chipless RFID tag shows the potential to achieve much lower manufacturing costs compared to other chipped and chipless RFID tags [1, 3].

## 2.2   Multiresonator Based Chipless RFID Tag

Multiresonator based chipless RFID tags are different from other chipped and chipless RFID tags in following ways:

**Table 2.1**   Comparison of Different Chipless RFID Tags

| | Division Based on | Types | Data Density (no. of bits/ area in $cm^2$) | Printability | 64/96-Bit Tag Possible within Credit Card Size? | Increment in Size with No. of Bits |
|---|---|---|---|---|---|---|
| Chipless RFID Tags | Time Domain | SAW | >1.0 bits/cm$^2$ | No | Yes | Yes |
| | | Delay line | 0.17 bit/cm$^2$ | Yes | No | Yes |
| | Frequency Domain | Dipole barcode | 1.01 bits/cm$^2$ | Yes | No | Yes |
| | | Multiband dipole | 0.81 bit/cm$^2$ | Yes | No | Yes |
| | | SRR array | 2.8 bits/cm$^2$ | Yes | No | Yes |
| | | Coplanar strip | 3.3 bits/cm$^2$ | Yes | No | NA |
| | | Letter ID | 4.6 bits/cm$^2$ | Yes | No | NA |
| | | MRDA | 0.16 bit/cm$^2$ | Yes (two sides) | No | Yes |
| | | Fractal | 0.03 bit/cm$^2$ | Yes (2 Side) | No | Yes |
| | | Slotted monopole | 1.0 bits/cm$^2$ | Yes (two sides) | No | No |
| | | Chemical | 1.76 bits/cm$^2$ | Yes | Yes | No |
| | | Space-filling curve | 0.55 bit/cm$^2$ | Yes | No | Yes |
| | | LC resonant | 1.0 bit/cm$^2$ | Yes | No | Yes |
| | | Spiral resonator | 0.61 bit/cm$^2$ | Yes | No | Yes |
| | | Elliptical dipole | 0.36 bit/cm$^2$ | Yes | No | No |
| | Phase | SLMPA | 0.18 bit/cm$^2$ | Yes | No | Yes |
| | | Multilayer patch | 0.61 bit/cm$^2$ | No | No | Yes |
| | | Complex impedance | >0.1 bit/cm$^2$ | Yes | No | Yes |
| | Image | SAR | >1.0 bits/cm$^2$ | Yes | Yes | Yes |
| | | Ink tattoo | >1.0 bits/cm$^2$ | Yes | Yes | No |

From Md. A. Islam, *Compact Printable Chipless RFID System Using Polarization Diversity*, unpublished confirmation to PhD Report, Electrical and Computer Systems Engineering Department, Monash University, Melbourne, Australia, October 2011.

1.  Operating frequency;
2.  Data encoding technique;
3.  Communication protocol used to communicate with the tag.

In terms of operating frequency, other active and passive chipped tags mostly work on the LF, HF, UHF, and ISM frequency bands and in a very narrow bandwidth [17]. In contrast, this chipless RFID system works in a very wide bandwidth, greater than 500 MHz [29]. In conventional chipped tags, amplitude shift keying (ASK) and binary phase shift keying (BPSK) are mostly used as modulation techniques and time domain reflectometry for transmitting and reception of data. On the other hand, this system uses amplitude attenuation and phase jumps in a continuous wave (CW) for encoding data. In terms of a communications protocol used to establish data communication between the tags and their reader, chipped tags use a handshake algorithm by responding to the reader's polling signal. However, this chipless RFID system is fully passive and no handshake algorithms are possible.

The tag is a planar microwave resonance structure consisting of spiral resonators, split-ring resonators, and so forth. They resonate at different frequencies of the operating frequency bandwidth. Each resonator is a bandstop filter. Therefore, the tag could be modeled as a series of bandstop filters connected in series with two cross-polarized UWB monopole antennas [29]. A block diagram and a photograph of a prototype of the tag are shown in Figures 2.9.

Spiral resonators are gap coupled to a 50 microstrip transmission line. A single spiral resonator coupled to a microstrip line can be modeled with a parallel RLC network. The equivalent circuit of the tag derived based on this model is shown in Figure 2.10. Each parallel RLC network generates a high-Q stopband at a designated frequency [51]. Therefore, $N$ spirals coupled to the microstrip line generate

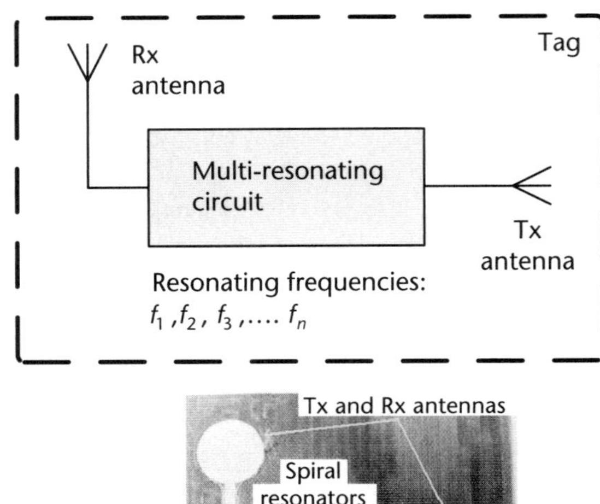


**Figure 2.9**  Schematic diagram of a multiresonator based chipless RFID tag and a photo of a tag made on Taconic TLX-0 ($\varepsilon_r$ = 2.45, tan $\delta$ = 0.0019, h = 0.7874 mm) substrate.

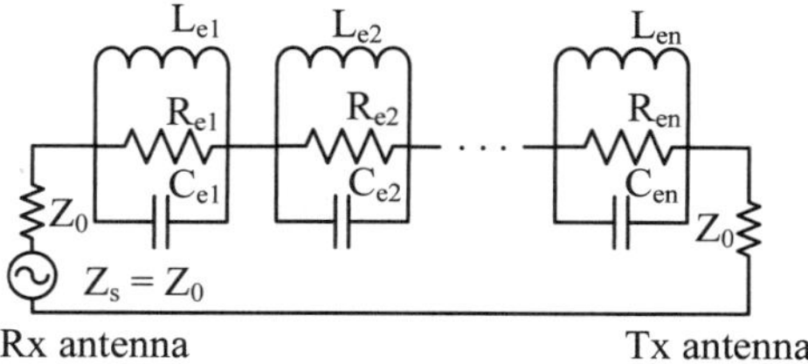

**Figure 2.10**   Equivalent circuit of a chipless RFID tag with transmitting and receiving antennas.

$N$ notching in the swept-frequency band [51–53]. Here, $L_e$ and $C_e$ are the total equivalent inductance and capacitance of the bandstop filters; $R_e$ is the equivalent resistive loss of the filter. The equivalent lumped components can be derived from the quasi-static analysis of the planar structure [54].

The resonance frequency of the generic $n$th RLC circuit () of the $N$-order multiresonator circuit can be expressed as:

$$f_{r_n} = \frac{1}{\sqrt{L_{e_n} C_{e_n}}}$$

(2.1)

If the response of the filter in the frequency domain is $H(f)$, then the amplitude response, $A(f)$ and phase response, $\phi(f)$ of an N-bit tag can be expressed respectively as:

$$A(f) = \prod_{n=1}^{N} |H_n(f)|$$

(2.2)

and

$$\phi(f) = \sum_{n=1}^{N} \angle H_n(f)$$

(2.3)

Using (2.2) and (2.3) it is possible to describe the frequency response of any multiresonator based chipless RFID tag.

### 2.2.1   Operating Principle for Reading of Multiresonator Based Chipless RFID Tags

Since the chipless tag is a fully passive microwave structure, an external source of electromagnetic signals is needed for the interrogation. When the tag is illuminated with a sufficiently wideband signal to cover all the frequencies of the multiresonating structure, it creates a spectral signature, which can be converted into a series of bits (1's and 0's). Figure 2.11 explains the operation of the tag. When the tag is illuminated with a wideband signal as shown in the figure, due to the captured signal by the receive (Rx) antenna of the tag, resonators start resonating at the designed frequencies and create attenuations in the amplitude. Further, it creates phase jumps at each resonating frequency since each resonator acts as a bandstop filter. This signal is then retransmitted back from the transmit (Tx) antenna of the tag. The presence of a resonator creates attenuation in amplitude of the frequency spectrum

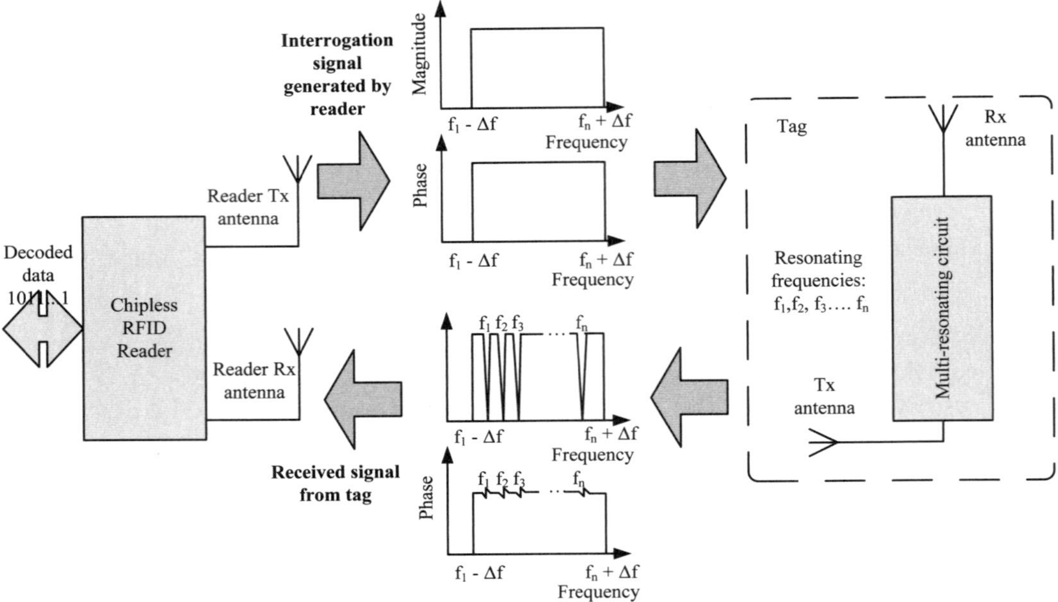


**Figure 2.11**   Operating principle of the multiresonator based chipless RFID system.

and creates phase jumps, whereas the absence of a resonator does not create either amplitude attenuation or phase jumps. Ultimately, the resonator circuit creates a frequency signature in the received signal from the Rx antenna of the tag. This signal with its frequency signature is transmitted back through the transmit antenna of the tag. The frequency signature can be controlled by varying the resonance frequencies and using different combinations of resonance frequencies. This property is converted into bits by using a convention such as this: The presence of amplitude attenuation and phase jump is bit 0 and the absence of amplitude or phase jump at the predetermined frequency is bit 1.

The techniques and components used for generating the wideband signal, receiving the signals from the tags, and decoding the data by identifying the frequency signature are described in Chapters 3 and 4.

### 2.2.1.1   Operation of a Multiresonator Circuit

Figure 2.12 shows the measured results for a the multiresonator circuit. An Agilent E8361A performance network analyzer (PNA) was used to measure the tags. The forward transmission coefficient (S21) of the multiresonator circuit is measured to identify the resonances. The forward transmission coefficient of the multiresonator circuit contains two parts, amplitude and phase. The resonance can be observed in both the amplitude and phase of the transmission coefficient measurements as shown in Figures 2.12(a) and (b). The measured results show nine dips in the amplitude response and nine deviations from the linear phase variation. Each dip and phase variation corresponds to the resonance that occurs in the multiresonator circuit. Based on the notation used earlier in this chapter, the tags contain 9 bits. The

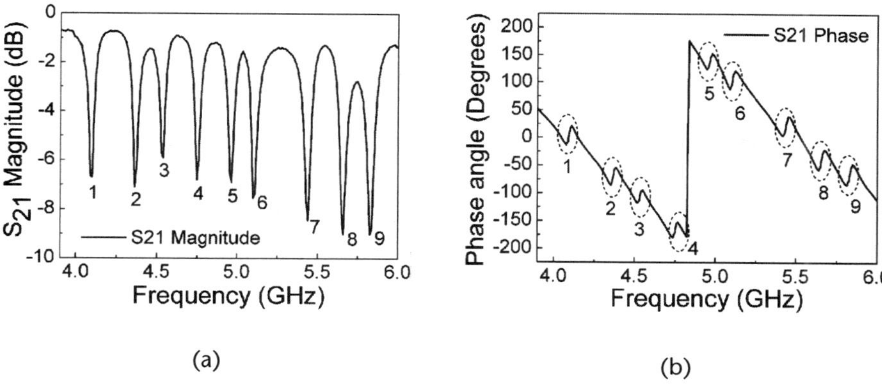

(a)                                                                              (b)

**Figure 2.12**   Measured (a) magnitude and (b) phase of the transmission coefficient of a 9-bit chipless RFID tag [55].

data embedded on the tags can be decoded as "000000000" and "100101110," respectively.

## 2.3   Methods for Reading RFID Tags

Many types of RFID readers are available for reading various tags with ASICs and chipless tags. Any RFID reader that is used to read chipped or chipless tags contains three major sections:

1. An antenna;
2. An RF section;
3. A control section.

A block diagram of an RFID reader is shown in Figure 2.13. One or more antennas are used to transmit and receive RF signals.

The RF section usually contains both the transmitter and receiver circuits in a section called the transceiver section. The outputs of the RF transceiver section are interfaced with the data and signal processing unit belonging to the control section. The processing of the outputs of the RF section and decoding data is done at the data and signal processing unit. The RF transceiver section is controlled by the control unit and usually implemented in the digital section of the reader. Even RFID tags with ASIC chips use many techniques such as ASK, BPSK, and FSK to

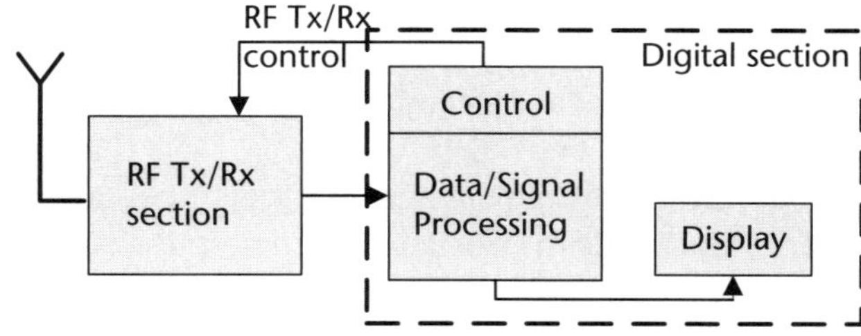


**Figure 2.13**   A block diagram of a typical RFID reader.

transmit data. Therefore, the RF section is customized to suit the communication technique used in the tag.

As discussed in this chapter, chipless RFID systems do not contain an RF transmitter or any active modulation scheme on the tag. Therefore, it is not possible to use conventional data communication techniques to extract data from chipless tags. Because many methods are being used to encode data in chipless tags, highly customized readers are required to extract the data from them. Although it would be preferable to use an off-the-shelf reader in chipless RFID systems, it is impossible due to the differences of the operating principles. The chipless readers are typically designed to read a particular type of tag and cannot be used with another type of tag. This section provides an overview of the reading processes of different types of chipless RFID tags.

### 2.3.1 Reading Time Domain Based Chipless RFID Tags

Time domain tags either use reflected echoes from the passive structure or a stream of reflected pulses from the tag to encode data. The time information of the received signals is important in these readers. Therefore, the reading process involves generation of the interrogation signal and recording of the incoming RF signal with accurate timing information. Typically high-speed sampling devices or envelope detectors are used based on the requirement of the signal processing technique. As an example, if the chipless tag uses the timing information of the reflected echo from the tag as mentioned in [56], a high-speed sampling device is needed for the reader. The sampled signal is analyzed and processed to recover data in the digital section. In a laboratory a high-speed sampling oscilloscope or some other high-speed sampler can be used [56].

### 2.3.2 Reading Frequency Domain Based Chipless RFID Tags

Frequency domain based chipless RFID systems require the measurement or identification of the frequency response of the tag. For most of the frequency domain chipless RFID tags, a similar process of characterization of the tag in the frequency domain is needed to extract the data. Therefore, the reader should be capable of generating and receiving the signals with sufficient bandwidth to cover the operating frequency range of the tag. Typically the operating range is several gigahertz in most of the frequency domain chipless RFID tags [3, 29]. Identification of the amplitude and phase changes between the transmitted and received RF signal is the key operating principle for most of the chipless tags. Therefore, most of the time, the timing information of the RF signals is not critical as long as they provide accurate frequency and sufficient bandwidth. RF imaging based identification also can be considered a frequency domain based reading method. The SAR based chipless RFID tag uses a small radar system and signal processing method to identify the data encoded in the tag [46, 47].

Typical laboratory equipment that is used to characterize the RF components in the frequency domain can be used to read chipless RFID tags. A vector network analyzer is a good example of such laboratory equipment that can be used to read some of the chipless RFID tags. The PNA works as both the signal source and the receiver and displays the frequency response of the tag [29]. In addition to that, using a signal analyzer and a signal generator, the amplitude response of the chipless RFID tag can be measured. Use of laboratory equipment to read the chipless RFID tags is discussed in Chapters 3 and 4.

### 2.3.3   Reading Hybrid Domain Based Chipless RFID Tags

Hybrid domain based chipless RFID tag reading requires accurate generation and identification of timing information of the transmitted and received signals. In addition to the recording of timing information, the reader should be capable of analyzing the signal in the frequency domain to identify the spectral features of the incoming RF signal. Combining time and frequency domains increases the data carrying capability of the chipless tag. However, it also increases the complexity and the cost of the reader. Figure 2.14 shows a proposed reading method for a hybrid domain chipless tag. A UWB RF pulse is transmitted to the receive antenna of the chipless tag. The two outputs of the tag are transmitted back to the reader and it records the two pulses in the time domain by using a high-speed sampler. Then, with a digital signal processing technique, the frequency spectrum is obtained to identify the resonances introduced by the ID generation circuit [50]. Together with the identified resonances, the timing information in the recorded signal can be used to identify the encoded data bits in the tag.

In a laboratory a high-speed sampling oscilloscope can be used to record the signals and signal processing software used to analyze the frequency response.

### 2.3.4   SAR Based Reading Process

SAR principles require the relative motion between the antenna and the target. Figure 2.15 shows such a reading technique [46]. Usually the antenna is mounted on a moving platform and by using a single beam-forming antenna the target is illuminated repeatedly with short bursts of RF waves. Alternately, a semicircular grid switched beam antenna array can be used to avoid the physical movement of

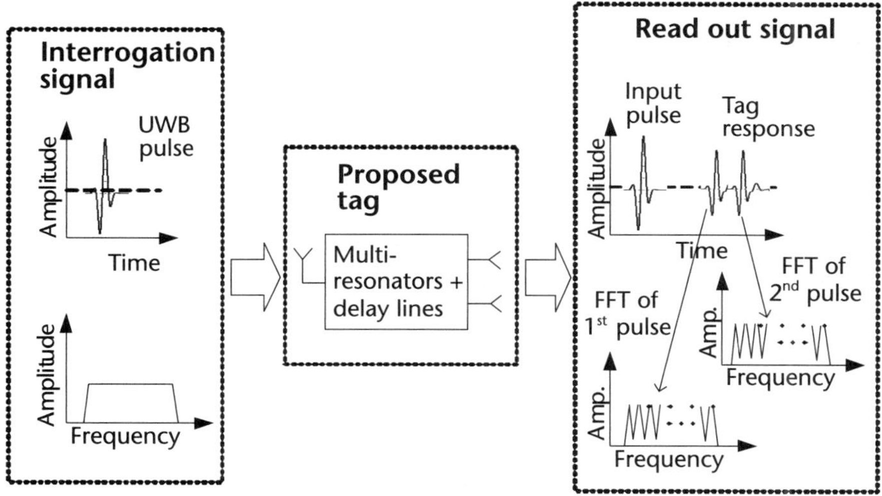


**Figure 2.14**   The reading of a hybrid domain chipless RFID tag.

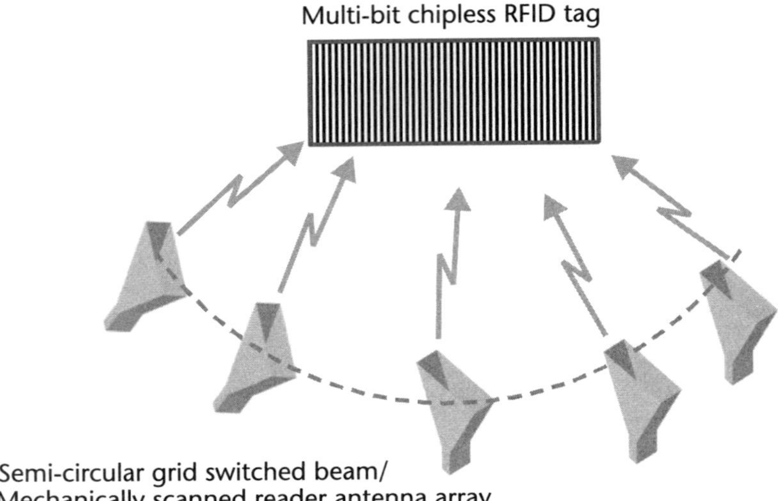


**Figure 2.15**   The reading of a SAR based chipless RFID tag [46].

the reader antenna. Detailed technical descriptions of switched beam antennas can be found in [57–61].

The reflected signals are received successively at different antenna positions and coherently detected and processed together to form an image of the illuminated target. The working principle of the SAR based chipless RFID tags is very similar to the conventional SAR implementation in microwave imaging systems. Sometimes a switching set of antennas is used instead of the relative movement to scan the full surface of the SAR based chipless tags. Then the received signals are processed to obtain an image of the tag, which is converted into a stream of bits using a signal processing algorithm. The resolution of the image depends on the frequency of the RF signal used to illuminate the tag. Millimeter-wave RF signal provides resolution of a few millimeters. Therefore, a high data capacity is possible with smaller size tags.

## 2.4   Conclusion

This chapter presented a comprehensive review of chipless RFID tags. Time domain and frequency domain tags were introduced as the two main categories of chipless tags. Time domain based tags use the timing information of the RF signals to encode data. Frequency domain tags use spectral features to encode data. The new term hybrid domain was introduced to categorize the tags that use both time domain and frequency domain principles to encode data. Detailed descriptions of the operating principle of multiresonator based chipless RFID tags act as the foundation for Chapters 3 through 6 of this book. In this chapter a brief introduction to various reading methods has been provided for different types of chipless tags. Whatever techniques are implemented in the chipless RFID systems, the common goal is to increase the data capacity, compactness, and compatibility with existing systems and, finally, to achieve sub-cent tag prices.

## Questions

1. What is the main motivation for driving research on chipless RFID systems?
2. What does *chipless RFID* mean?
3. What are the main categories of chipless RFID tags?
4. How can a time domain based chipless tag be interrogated?
5. What are the recent developments in SAW tags?
6. How does a TDR tag work?
7. What is the main drawback of left-handed negative refraction index transmission line sections for TDR tags?
8. How are data encoded into a frequency domain tag?
9. What are the main types of image based chipless RFID tags?
10. What can be considered as a hybrid domain chipless RFID tag?
11. How is a multiresonator based chipless RFID tag different from other tags?
12. How can a multiresonator based chipless RFID tag be modeled using a lumped component equivalent circuit?
13. What is the basic operating principle of a multiresonator based chipless RFID tag?
14. What are the main components in an RFID reader system?

## References

[1]   S. A. Ahson and M. Ilyas, *RFID Handbook: Applications, Technology, Security, and Privacy*, Boca Raton, FL: CRC Press, 2008.

[2]   C. S. Hartmann, "A Global SAW ID Tag with Large Data Capacity," in *Proc. IEEE Ultrasonics Symposium*, Munich, Germany, October 2002, pp. 65–69.

[3]   S. Preradovic, N. Karmakar, and I. Balbin, "RFID Transponders," *IEEE Microwave Mag.*, Vol. 9, No. 5, pp. 90–103, October 2008.

[4]   R. Das, *Chip-less RFID—The End Game*, IDTechEx, http://www.idtechex.com/products/en/articles/00000435.asp (February 2006).

[5]   J. R. Baker, H. W. Li, and D. E. Boyce, "CMOS Circuit Design, Layout, and Simulation, 2nd Edition," *IEEE Circuits and Devices Magazine*, Vol. 22, No. 3, p. 37, 2006.

[6]   J. McVay, A. Hoorfar, and N. Engheta, "Theory and Experiments on Peano and Hilbert Curve RFID Tags," *Proc. SPIE*, Vol. 6248, No. 1, 2006, pp. 624808-1-1.

[7]   S. Harma et al., "Inline SAW RFID Tag Using Time Position and Phase Encoding," *IEEE Trans. on Ultrasonics Ferroelectrics and Frequency Control*, Vol. 55, pp. 1840–1846, August 2008.

[8]   V. Plessky and L. Reindl, "Review on SAW RFID Tags," *IEEE Trans. on Ultrasonics, Ferroelectrics and Frequency Control*, Vol. 57, pp. 654–668, 2010.

[9]   S. Härma, *Surface Acoustic Wave RFID Tags: Ideas, Developments, and Experiments*, PhD thesis, http://lib.tkk.fi/Diss/2009/isbn9789512297436/isbn9789512297436.pdf.

[10]  C. Hartmann et al., "Anti-Collision Methods for Global SAW RFID Tag Systems," in *2004 IEEE Ultrasonics Symposium*, 2004, Vols. 1–3, pp. 805–808.

[11]  T. Zhijun and H. Yigang, "Research of Multi-Access and Anti-Collision Protocols in RFID Systems," in *2007 IEEE International Workshop on Anti-counterfeiting, Security, Identification*, 2007, pp. 377–380.

[12]   B. Shao, Y. Amin, Q. Chen, R. Liv and R. Zheng, "Design of fully printable and configurable chipless RFID tag on flexible substrate," *Mircowave and Optical Technology Letters,* Vol. 54, Issue 1, Jan. 2012, pp. 226–230.

[13]   L. M. Reindl and I. M. Shrena, "Wireless Measurement of Temperature Using Surface Acoustic Waves Sensors," *IEEE Trans. on Ultrasonics, Ferroelectrics and Frequency Control,* Vol. 51, pp. 1457–1463, 2004.

[14]   D. C. Malocha, D. Puccio, and D. Gallagher, "Orthogonal Frequency Coding for SAW Device Applications," in *Proc. 2004 IEEE Ultrasonics Symposium,* Vol. 2, pp. 1082–1085.

[15]   A. Stelzer et al., "Multi Reader/Multi-Tag SAW RFID Systems Combining Tagging, Sensing, and Ranging for Industrial Applications," in *2008 IEEE International Frequency Control Symposium,* Vols. 1 and 2, pp. 263–272, 2008.

[16]   B. Wolf, FYP thesis, Dept. Electrical and Computer Systems. Eng., Monash University, Melbourne, Australia, 2011.

[17]   D. M. Dobkin, *The RF in RFID: Passive UHF RFID in Practice,* Boston: Newnes, 2007.

[18]   N. C. Karmakar et al., "Development of Low-Cost Active RFID Tag at 2.4 GHz," in *36th European Microwave Conference,* 2006, pp. 1602–1605.

[19]   I. Jalaly and I. D. Robertson, "Capacitively-Tuned Split Microstrip Resonators for RFID Barcodes," in *European Microwave Conference (EuMC 2005),* Paris, France, 2005, pp. 1–4.

[20]   V. Deepu et al., "New RF Identification Technology for Secure Applications," in *2010 IEEE International Conference on RFID-Technology and Applications (RFID-TA),* 2010, pp. 159–163.

[21]   A. Blischak and M. Manteghi, "Pole Residue Techniques for Chipless RFID Detection," in *IEEE Antennas and Propagation Society International Symposium (APSURSI'09),* June 2009, pp. 1–4.

[22]   K. Finkenzeller, *RFID Handbook: Fundamentals and Applications in Contactless Smart Cards and Identification,* 2nd ed. Chichester: John Wiley Sons Ltd., 2003.

[23]   J. Hyeong-Seok et al., "Design of low-cost chipless system using printable chipless tag with electromagnetic code," *IEEE Microwave and Wireless Components Letters,* Vol. 20, pp. 640–642, 2010.

[24]   A. Vena, E. Perret, and S. Tedjini, "Novel Compact RFID Chipless Tag," in *Progress In Electromagnetics Research Symposium Proceedings,* Marrakesh, Morocco, 2011, pp. 1062–1066.

[25]   T. Singh et al., "A Frequency Signature Based Method for the RF Identification of Letters," in *2011 IEEE International Conference on RFID,* 2011, pp. 1–5.

[26]   S. Preradovic et al., "Multiresonator-Based Chipless RFID System for Low-Cost Item Tracking," *IEEE Trans. Microw. Theory Tech.,* Vol. 57, No. 5, pp. 1411–1419, May 2009.

[27]   S. Preradovic, I. Balbin, and N. Karmakar, "The Development and Design of a Novel Chipless RFID System for Low-Cost Item Tracking," in *Asia-Pacific 2008 Microwave Conference,* pp. 1–4.

[28]   S. Preradovic et al., "A Novel Chipless RFID System Based on Planar Multiresonators for Barcode Replacement," in *2008 IEEE International Conference on RFID,* pp. 289–296.

[29]   S. Preradovic et al., "Chipless Frequency Signature Based RFID transponders," in *38th European Microwave Conference (EuMC 2008),* pp. 1723–1726.

[30]   S. Preradovic et al., "Multiresonator-Based Chipless RFID System for Low-Cost Item Tracking," *IEEE Trans. on Microwave Theory and Techniques,* Vol. 57, pp. 1411–1419, 2009.

[31]   S. Preradovic and N. Karmakar, "Chipless RFID Tag with Integrated Sensor," in *2010 IEEE Sensors,* pp. 1277–1281.

[32]   S. Preradovic and N. Karmakar, "4th Generation Multiresonator-Based Chipless RFID Tag Utilizing Spiral EBGs," in *2010 European, Wireless Technology Conference (EuWIT),* pp. 269–272.

[33]  S. Preradovic, N. Karmakar, and I. Balbin, "RFID Transponders," *IEEE Microwave Magazine,* Vol. 9, pp. 90–103, 2008.

[34]  S. Preradovic, N. Karmakar, and M. Zenere, "UWB Chipless Tag RFID Reader Design," in *2010 IEEE International Conference on RFID—Technology and Applications (RFID-TA),* pp. 257–262.

[35]  S. Preradovic and N. C. Karmakar, "Design of Fully Printable Chipless RFID Tag on Flexible Substrate for Secure Banknote Applications," in *ASID 3rd International Conference on Anti-counterfeiting, Security, and Identification in Communication,* pp. 206–210.

[36]  S. Preradovic and N. C. Karmakar, "Design of Chipless RFID Tag for Operation on Flexible Laminates," *IEEE Antennas and Wireless Propagation Letters,* Vol. 9, pp. 207–210, 2010.

[37]  S. Preradovic and N. C. Karmakar, "Design of Short Range Chipless RFID reader Prototype," in *5th International Conference on Intelligent Sensors, Sensor Networks and Information Processing (ISSNIP),* pp. 307–312.

[38]  S. Preradovic and N. C. Karmakar, "Chipless RFID: Bar code of the Future," *IEEE Microwave Magazine,* Vol. 11, pp. 87–97, 2010.

[39]  I. Balbin and N. Karmakar, "Novel Chipless RFID Tag for Conveyor Belt Tracking Using Multi-Resonant Dipole Antenna," in *European Microwave Conference (EuMC 2009),* pp. 1109–1112.

[40]  S. Preradovic et al., "Chipless Frequency Signature Based RFID Transponders," in *European Conference on Wireless Technology (EuWiT 2008),* pp. 302–305.

[41]  I. Balbin and N. Karmakar, "Radio Frequency Transponder System," Australian Provisional Patent, DCC Ref: 30684143/DBW, October 20, 2008.

[42]  I. Balbin and N. C. Karmakar, "Phase-Encoded Chipless RFID Transponder For Large-Scale Low-Cost Applications," *IEEE Microwave and Wireless Components Letters,* Vol. 19, pp. 509–511, 2009.

[43]  S. Mukherjee and G. Chakraborty, "Chipless RFID Using Stacked Multilayer Patches," in *2009 Applied Electromagnetics Conference (AEMC),* pp. 1–4.

[44]  S. Mukherjee, "Chipless Radio Frequency Identification by Remote Measurement of Complex Impedance," in *2007 European Conference on Wireless Technologies,* pp. 249–252

[45]  G. Chakraborty, S. Mukherjee, and K. Chiba, "Synthesis of Passive RFID from Backscatter Using Soft-Computing Techniques," in *Second International Conference on Emerging Applications of Information Technology (EAIT),* pp. 325–328.

[46]  "SAR code identification," InkSure, http://www.inksure.com/images/stories/presentations/ SARCodeIntroduction%202.09.pdf (accessed March 6, 2011).

[47]  R. M. Mays and A. M. Grishin, "Microwave Readable Dielectric Barcode," US Patent 20060125491, June 15, 2006.

[48]  A. M. Grishin and R. M. Mays, "Bar Code Interrogation System," US Patent 7221168, May 22, 2007.

[49]  S. Preradovic and N. Karmakar, "Fully Printable Chipless RFID Tag," in *Advanced Radio Frequency Identification Design and Applications,* S. Preradovic (Ed.), http://www.intechopen.com/articles/show/title/fully-printable-chipless-rfid-tag.

[50]  M. S. Bhuiyan, R. Azim, and N. Karmakar, "A Novel Frequency Reuse Based ID Generation Circuit for Chipless RFID Applications," in *2011 Asia-Pacific Microwave Conference Proceedings (APMC),* pp. 1470–1473.

[51]  H. Lim et al., "A Novel Compact Microstrip Bandstop Filter Based on Spiral Resonators," in *Asia-Pacific Microwave Conference (APMC 2007),* Bangkok, Thailand, 2007, pp. 1–4.

[52]  K. Yoon et al., "Design of a High-Q Resonator for Satellite Broadcasting Application," in *IEEE Antennas and Propagation Society International Symposium (AP-S 2008),* San Diego, CA, 2008, pp. 1–4.

[53]  Y. Lee et al., "A compact-Size Microstrip Spiral Resonator and Its Application to Microwave Oscillator," *IEEE Microwave and Wireless Components Letters,* Vol. 12, No. 10, pp. 375–377, December 2002.

[54]    N. C. Karmakar, S. M. Roy, and I. Balbin, "Quasi-Static Modeling of Defected Ground Structure," *IEEE Trans. on Microwave Theory and Techniques*, Vol. 54, No. 5, pp. 2160–2168, May 2006.

[55]    R. V. Koswatta and N. C. Karmakar, "Time Domain Response of a UWB Dipole Array for Impulse Based Chipless RFID Reader," in *2011 Asia-Pacific Microwave Conference Proceedings (APMC)*, pp. 1858–1861.

[56]    A. Lazaro et al., "Chipless UWB RFID Tag Detection Using Continuous Wavelet Transform," *IEEE Antennas and Wireless Propagation Letters*, Vol. 10, pp. 520–523, May 2011.

[57]    N. C. Karmakar and M. E. Bialkowski, "A Beam-Forming Network for a Circular Switched-Beam Phased Array Antenna," *IEEE Microwave and Guided Wave Letters*, Vol. 11, No.1, pp. 1–3, January 2001.

[58]    N. C. Karmakar And M. E. Bialkowski, "A Compact Switched-Beam Array Antenna For Mobile satellite communications," *Microwave and Optical Technology Letters*, Vol. 21, No. 3, pp. 186–191, May 5, 1999.

[59]    N. C. Karmakar and M. E. Bialkowski, "An 8-Element Switched Beam Array for Mobile Satellite Communications," in *Proc. TENCON'98*, Delhi, India, December 17–19, 1998, pp. 241–244.

[60]    N. C. Karmakar and M. E. Bialkowski, "Design and Development of Low Cost Components and Sub-System for L-Band Switched Beam and Phased Array Antennas," in *Proc. IEEE APCC'98/ICCS'98*, Singapore, November 23–27, 1998, pp. 423–427.

[61]    N. C. Karmakar and M. E. Bialkowski, "A Low Cost Switched Beam Array Antenna for L-Band Land Mobile Satellite Communications in Australia," in *Digest of the 1997 IEEE AP-S International Symposium*, Montreal, Canada, July 13–18, 1997, pp. 2226–2229.

## CHAPTER 3

# Chipless RFID Readers

## 3.1 Introduction to Chipless RFID Readers

The preceding chapter explained the three main types of chipless RFID tags: frequency, time, and hybrid domain based tags and their operating principles. This chapter presents various types of reader architectures and the operating principles used to successfully read those tags. The main theme of this book is chipless RFID tag reader architectures. Therefore, this chapter lays the foundation for the whole theme of the book. After reading the first three chapters, the readers of the book will have learned background information and the fundamental principles of chipless RFID technology on the whole. These three chapters present the state of the art of the chipless RFID systems. Then the reader will gain more specific information on technical specifications, design details, and niche areas of chipless RFID reader designs in later chapters.

A *reader* is the device that performs the interrogation of any RFID system. In technical nomenclature, it is sometimes called the *interrogator*. In chipped RFID systems, the architecture of the reader is almost the same even for different types of tags, because common data communication techniques are used for transferring data between the tag and its reader [1]. However, in chipless RFID systems, due to the differences discussed in Chapter 2, the architectures of the readers need to be different. The signal processing techniques that are used in the readers also need to be unique based on the technique used in the tag to encode data.

As discussed in Chapter 2 chipless RFID tags are passive devices and have no ability to generate their own signal for communication purposes. Without a signal generated by an external source, chipless tags cannot send information to the reader. Once the tag is illuminated with a suitable signal generated by a reader, the received signal from the tag is captured. Then it is processed in the software (sometimes it is called the *middleware*) and converted into useful information typically as data bits [2].

This chapter provides information on different types of chipless RFID reader systems with a comprehensive review of commercially available and reported reader systems in the open literature. Also, this chapter provides a foundation for understanding the different principles and techniques used in chipless RFID tag interrogation in the rest of the book.

Next we take a look at the details of chipless RFID reader system architectures.

## 3.2  Chipless RFID Reader System Architectures

RFID reader technology is a combination of several disciplines such as microwave or RF engineering, digital systems design, antenna engineering, digital signal processing, and computer engineering. To understand the architecture of an RFID reader, a basic understanding of the technologies and disciplines used in the design is required due to the multidisciplinary nature of RFID technology.

As discussed in Chapter 2 three major subsystems can be identified in any chipped or chipless RFID system, although there are many differences between these systems. They are:

- Antenna;
- RF section;
- Digital control and data processing section.

Figure 2.13 shows a block diagram of a general RFID reader system that illustrates the above three units. Although there are three common subsystems in every RFID system, the architecture of each subsystem is different. Due to these differences, the controlling and signal processing algorithms are different from one reader system to another.

The following section provides an overview of a chipless RFID reader system architecture and also details for each subsystem.

### 3.2.1  General Overview of Chipless RFID Reader Architecture

The architecture of a chipless RFID reader is described using an existing chipless RFID reader design. Figure 3.1 shows the design of a chipless RFID reader used for multiresonator based chipless RFID tag reading [3, 4]. This reader design is used only as an example to introduce the functionality and the typical components used to build up each subsystem. It is important to note again that the architecture and the operation of the reader are different for different types of chipless RFID tags.

#### 3.2.1.1  Antennas

Antennas work as the sensory organs ( eyes, ears, and mouth) of the RFID system. All interactions between the reader and the tags pass through the antennas in the form of electromagnetic (EM) waves. The system can have a single or multiple antennas depending on the requirements of the application and technology used in the RFID tags. Irrespective of the number of antennas, the basic function of an antenna is to transmit and receive EM waves to establish communication between the tag and the reader.

The function of antennas can be understood by comparing the communication of a human body. The antennas are illuminating an area of the surrounding space, similar to looking in one direction with the eyes to pick up one or a few objects

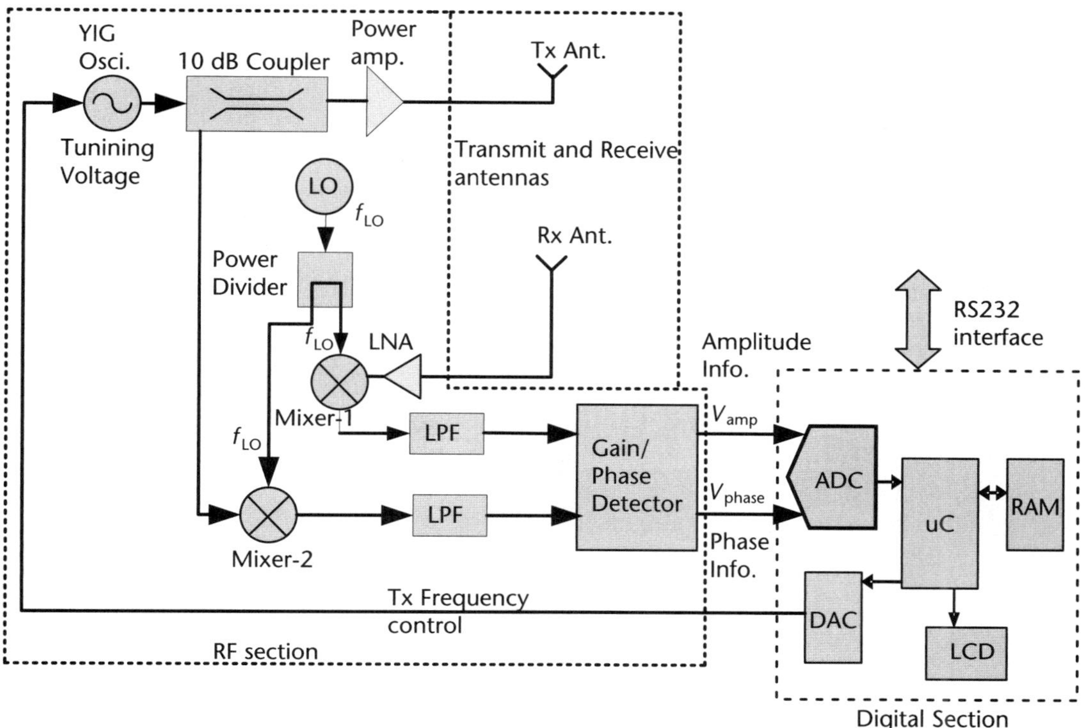


**Figure 3.1**   A complete chipless RFID reader design for reading multiresonator based chipless RFID tags [3].

from a large set. This process is also called *spatial filtering*. Spatial filtering reduces the collision of communication with tags [4]. Similar to hearing sounds with ears, antennas "listen" to EM waves and this process is called *reception of EM waves*. The speaking (function of the mouth) of a communication process can be identified as transmitting EM waves toward a certain direction. Chapter 7 discusses various antennas for RFID reader systems.

The use of smart antennas increases the capabilities of the antennas and enhances the performance of the RFID systems. A comprehensive description on the use of smart antennas in RFID systems is presented in another publication by the author [5].

### 3.2.1.2   RF Section

The RF section is responsible for generating the interrogation signal and receiving and demodulating the information carrying signal from the tag. The RF section is an interconnected collection of different RF components. The typical components are an RF signal source such as an oscillator; frequency conversion mixers, power amplifiers, low-noise amplifiers, lowpass/bandpass filters; and power dividers. Depending on the technology used in the reader architecture, there may be components such as a gain/phase detector (GPD) as shown in Figure 3.1 and power detectors, which are not common in other RF receiver or RFID reader units.

There are two different signal paths in the RF section: the receiver path and the transmitter path. The transmitter path usually contains strong RF signals compared to the receiver path. The signal source or sources generate the RF signal to be transmitted using an oscillator. It is a usual practice to use a VCO and tune it to obtain the desired frequency in wideband or multiband RF systems [6]. For narrowband systems, oscillators that are tuned to a fixed frequency are used. The type of oscillator or signal source basically depends on the operating frequency and the desired frequency band of operation. In the design shown in Figure 3.1 a YIG oscillator [6] is used and the operating frequency is 3 to 6 GHz. The output frequency is controlled using a voltage input as shown in the figure.

There might be a coupler to extract a small portion of the transmitted signal, as a reference signal to be used in the demodulation process in the receiver path. The presented reader design as shown in Figure 3.1 has an asymmetric coupled line coupler that extracts a small portion of the transmitted signal, also called a pilot signal or a reference signal. There is an RF power amplifier for amplifying the RF output of the oscillator if the required power level is not generated from the oscillator. It is important to be aware of the input power level to the power amplifier; otherwise, higher power level inputs might drive the RF power amplifier into saturation.

Received signals are weak in comparison to the transmitted signals and require amplification. LNAs are used to amplify such weak signals in the received path as shown in Figure 3.1. The two mixers in the reader design down-convert the incoming RF signal and reference signal into an intermediate frequency (IF) signal for demodulation. In this design only the frequency down-conversion is done. However in some reader architectures, a frequency up-conversion might happen using mixers. Lowpass filters (LPFs) or sometimes bandpass filters (BPFs) are used to filter out the unwanted frequencies created during the frequency up-conversion or down-conversion processes. Sometimes, a bandpass filter is applied immediately after the receiver antenna in the receiver path.

The gain/phase detector (GPD) is a specialized RF component that compares the amplitude and phase of two RF signals [7]. In this design, a portion of the transmitted signal and the received signals is down-converted and compared with the GPD to identify the features of the frequency signature. More information on the operation of the GPD and the reader is presented in Chapter 4. The information generated by the GPD output is acquired by the digital section for further processing.

This reader design example is presented to help this book's readers understand the general overview of the reader architecture as well as to emphasize the use of unique features and components that are not commonly used in typical RFID readers.

### 3.2.1.3  Digital Section

A block diagram of the digital section is shown in Figure 3.1. The control of the reader is implemented on the middleware of the digital section. The control of the transmission of the interrogation signal, acquiring the down-converted or demodulated signals from the RF section, and, if applicable, any additional signal processing algorithms or anticollision algorithms are also implemented in this section. The primary function of the digital section of the RFID reader is to perform digital signal processing and recover the received data from the RFID transponder. Also, the control section enables the reader to communicate with the transponders wirelessly by performing modulation and decoding the received data from the tags. These data are usually used to interrogate tags (read) or to reprogram the tag (write) in case of chipped tags. However, there are no reported chipless tags, which have a wireless reconfiguration capability. Therefore, for chipless RFID tags it can be said that the data are only used in reading the tag. This section usually consists of a microprocessor, a memory block, a few analog-to-digital converters, and a communication block for the software application. The typical digital section is based on a microprocessor or a field programmable gate array (FPGA). Chapter 9 presents the digital control section of the chipless RFID reader.

## 3.3   Chipless RFID Readers and Tag Reading Techniques

In Chapter 2, three major categories of chipless RFID tags were identified and their operation and techniques used in encoding data without using a chip were discussed. The identified three types of tags are:

- Time domain (TD) chipless tags;
- Frequency domain chipless tags;
- Hybrid domain chipless tags.

A short description of the techniques used for reading the tag ID or encoded data on the tags was given at the end of Chapter 2. To read the three types of chipless RFID tags, four types of reading techniques were also introduced in the last section of Chapter 2. Based on the four types of techniques used to read the tags, chipless RFID readers can be categorized into four types:

1. Time domain chipless RFID readers;
2. Frequency domain chipless RFID readers;
3. Hybrid domain chipless RFID readers;
4. SAR based chipless RFID readers.

This section provides a review of reader architectures and reading methods used in commercially available chipless RFID systems and reported chipless RFID systems in the open scientific literature. More focus is being given to the reader architecture and operating principle of the reader system since the operating principles of tags are discussed in Chapter 2. The construction of the reader system and the details of the functional blocks of the reader are explained in detail. The block diagrams of the reader systems discussed in this section are also provided whenever possible.

Since the chipless tag is a fully passive device, it is required to have two common features in all four types of chipless RFID readers. They are:

1. The generation of an interrogation signal;
2. All the signal processing related to tag ID extraction.

Therefore, all RFID readers are required to have a unit to generate a suitable interrogation signal and a unit to process the received signal and decode the tag ID.

### 3.3.1    Time Domain Readers and Tag Reading Techniques

Time domain based chipless RFID tags use reflected echoes as a stream of pulses reflected from the tag to encode the data. The tag reading process involves generating an interrogation signal, capturing the signals received from the tag, processing the signals, and decoding the tag ID or data bits encoded in the tag. This section describes few selected chipless RFID readers that work on time domain based operating principles and the different methods used to interrogate and decode the data encoded in the tags.

#### 3.3.1.1    SAW Time Domain Reader

A SAW based RFID tag is one of the successful commercial chipless tags [2]. The operating principles of SAW tags are described in detail in Chapter 2. The SAW tag is categorized as a time domain based chipless tag and uses short RF pulses as interrogation signals [8]. However, frequency domain based SAW tag reading processes have also been reported [9, 10]. Figure 3.2 shows a block diagram of a reader that works on the time domain based principle.

As shown in Figure 3.2, an RF oscillator is used as the RF signal source. Using rapid HF switches, a short RF pulse is generated. The RF signal (pulse) is amplified with a power amplifier if the strength of the output of the RF signal source is not sufficient. The international scientific and medical (ISM) frequency band (typically 2.45 GHz) is used in most of the SAW tags since the unlicensed usage of the frequency band is possible. The bandwidth of the signal is usually kept around 100 MHz. The pulse created using rapid HF switches is applied to the antenna by placing the RF switch in the proper position. If a SAW chipless tag is in the interrogation range of the reader, the transmitted short pulse is received by the antenna of the tag.

The tag converts the short pulse to a surface wave [8], and the reflective structures of the tag start reflecting the generated surface wave. Once the reflected pulses reach the reader antennas of the SAW tag, they are retransmitted with the tag antenna elements. The usual return time of the reflected pulses from the tag is a few microseconds. Due to this comparatively long propagation delay compared to other electromagnetic reflections caused by the surrounding environment, it is easy to separate the reflections of the signals from the surrounding objects. The received signals are passed through the receiver path of the reader by choosing it with the HF switch (also called a duplex switch). The signal is amplified with an LNA and demodulated with a quadrature demodulator [11]. The decoded in-phase (I) and

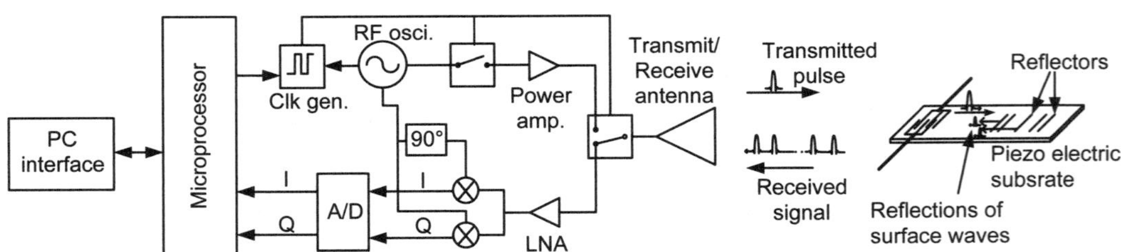


**Figure 3.2** SAW time domain chipless RFID reader. It detects the time delay or phase delay between the two pulses received from the tag [11].

quadrature-phase (Q) components are sampled with an analog-to-digital converter (ADC). The two orthogonal components of the received signals can be used to identify the time delay between the received pulses or the phase angle between those. The time delays are then converted into a sequence of bits according to the protocol used to encode data. Usually a group of bits is represented with a number of pulses (e.g., 16 bits are represented with four pulses [12]) and the data bits are represented according to the position of the pulse in a specific time window. A photo of a RFSAW reader system and an antenna is shown in Figure 3.3.

### 3.3.1.2   UWB-IR Time Domain Chipless RFID Readers

Ultra-wideband impulse radio (UWB-IR) systems have a distinguishing feature such as extremely large signal bandwidth. This feature enables a number of properties unique to UWB-IR systems: high data rate with low transmitting power, good signal penetration, little fading of the signal amplitude at the receiver, and the ability for precise localization. Much research has been conducted on data communication, medical applications, sensors, radar, and positioning.

More recently, increasing interest is being paid to UWB technology based chipless RFID systems. This section provides an introduction to chipless RFID reader systems recently reported about in the scientific literature [13–18]. To the authors' best of knowledge, there are no reported UWB-IR based commercial chipless RFID systems as yet.

Based on published work, the block diagram of a UWB-IR based reader is shown in Figure 3.4. There is a UWB impulse generator that generates a short RF pulse. The width of the RF pulse is in the range of a few hundred picoseconds to a few nanoseconds [16, 17] based on the required bandwidth. The chipless RFID tag may operate on different principles and most of the published works are based on time domain reflectometry [14, 16, 17]. The receiver block contains a high-speed sampling device. The experimental work has been carried out using either a PC based high-speed sampler [16] or a high-speed digital storage oscilloscope (DSO) [13, 17].

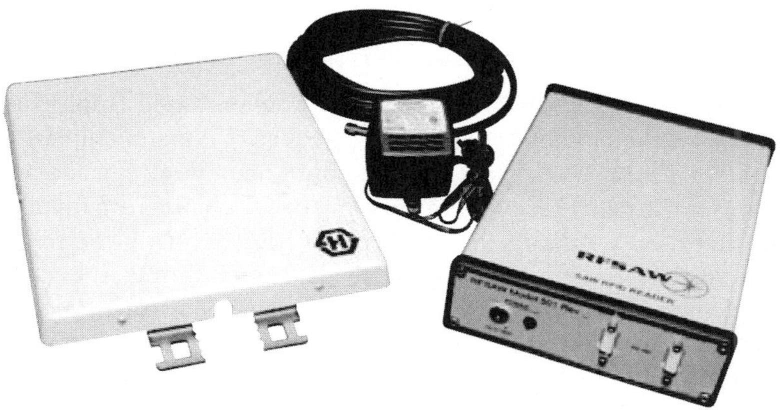

**Figure 3.3**   A SAW chipless RFID reader and antenna. (Courtesy of RFSAW Inc.)

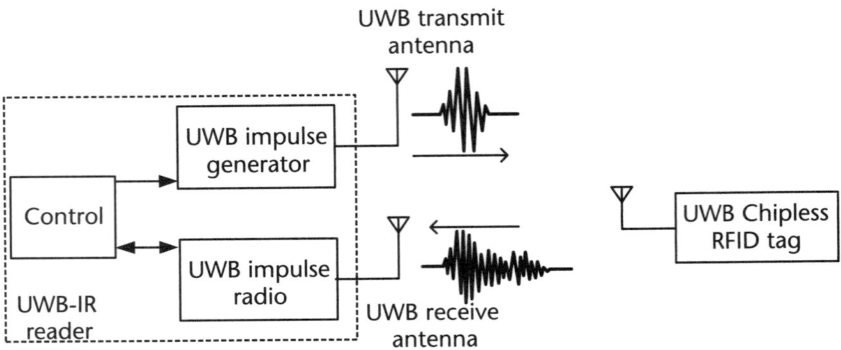


**Figure 3.4**  A block diagram of a UWB-IR based chipless RFID reader system.

The sampled signal is processed with different signal processing techniques. Lazaro et al. [16] used a continuous wavelet based signal processing technique to detect a chipless RFID tag based on time domain reflectometry (TDR). A time-coded tag with terminated transmission line is used in this experiment. Hu et al. [17] analyze the antenna mode scattering of a UWB pulse of the tag. The tag has a simple passive structure with one antenna and terminated transmission line. Kalansuriya and Karmakar [14] proposed a different tag design with a spiral resonator loaded transmission line and a UWB monopole antenna. They use a technique similar to that proposed for identifying the signature of the tag [14]. Another TDR based chipless tag is reported by Zheng et al. [15]. A transmission line with capacitive discontinuities is used to encode the data. By analyzing the reflected signal recorded with a high-speed DSO, the bits are identified. A pulse position modulation (PPM) based UWB chipless RFID tag has been proposed by Gupta et al. [13]. The chipless tag contains two cross-polarized antennas as the receive and transmit antennas. A dispersive delay structure (DDS) is used to obtain different group delays for different frequencies. The retransmitted pulses are recorded with a high-speed DSO and analyzed for obtaining the encoded data.

The next section presents information about existing frequency domain reader designs.

### 3.3.2  Frequency Domain Readers and Tag Reading Techniques

A comprehensive review of frequency domain based chipless RFID tags was presented in the previous chapter. The frequency domain chipless tags use the amplitude or phase of the frequency response or both amplitude and phase to encode data on the tag. Therefore, the reading of the chipless tag involves a process similar to that used for characterizing a passive RF device in the frequency domain. According to the technique used to encode data on the tag, the reading process and the reader architecture are different from one another. However, the sequence of operation introduced in time domain readers (generate the interrogation signal, transmit and receive it, process and decode data) is also valid for frequency domain readers. In this section, a selected set of chipless RFID readers that work in the frequency domain is presented. The selection of readers is done based on the different techniques carried out in the reading process. The selected set of readers covers the frequency range of the 1- to 10-GHz frequency band and shows the diversity of frequency domain reader operating principles and operating frequency bands.

### 3.3.2.1   Tagsense Chipless RFID Readers

Two chipless RFID systems reported on in [19–21] have been developed by Tagsense. They work in the 1- to 50-MHz frequency range. The reader reported in [20, 22] is commercially available. The block diagram of the reader system is shown in Figure 3.5.

The operation of the chipless RFID tag used with the readers is explained in [20]. The frequency signature depends on the shape of the conductive tracks of the tag. With different track shapes and modifications done by shorting and so on, it is possible to achieve multiple resonances. The multiple resonance frequencies are used to represent multibit data on the chipless RFID tag.

The reader system used to read these tags is shown in Figure 3.6. It contains a swept-frequency source, a signal detector, a directional coupler, an antenna, and a data and signal processing unit. One loop antenna is used as the reader antenna. The directional coupler transfers the signal generated by the source to the reader antenna and this signal is used as the interrogation signal. The frequency is swept in the range of 1 to 50 MHz. The antenna transmits the signals to the tags within its effective range. Then the antenna receives the response signal from the tag with frequency signatures. The received signal travels in the reverse direction, which is the opposite direction of the interrogation signal as shown in the figure. This reverse signal does not travel to the RF source. It passes to the third port, which is connected to the detector due to the operation of the directional coupler. The detector samples the received signal and senses the power of the signal. It is possible to use a threshold based technique to identify the resonance peaks by observing the amplitude of the received signal. Other techniques that are used to convert the signal into the frequency domain use either the fast Fourier transform (FFT) or a SAW transducer. Then, by determining the number of resonance peaks of the received signal, it is possible to identify the encoded data by analyzing the frequency

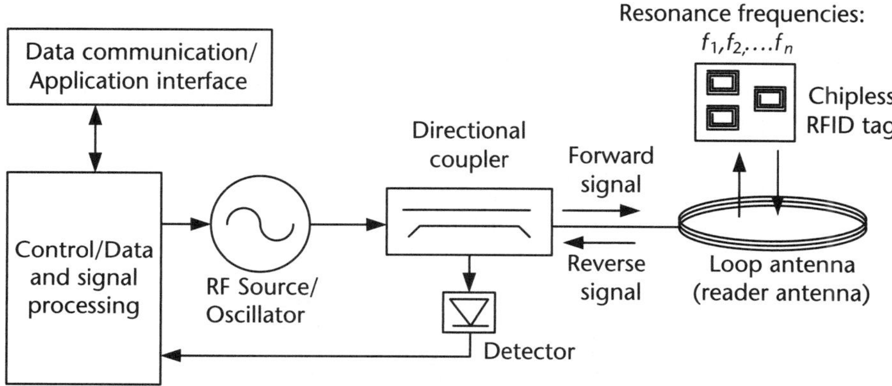


**Figure 3.5**   Tagsense LC series portable chipless RFID reader. It detects the tag by identifying the tag's resonance peaks [20].

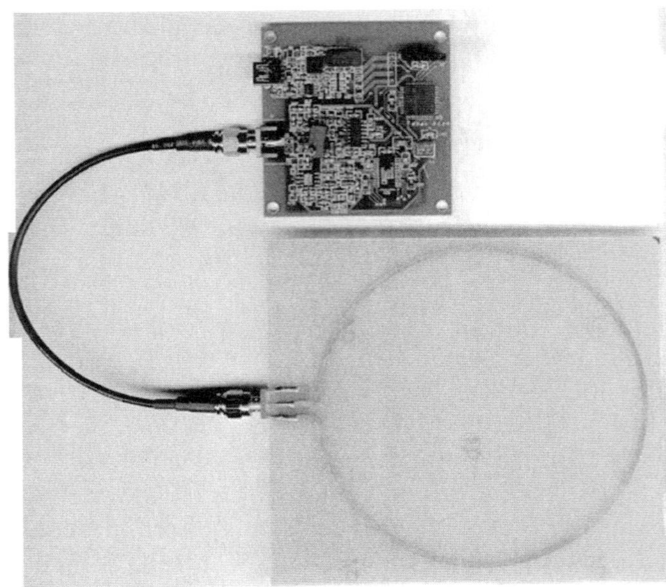

**Figure 3.6**  Tagsense LC-10 portable chipless RFID reader and antenna [22]. (Courtesy of Tagsense Inc.)

response of the tag. The tag reader system provides encoding of up to 32 bits of data within the operating frequency range of 1 to 50 MHz.

The other chipless RFID system [21] developed by Tagsense works on a different principle. The block diagram of the reader system is shown in Figure 3.7. Some of the materials in the tag have high permittivity and permeability. As shown in the figure, the reader system contains a tunable oscillator connected to a coil antenna. The coil antenna works as one tuning element of the oscillator. When there are no tags in the range of the reader antenna coil, the tunable oscillator oscillates at a certain frequency. This frequency can be called the *characteristic frequency*. In the presence of a tag in the range of the reader antenna coil, the characteristics of the coil change (inductance of the coil). This change results in a different operating frequency in the tunable oscillator. By counting the frequency of the oscillator, it is possible to identify different sets of objects. However, the number of objects that can be used with this system (i.e., the number of bits that can be encoded) is less than that for the previous system, the LC-10 reader system [20]. The two reader systems have a limited reading range and operating frequency range.

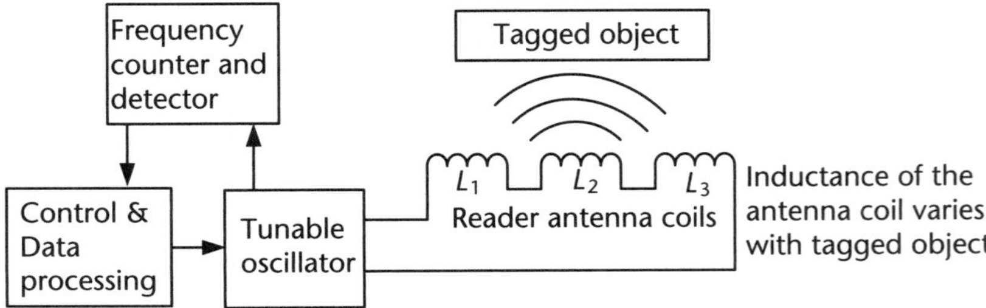


**Figure 3.7**  Tagsense chipless RFID reader. It detects the frequency shift of the characteristic oscillating frequency of a coil due to the materials used in chipless tag [21].

### 3.3.2.2 Surface Acoustic Wave (SAW) Frequency Domain Reader

The SAW chipless RFID tags belong to the time domain based category of tags. The operating principle of a reader that works on time domain principles was discussed in Section 3.3.1. However, techniques that were developed based on frequency domain principles are available [9, 10] for interrogating SAW chipless RFID tags. A reader based on the frequency modulated continuous wave (FMCW) radar principle is shown in Figure 3.8.

As shown in Figure 3.8, the RF source generates a chirp signal and it is passed through the directional coupler to the reader antenna. The antenna does the signal transmission and the reception. The retransmitted signals from the SAW tag reach the antenna, travel through the directional coupler, and then pass through the directional coupler to its port 3. The transmitted signal is coupled to port 4. Homodyne reception is achieved by mixing the two signals obtained from ports 3 and 4 of the directional coupler. The output of the mixer is passed through a lowpass filter (LPF) and sampled with an ADC. For the interrogation of SAW chipless RFID tags, the standard practice is to use either a linear frequency modulated continuous wave (LFMCW) or linear frequency stepped continuous wave (LFSCW) interrogation signal [10]. However, most of the VCO based LFMCW or LFSCW generation is a slow process due to the stability of the oscillator frequencies. Therefore, in [10] a direct digital synthesis (DDS) based FMCW signal generation scheme is used that is faster than other methods. However, this FMCW signal generation scheme generates a nonlinear signal. The effect of nonlinearity is compensated by rearranging the sampled signal in a timescale to obtain a signal similar to the one received from a LFMCW signal.

As shown in Figure 3.8, the transmitted signal is an up-chirp signal (frequency of the chip signal increases with time). Due to the operation of the SAW transducer, the retransmitted signal from the tag is the time-reverse of the interrogation signal [9]. Therefore, the response of the tag becomes a down-chirp signal (frequency of the chirp signal decreases with time). However, the electromagnetic reflections that might occur due to the surrounding environment have an up-chirp frequency profile similar to that of the transmitted signal. Therefore, it is easy to separate the response of the tag from surrounding reflections with signal processing. After the rearrangement of samples to remove the effect of the nonlinear interrogation chirp signal, the FFT can be used to recover the frequency response of the SAW tag and decode the ID.

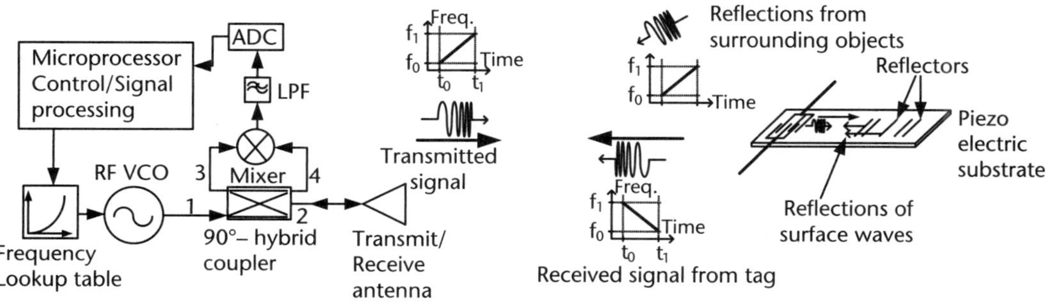


**Figure 3.8** SAW chipless RFID reader. It uses a FMCW signal for interrogating the tags [10].

### 3.3.2.3  Chipless RFID Readers for Multiresonator Based Chipless Tag Interrogation

Three generations of chipless RFID readers for interrogating multiresonator based chipless RFID tags have been reported by the authors' research group [3, 4, 23, 24]. The block diagram for a first-generation reader is shown in Figure 3.9. The reader operates over a 500-MHz frequency band from 2.0 to 2.5 GHz. A DAC is used to generate a swept-frequency signal with a VCO. The generated signal is passed through the tag as shown in the figure and the retransmitted signal is captured by the receiving antenna. The received signal is passed through a bandpass filter and amplified with a LNA. The amplified signal is fed to a diode detector to capture the envelope of the received signal. The resonances that occur in the tag make local minima in the envelope. If there is a resonance, it is considered to be a logic 1; otherwise, it is considered to be a logic 0. For the best results, a calibration tag is required to differentiate between the 1's and 0's. With envelope detection, only the amplitude response of the frequency signature can be captured with a first-generation reader.

The block diagram of the second-generation reader design is shown in Figure 3.10. The operating frequency is as same as for the first-generation reader. However, with the use of a gain/phase detector (GPD) it is possible to extract both the amplitude and phase responses of the frequency signature of the chipless tag. The swept frequency is generated with the same technique used in the first-generation reader. A reference signal to the GPD is extracted from the output of the VCO. The signal received from the tag is amplified and compared with the reference signal to identify the amplitude differences and phase jumps that occur due to resonators of the chipless tag. The ability to recover both amplitude and phase increases the accuracy and reliability of the readings.

The block diagram of the UWB reader, a third-generation reader, is shown in Figure 3.1 [4, 23]. The components of the reader were explained in the previous section. This reader design uses a GPD to capture the amplitude and phase responses of the frequency signature of the chipless tag, similar to the second-generation version. However, the VCO of this reader generates higher frequencies than the operating frequency range of the GPD. Therefore, to compare the reference and received signals from the tag for the amplitude and phase differences, it is required to down-convert to a suitable frequency. The two mixers are used to down-convert the reference signal extracted from the transmitted signal and received signal from the tag along with a local oscillator (LO). The LO is kept at a fixed frequency in

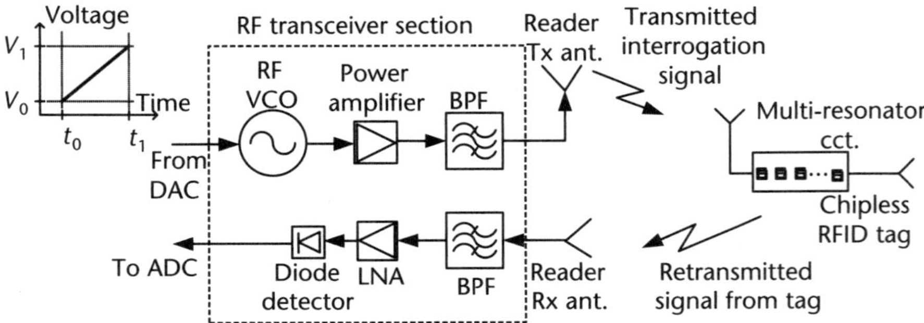


**Figure 3.9**  First-generation chipless RFID reader for interrogating multiresonator based chipless tags. It detects only the amplitude response of the frequency signature [4].

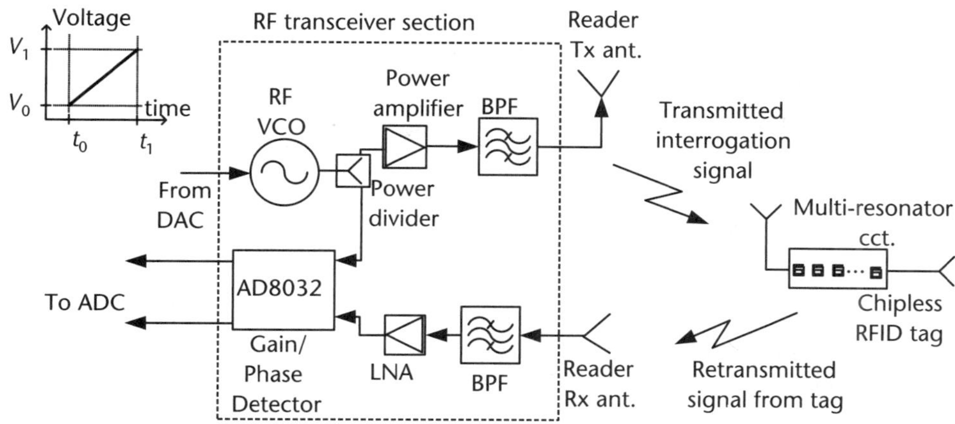



**Figure 3.10**   Second-generation chipless RFID reader for interrogating multiresonator based chipless tags. It detects both the amplitude and phase responses of the frequency signature [4].

this design. This introduces a limitation on the operating bandwidth of the reader. Because the local oscillator is kept at a fixed frequency, the intermediate frequency (IF) of the two mixers varies with the transmitted frequency. Therefore, the IF can have a maximum swing of 2.6 GHz (operating range of the gain phase detector: 0.1 to 2.7 GHz). Due to the varying IF frequency, the extraction of phase information process is also poor in this reader. A photo of the reader RF section is shown in Figure 3.11.

### 3.3.2.4   Nicanti Swipe Reader for NiCode Chipless Tag Reading

A printable chipless RFID system developed by Nicanti [25–27] uses a reader that can be considered as working with frequency domain based principles. The block diagram of the reader, a cross section of a printed tag and reading process, is shown in Figure 3.12. The tag, called NiCode, is printed on any substrate used for packaging

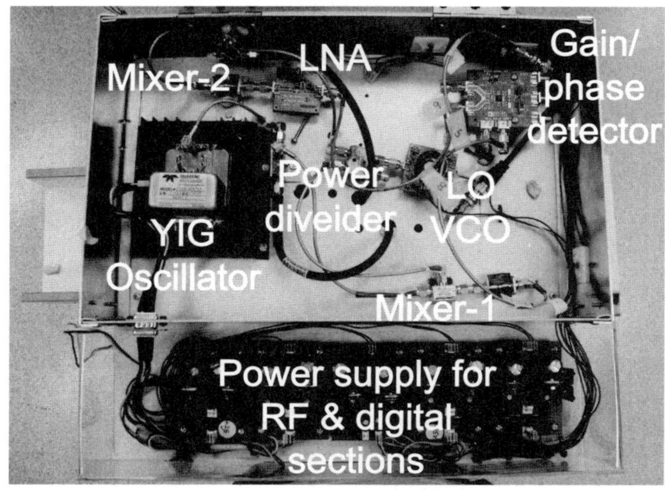



**Figure 3.11**   The UWB reader RF section [23].

goods such as paper or plastics. A combination of proprietary conductive inks is used to print patterns to form a chipless tag. Printing techniques such as ink-jet or screen printing can be used for small-scale applications and a thermal transfer based technique is proposed for large-scale printing applications [27]. The printed pattern is protected with an insulating layer. On top of this insulation layer, anything can be printed using standard nonconductive inks as shown in Figure 3.12.

The block diagram of the reader device is also shown in Figure 3.12. An AC bridge circuit is formed by the section of the tag covered with electrodes. An RF signal is applied to the formed bridge circuit by the reader unit RF generator. The tag section covered by the electrodes can be modeled as a network of resistors and capacitors [25]. The set of electrodes does not directly contact the conductive section of the tag. Only placing the reader electrodes over the insulating layer is sufficient to make a capacitive coupling from the bridge circuit for the reading. Therefore, this method can be considered to be a noncontact reading technique and can be considered to be a type of chipless RFID system. An RF signal of 50 MHz is generated by the reader and injected to the AC bridge circuit. The current flowing through the electrodes is measured and the impedance of the tag section is calculated. The real and imaginary components of the impedance are used to encode the data of the tag. The size of the tag depends on the number of data bits encoded in the tag. When the size of the tag increases, the reader is swiped over the tag area to cover the whole tag. During the swiping process, the current flowing through the electrodes is recorded. Using signal processing techniques, the impedance at different sections is calculated from the recorded signal. The reader is interfaced to the PC with a cable or a wireless technology such as Bluetooth or Wi-Fi. Secure data transfer protocols are used to transfer data from reader to PC when using the reader with security-sensitive work. A photograph of the reader is shown in Figure 3.13.

### 3.3.3  Hybrid Domain Readers and Tag Reading Techniques

The hybrid domain tag design [28] reported by the authors' research group and a reading technique were discussed in Chapter 2. A comprehensive description of the design of the readers for hybrid domain tags is presented in Chapter 6.

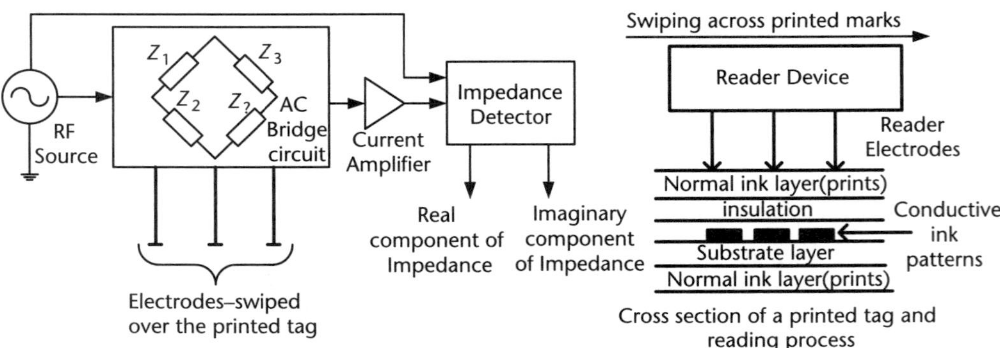


**Figure 3.12**  Chipless RFID reader developed by Nicanti for reading NiCode chipless tags—cross section of a printed tag and reading process [25–27].

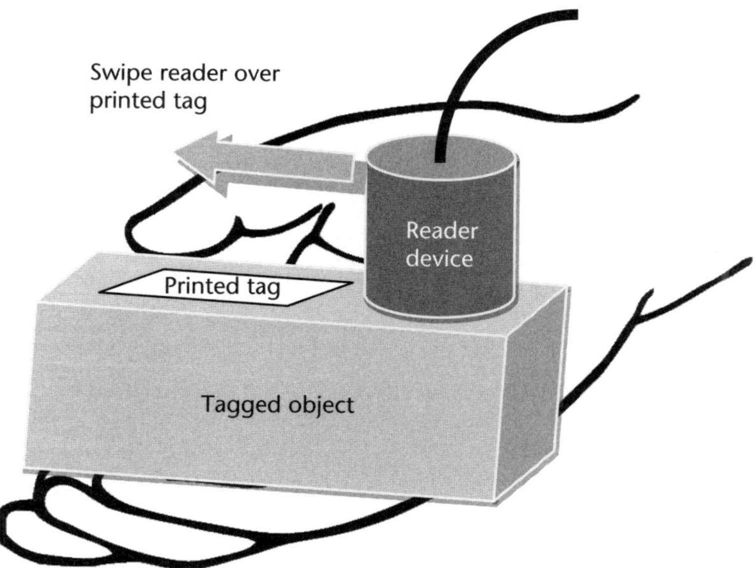


**Figure 3.13**   Reading process of Nicanti NiCode chipless tag with RFID readers (adapted from http://www.vttventures.fi/portfolio-companies/nicanti).

### 3.3.4   SAR Based Readers and Tag Reading Techniques

This section discusses three commercial chipless RFID system tags that use SAR imaging techniques. The reader architectures are different from each other and they operate on different working principles.

#### 3.3.4.1   Printed Chipless RFID Tag Reader System Developed by Vubiq

As shown in Figure 3.14, the reader system developed by Vubiq [29, 30] utilizes radar technology to identify and decode chipless tags. The chipless tag is printed using conductive inks. Typically an array of antenna shapes is printed on a substrate. The array of antennas is organized to be resonating at different frequencies and scatter the received RF signals with different polarizations. The data of these tags are encoded in phase and polarization of these antenna elements. The reader transmit antenna polarizations are controlled using a control unit as shown in Figure 3.14. Using this arrangement it is possible to transmit and receive signals with two different polarizations simultaneously. The reader illuminates the tag area with a millimeter-wave RF signal in the frequency range of the resonant frequency of the antenna elements on the tag.

The beam scanning is achieved with an antenna and a control unit. It utilizes the scanning of the tag area to support the SAR imaging. The receive antenna captures the reradiated signals from the tag and the captured signals are stored in the memory of the reader device. Using signal processing algorithms based on SAR, the reader develops an image of the reader system. Mathematical focus algorithms used in radar imaging are used to obtain high-resolution images of the scanned area. The antenna orientation and sizes are obtained from the image by analyzing the phase and polarization data captured in the reading process. The identified

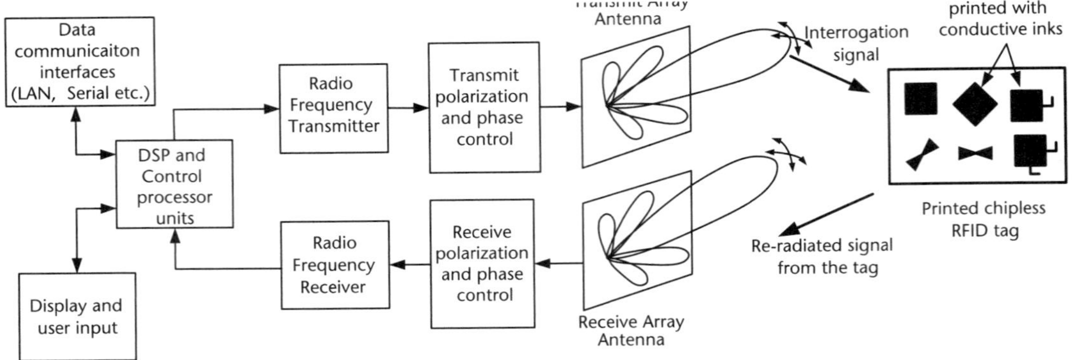



**Figure 3.14** Chipless RFID reader developed by Vubiq to read the printed radar array type chipless RFID tag. It obtains a 2D image of the tag with a SAR technique to obtain the tag ID [29, 30].

phase and polarization information are converted back to the data or as the ID of the tag.

The reader device contains an RF transmitter and receiver, a digital signal processor (DSP), a control processor, and control units to drive the transmit and receive antennas. The reader device needs to perform complex signal processing using the DSP to develop the image of the scanned tag. Another processor (control processor) is used to control and synchronize the operation of other units of the reader. The reader is utilized with wired or wireless networks to support industry standard data communication interfaces.

### 3.3.4.2  Chipless RFID Tag Reader System Developed by InkSure

Another SAR based chipless RFID tag reading system has been developed by InkSure as reported in [31]. The block diagram of the reader and tag reading process is shown in Figure 3.15.

The tag comprises a substrate layer patterned with several diffraction elements having the dimension of the order of the wavelength of the illuminated RF signal. The frequency of the signal used to interrogate the tag is in the millimeter-wave range (typically 30 to 60 GHz). The elements are printed using conductive inks.

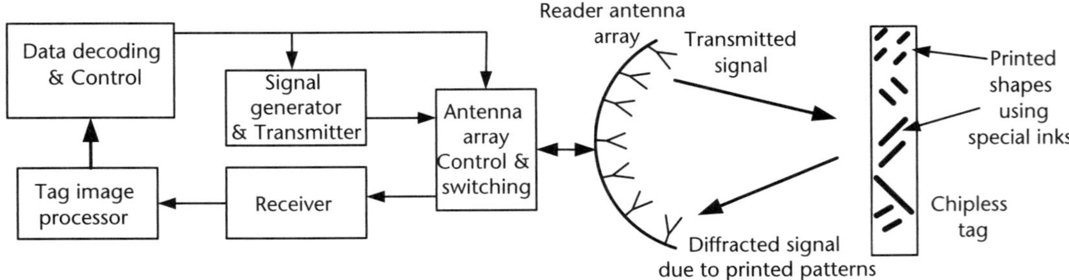



**Figure 3.15** Chipless RFID reader developed by InkSure to read the SARCode chipless RFID tag. It obtains a 2D image of the tag. The obtained image is used to generate the code corresponding to the tag [31].

The printed patterns diffract the illuminated RF signals in different power levels. The operation of the tag in lower frequencies is also possible. However, the size of the tag becomes excessively large, which makes the tag not suitable for tagging applications.

The reader generates an RF signal in the millimeter-wave range and transmits using the array of antennas as shown in Figure 3.15. The antenna contains a large number of antenna elements. The antennas are switched to scan the interrogation beam across the tag. The millimeter-wave RF signals generated by the RF source are transmitted toward the tag with the antennas. The diffracted signals from the tag are captured with the same antenna array. The received signals by the receiver are processed and an image is developed using SAR signal processing techniques. The image is processed with a DSP assisted unit in the reader. The image represents the amount of diffraction of RF signals done by the tag. The data are extracted by considering the amount of diffraction and position of the diffraction of the tag area. The control unit provides the control of the operation of the reader and interfaces external communications with the reader.

This system is different from the Vubiq in two ways although it uses SAR imaging techniques. First, the operating principle of the tag and reader architecture is different from that of the Vubiq tag reader system. Second, the type of materials used in the tag and reader architecture is different from InkSure's chipless tags and readers.

### 3.3.4.3  Somark Ink Tattoo Chipless RFID Tag Reader System

A chipless RFID system focused on animal tagging applications has been developed by Somark [32]. The chipless tag is printed in the form of a tattoo by using a special ink on the animal's skin. The tattoo is transferred to the animal's skin using a technique similar to that used to print tattoos. However, it can be implemented on other substrates and the tag is not limited to tattoos. The tattoo ink contains dielectric materials that scatter the RF signals. The tattoo contains a series of bars with different widths similar to a barcode. The data are encoded by using a number of bars having different widths [32].

The block diagram of the reader system and a model of the tag developed by Somark [32] are shown in Figure 3.16. The tag design is done for millimeter-wave frequencies to reduce the size of the tags. The reader transmits a signal with the transmitter antenna. The antenna is either mechanically steered to cover the tag area or a beam is steered. As shown in Figure 3.16, the ink bars with smaller widths scatter the incoming RF signal in a smaller amount than the lines with higher widths. The reader antenna receives the scattered signals from the ink bars and records the strength of the scattered signal toward it. The information on the scattered signal strength and the corresponding beam position are combined to form a 2D image of the tattoo with SAR image processing algorithms. The constructed image is used to identify the data encoded in the tag.

Although the three systems reported above use SAR techniques in signal processing, they use entirely different operating principles and hardware configurations in the physical layers.

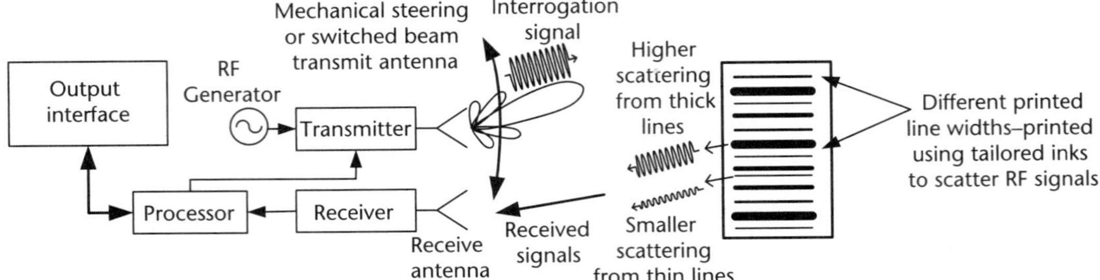



**Figure 3.16** Chipless RFID reader developed by Somark to read the ink tattoo chipless tag. The tag structure is similar to that of a barcode. Bars with different widths are printed with a dielectric ink to obtain a unique code. A 2D image is developed to obtain the information coded in the tag using a similar technique to SAR [32].

## 3.4 Limitations and Issues with Current Chipless RFID Readers

As discussed earlier in the literature review, chipless RFID technology has attracted strong interest from many industry sectors. The chipless RFID is promising for many application areas for not only its low-cost characteristic but also its special properties that conventional chipped tags cannot offer. For conventional chipped RFID technology, many international (ISO) standards have been published that allow interoperability of different devices from different sources. For chipless RFID tags, some tags [8, 31] produce industry standard data formats for tagged items (e.g., SAW tags). However, many chipless RFID systems are still in their inception\ experimental stages for practical applications. Therefore, many challenges have to be overcome to enable the technology to be more attractive commercially. Figure 3.17 shows some of the challenges that chipless RFID systems have been encountering. This section discusses the limitations and challenging issues related to existing chipless RFID reader systems.

### 3.4.1 Cost of Readers

In general, the function of the chipless tag is to produce a signature that can be identified by the reader. By removing the chip from the tag, all the burden of signal

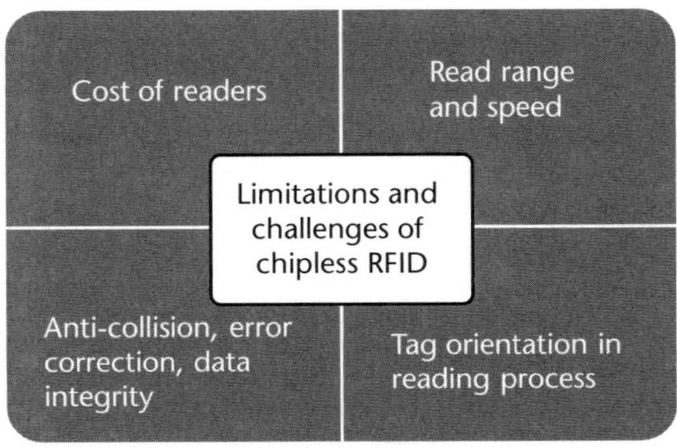



**Figure 3.17** Limitations and challenges of chipless RFID systems.

processing is transferred to the reader. Therefore, the complexity of the reader is comparatively higher than that for conventional RFID readers. The increased complexity in the physical layer design and its processing techniques generally increases the cost of the reader system. For example, the chipless systems that use SAR techniques [29, 31, 32] require high-performance hardware platforms to meet the requirements of their complex signal processing schemes. Although the cost of the tag is reduced, the cost of the reader increases due to the expensive components and signal processing chips[1]. However, when the technology matures, the manufacturing costs will decrease and the low-cost components and services required for system development will be available. IDTechEx advocates that the cost of the chipless tag reader and its peripherals will be comparable to that for the chipped tag reader [33].

### 3.4.2   Read Range

One of the main reasons RFID technology became popular is its reading flexibility and long-distance capabilities, which mean that less human intervention is required. With chipped active and passive RFIDs, achieving long operating distances compared to barcode systems is not difficult due to the on-chip transmitter of the tag. However, with passive chipless RFID tags, achieving long operating ranges is challenging due to the unavailability of the signal source on the tag. The developed chipless RFID tags operate over UWB frequency ranges from 3 to 11, 22 to 26, and around 60 GHz. There are stringent regulatory issues surrounding the transmitting of higher power in similar orders of magnitudes for the conventional chipped tags. For example, for UWB chipless tags only a few milliwatts of transmission power is permissible to comply with stringent regulations and not to interfere with existing telecommunications and weather radars. Contrary to this, the conventional RFID systems can transmit power on the order of a few watts. Therefore, reading distances for the chipless tags could be an issue.

A few tags having long operating ranges of up to about 100 feet in line-of-sight conditions [29, 30]. However, with multiple reflections (multipath propagation), it is not possible to achieve such a long distance in non-line-of-sight (NLOS) operation. The NLOS characteristic is one of the main reasons why RFID technology became popular, and it is becoming a challenging issue in chipless RFID technology today.

### 3.4.3   Tag Reading Speed

As discussed in Chapter 2 and in this chapter, most of the chipless RFID systems operate in a very wide bandwidth. In particular, those chipless RFID tags that use frequency signatures to encode data [3, 34, 35] use up to several gigahertz of bandwidth. Reading of these tags requires generation of UWB signals. The widely used techniques for generating wideband signals with a several gigahertz frequency band use swept-frequency signals (frequency sweep) and narrow UWB RF pulses. The first technique swept-frequency signal generation is usually done with a wideband VCO or similar synthesized source. These wideband oscillators require a certain

---

1.   SAR techniques use millimeter-wave technology at 60 GHz and above. Hence, the hardware is very expensive.

settling time to generate linear swept frequencies. They take significant time to cover the whole bandwidth using linear chirp. The second technique narrow UWB RF pulses can generate the UWB pulses within a very short period of time (typically several hundreds of picoseconds to several nanoseconds) [13, 14, 17]. The time required for signal generation is reduced with the use of short RF pulses. However, the main issue is the reception of short UWB pulses. To achieve the full benefit of the nanosecond range of signal generation time, it is required to sample the received signal with ultra-high-speed ADCs. The use of ultra-high-speed ADCs increases the cost of the reader system significantly.

Equivalent time sampling (ETS) is another standard technique that is used for sampling the signals in very high speed DSOs [36] to obtain the trace of subnanosecond signals. However, to use ETS, the input signal should be repeated a number of times to obtain enough samples to reconstruct the original signal with signal processing techniques [36, 37]. The requirement for the repetition of the signal means that the full benefit of the very small signal generation time is lost because signal reception requires much longer times. In addition to the longer reception time, signal processing techniques have to be applied to reconstruct the original received signal [38, 39]. The reconstruction process leads to increased signal processing complexity and increased time is required for the reading process. Therefore, achieving a high tag reading speed is another challenge for most chipless RFID systems.

### 3.4.4 Anticollision, Error Correction, and Data Integrity

The vision for chipless RFID tag development is item-level tagging in mass deployment. Therefore, multiple tag reading in proximity will be a common feature of chipless RFID applications. Because the chipless tag is a fully passive device, it cannot control the reception and retransmission of the incoming interrogation signal. If the tag receives a signal, the retransmission or backscattering happens automatically. If there are multiple tags in the vicinity of a reader, all the tags will respond to the interrogation signal at the same time. The simultaneous responses of the tags introduce collisions in the tag-reader communication process. Therefore, reading multiple tags or implementation of anticollision algorithms should be done entirely in the reader without any assistance from the tag. A few anticollision techniques for SAW chipless RFID tags have been reported on in [12, 40]. Anticollision techniques based on the signal strength of backscatterers, time separation, spatial separation, and code division based schemes are proposed in [12]. A linear block code based anticollision technique is proposed in [40] for RF-SAW tags. Azim and Karmakar [41] proposed an anticollision method for multiresonator based chipless RFID tags [24]. The technique uses a signal processing technique based on the fractional Fourier transform (FrFT) to separate the frequency signatures of multiple tags. The SAR technique based reading methods also report multiple tag reading with high-end signal processing schemes [29, 31]. However, there is the demanding requirement of implementing multiple tag reading in chipless RFID systems if they are to be used in real-world applications. A companion book, *Chipless RFID Reader Signal Processing,* by the lead author of this book presents a comprehensive literature review and proposed methods for anticollision, error correction, and data integrity for chipless RFID tags.

### 3.4.5 Orientation of the Tag

Readability of existing chipless RFID tags is sensitive to the orientation of the tag with respect to the reader. For example, the tags based on SAR principles [29, 31, 32] should be kept in a way such that the tag backscatters the interrogating RF signal toward the reader device. Transmit and receive antennas should be polarization matched with the reader antennas to achieve proper operation of the chipless RFID tags [24]. The polarization diversity in tags is used to achieve a higher bit density in a smaller area as reported in [34, 42]. Proper reading of these tags is only achieved by means of a fixed orientation with the reader's antennas. The reading of the retransmission chipless tag with two cross-polarized antennas requires a UWB RF pulse and is also sensitive to the reader antenna's orientation [13]. None of these tags can scatter a signal toward the reader if the plane of the tag is kept in parallel to the direction of propagation of the interrogating RF signal. In general, almost all of the chipless RFID systems are sensitive to the orientation of the tag—without the proper orientation, it is challenging to achieve a proper reading. Therefore, a totally new set of chipless RFID tags and reader antennas need to be specified to have orientation-insensitive tag reading. Chapter 7 presents many aspects of antenna design and propagation issues. The rest of the book provides information to overcome the identified limitations, at least up to a certain extent.

## 3.5 Conclusion

Chipless RFID is an emerging RFID technology that transforms the identification, asset tracking, security surveillance, anticounterfeiting, and many other sectors into lower cost solutions than conventional chipped RFID technology. The off-the-shelf RFID readers cannot be used with chipless RFID tags. The operation of the reader depends on the operating principle of the chipless tag. Therefore, different RFID tags require different reader systems.

This chapter presented a comprehensive review of the existing commercial and proof-of-concept prototype chipless RFID readers that have been reported on in the scientific literature. The review shows that the chipless RFID system is a combination of several disciplines such as RF/microwave engineering, antenna engineering, digital systems design, and digital signal processing. The interrogation is completely handled by the reader and the tag has no control over the communication process with the tag. In most of the chipless RFID systems, each and every component is tailored to the specific system. The review of reader architectures and reading techniques presented in this chapter and review of chipless tags presented in Chapter 2 will help build a foundation for understanding the rest of this book.

## Questions

1. Provide an overview of the architecture for a typical chipless RFID reader.
2. What different chipless tag reading techniques are available?
3. Briefly explain each of the techniques listed in Question 2 using an example.
4. Name some of the limitations with the current chipless RFID readers.

5.  What is the reading range for a chipless RFID reader?

6.  Name the operating frequency for each different type of reader.

7.  What is the approach for differentiating noise from data in a chipless RFID reader?

8.  How does a chipless RFID reader process information from the encoded chipless RFID tag?

9.  What factors contribute to the cost of a chipless RFID reader?

10. What approaches are being used to increase the reading speed of a chipless RFID reader?

11. How does orientation of a tag affect the chipless RFID reader's reading?

# References


[1]  D. M. Dobkin, *The RF in RFID: Passive UHF RFID in Practice*, Boston: Newnes, 2007.

[2]  S. Preradovic and N. Karmakar, "Chipless RFID: Bar Code of the Future," *IEEE Microwave Magazine*, Vol. 11, No. 7, pp. 87–97, 2010.

[3]  S. Preradovic and N. Karmakar, "Design of Short Range Chipless RFID Reader Prototype," in *5th International Conference on Intelligent Sensors, Sensor Networks and Information Processing (ISSNIP 2009)*, Melbourne, Australia, 2009, pp. 307–312.

[4]  S. Preradovic, *A Chipless RFID System for Barcode Replacement*, Ph.D. thesis, Monash University, Australia, 2009.

[5]  N. C. Karmakar, *Handbook of Smart Antennas for RFID Systems*, Hoboken, NJ: Wiley-Interscience, 2010.

[6]  U. L. Rohde, A. K. Poddar, and G. Barck, *The Design of Modern Microwave Oscillators for Wireless Applications: Theory and Optimization*, Hoboken, NJ: Wiley-Interscience, 2005.

[7]  Analog Devices Inc., "AD8302: 2.7GHz RF/IF Gain Phase Detector," June 2002, http://www.analog.com/static/imported-files/data sheets/AD8302.pdf (accessed June 12, 2012).

[8]  V. Plessky and L. Reindl, "Review on SAW RFID Tags," *IEEE Trans. on Ultrasonics, Ferroelectrics and Frequency Control*, Vol. 57, No. 3, pp. 654–668, 2010.

[9]  S. Harma, "Surface Acoustic Wave RFID tags: Ideas, Developments, and Experiments," Ph.D. thesis, Helsinki University of Technology, Finland, 2009.

[10] S. Scheiblhofer, S. Schuster, and A. Stelzer, "Signal Model and Linearization for Nonlinear Chirps in FMCW Radar SAW-ID Tag Request," *IEEE Trans. on Microwave Theory and Techniques*, Vol. 54, No. 4, pp. 1477–1483, June 2006.

[11] K. Finkenzeller, *RFID Handbook: Fundamentals and Applications in Contactless Smart Cards and Identification* (2nd Ed.). Hoboken, NJ: Wiley, 2003.

[12] C. Hartmann et al., "Anti-Collision Methods for Global SAWRFID Tag Systems," in *IEEE Ultrasonics Symposium 2004*, Vol. 2, pp. 805–808.

[13] S. Gupta, B. Nikfal, and C. Caloz, "Chipless RFID System Based on Group Delay Engineered Dispersive Delay Structures," *IEEE Antennas and Wireless Propagation Letters*, Vol. 10, pp. 1366–1368, 2011.

[14] P. Kalansuriya and N. Karmakar, "Time Domain Analysis of a Backscattering Frequency Signature Based Chipless RFID Tag," in *Asia-Pacific Microwave Conference Proceedings (APMC2011)*, December 2011, pp. 183–186.

[15] L. Zheng et al., "Design and Implementation of a Fully Reconfigurable Chipless RFID Tag Using Inkjet Printing Technology," in *IEEE International Symposium on Circuits and Systems, ISCAS 2008*, May 2008, pp. 1524 –1527.

[16] A. Lazaro et al., "Chipless UWB RFID Tag Detection Using Continuous Wavelet Transform," *IEEE Antennas and Wireless Propagation Letters*, Vol. 10, pp. 520–523, 2011.

[17]  S. Hu et al., "Study of a Uniplanar Monopole Antenna for Passive Chipless UWB-RFID Localization System," *IEEE Trans. on Antennas and Propagation*, Vol. 58, No. 2, pp. 271–278, February 2010.

[18]  S. Shrestha et al., "A Chipless RFID Sensor System for Cyber Centric Monitoring Applications," *IEEE Trans. on Microwave Theory and Techniques*, Vol. 57, No. 5, pp. 1303–1309, May 2009.

[19]  R. Fletcher and O. Omojola, "Methods and Apparatus for Counting and Positioning Using Resonant Tags," U.S. Patent 7,216,805, May 2007.

[20]  R. Fletcher and N. Gershenfeld, "Tunable Wireless Tags Using Spatially Inhomogeneous Structures," U.S. Patent 7,221,275, May 2009.

[21]  R. Fletcher, "Methods and Devices for Identifying, Sensing and Tracking Objects over a Surface," U.S. Patent 6,834,251, December 2004.

[22]  Tagsense Inc., "LC-10 Chipless Tag Reader," August 2006, http://www.tagsense.com/images/stories/products/chiplesstechnologies/LC-10-datasheet-v2.pdf (accessed June 12, 2012).

[23]  S. Preradovic and N. Karmakar, "Multi-Resonator Based Chipless RFID Tag and Dedicated RFID Reader," in *IEEE MTT-S International Microwave Symposium Digest*, Anaheim, CA, 2010, pp. 1520–1523.

[24]  S. Preradovic et al., "Multi-Resonator-Based Chipless RFID System for Low-Cost Item Tracking," *IEEE Trans. on Microwave Theory and Techniques*, Vol. 57, No. 5, pp. 1411–1419, May 2009.

[25]  H. Seppa et al., "Method and Device for Identifying an Electronic Code," International Patent WO 2009/138571, May 2009.

[26]  P. Wilkinson, "Method for Generating and Reading an Hybrid Code and Relative Hybrid Code," International Patent WO 2011/132160, April 2011.

[27]  P. Wilkinson, "Method for Applying onto a Substrate a Code Obtained by Printing Conductive Inks," International Patent WO 2012/059896, November 2011.

[28]  M. Bhuiyan, R. Azim, and N. Karmakar, "A Novel Frequency Reused Based ID Generation Circuit for Chipless RFID Applications," in *Asia-Pacific Microwave Conference Proceedings (APMC 2011)*, December 2011, pp. 1470 –1473.

[29]  M. G. Pettus, "RFID System Utilizing Parametric Reflective Technology," U.S. Patent 7,460, 014, Dec. 2005.

[30]  M. G. Pettus and J. R. A. Bardeen, "System and Method for Wireless Communication in a Backplane Fabric Architecture," U.S. Patent 2009/0028177, January 2009.

[31]  A. Yogev and S. Dukler, "Radio Frequency Data Carrier and Method and System for Reading Data Stored in the Data Carrier," U.S. Patent 2004/0159708, August 2004.

[32]  A. M. Grishin and R. M. Mays, "Radio Frequency Data Carrier and Method and System For Reading Data Stored in the Data Carrier," U.S. Patent 7,205,774, December 2005.

[33]  IDTechEx, *Printed and Chipless RFID Forecasts, Technologies & Players 2009–2029*, Executive Summary, September 2009.

[34]  M. A. Islam and N. C. Karmakar, "A Novel Compact Printable Dual-Polarized Chipless RFID System," *IEEE Trans. on Microwave Theory and Techniques*, Vol. 60, pp. 2142–2151, 2012.

[35]  A. Vena, E. Perret, and S. Tedjini, "Chipless RFID Tag Using Hybrid Coding Technique," *IEEE Trans. on Microwave Theory and Techniques*, Vol. 59, No. 12, pp. 3356–3364, December 2011.

[36]  Tektronix Inc., "Application Note: Real-Time Versus Equivalent-Time Sampling," January 2001, http://www.tek.com/application-note/real-time-versus-equivalent-time-sampling (accessed June 13, 2012).

[37]  A. Toya et al., "32 GS/s Ultra-High-Speed UWB Sampling Circuit for Portable Imaging System," *Electronics Letters*, Vol. 47, No. 3, pp. 165–167, 2011.

[38]    E. Moreno-Garcia, J. de la Rosa-Vazquez, and O. Alonzo-Larraga, "An Approach to the Equivalent-Time Sampling Technique for Pulse Transient Measurements," in *16th International Conference on Electronics, Communications and Computers, 2006. (CONIELE-COMP 2006)*, February 2006, p. 34.

[39]    Y. Jenq, "Perfect Reconstruction of Digital Spectrum from Nonuniformly Sampled Signals," *IEEE Trans. on Instrumentation and Measurement*, Vol. 46, No. 3, pp. 649–652, June 1997.

[40]    M. Brandl et al., "A New Anti-Collision Method for SAW Tags Using Linear Block Codes," in *IEEE 2008 International Frequency Control Symposium*, May 2008, pp. 284–289.

[41]    R. Azim and N. Karmakar, "A Collision Avoidance Methodology for Chipless RFID Tags," in *Asia-Pacific Microwave Conference Proceedings (APMC 2011)*, December 2011, pp. 1514–1517.

[42]    A. Vena, E. Perret, and S. Tedjini, "A Compact Chipless RFID Tag Using Polarization Diversity for Encoding and Sensing," in *IEEE International Conference on RFID (RFID 2012)*, April 2012, pp. 191–197.

## CHAPTER 4

# Frequency Domain Based RFID Reader Development

## 4.1  Introduction

The preceding two chapters presented comprehensive reviews of chipless RFID tags and readers, respectively. The review of available and reported chipless RFID transponders presented in Chapter 2 has demonstrated the need for dedicated reader designs for different types of chipless RFID systems. Therefore, the multiresonator based chipless RFID tags need readers that are tailored to the operating principles and requirements of the particular system. As discussed in Chapters 2 and 3, the architectures and signal processing techniques of the chipless RFID tag readers need to be different according to the principles of operations.

In laboratory settings, test equipment such as vector network analyzers (VNAs), high-performance signal analyzers, high-speed digital storage oscilloscopes (DSO), and high-speed samplers are used. However, it is not economical to use these expensive and bulky laboratory pieces of equipment for the real-world applications of chipless RFID systems. Therefore, while trying to reduce the cost of the tags by using chipless RFID technologies, it is essential to develop low-cost reader devices that can lead to the realization of low-cost chipless RFID systems.

This chapter describes the design of two chipless RFID readers based on frequency domain detection techniques. The frequency signature of the multiresonator based chipless RFID tag can be detected using either frequency domain or time domain techniques. A reader design based on the time domain based reading technique will be presented in Chapter 5.

### 4.1.1  Organization of This Chapter

The organization of this chapter is summarized in Figure 4.1. The frequency domain based reader architectures presented in this chapter can be divided into two categories based on the data decoding technique:

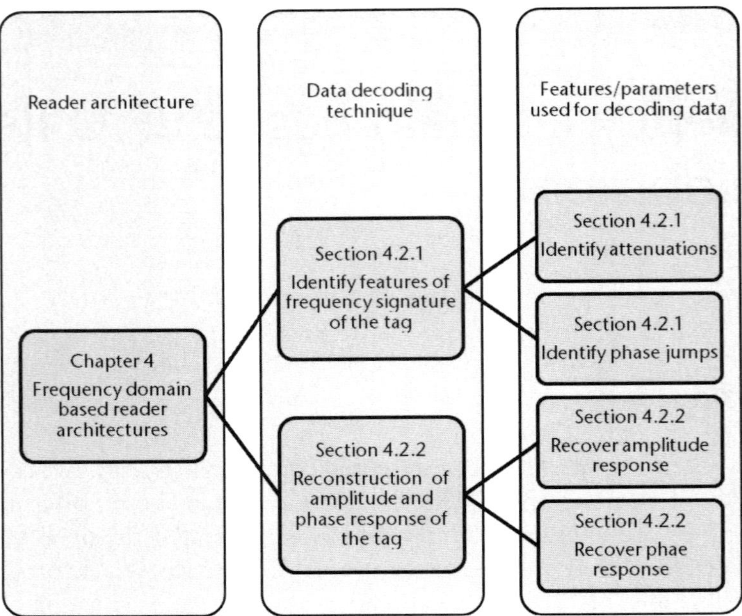


**Figure 4.1** Organization of the information presented in this chapter.

- Identification of the features of frequency signatures;
- Reconstruction (or recovering) of the frequency signature of a chipless RFID tag.

Two reader designs based on the above two techniques are described in this chapter along with their operating principles.

The first technique, the identification of the features of frequency signatures, can be divided into two subcategories as shown in the Figure 4.1. In this method, data can be identified using the attenuations that occur in the amplitude and deviations of the phase from the linear variation (also known as phase jumps) of the received signals from the tags. Each piece of information could be used to identify the data encoded in chipless tags. However, by using both pieces of information, amplitude and phase, the reliability of the decoded data is greater than when using only either amplitude or phase.

The second tag reading technique discussed in this chapter that is based on the frequency domain involves reconstructing the amplitude and phase information of the frequency signature of the chipless tag. The reconstruction of the amplitude and phase information is done using the signal received from the chipless tags. Using signal processing techniques, it is possible to reconstruct both the amplitude and phase information of a frequency signature based chipless RFID tag.

The following sections present information about the operating principles, design, simulation, and experimental results of the two readers based on the two techniques shown in Figure 4.1.

## 4.2   Operation of Frequency Domain Based Chipless RFID Readers

Before going into the details of the operating principles and designs of the readers, a brief description of the multiresonator based chipless RFID tag is presented. This will help in understanding the operating principles of chipless RFID readers that are presented in detail in later sections of this chapter.

As explained in Chapter 2, the chipless tag is a fully passive microwave structure. An external source of electromagnetic signals is needed for the interrogation of the tag. When the tag is illuminated with a sufficiently wideband signal to cover all frequencies of the multiresonating structure, it creates a spectral signature. The reader converts the spectral signature of the tag into a series of binary data bits (1's and 0's). Figure 4.2 explains the operation of the tag. When the tag is illuminated with a wideband signal as shown in the figure, with the captured signal by the receive (Rx) antenna of the tag, resonators start resonating at the designed frequencies and create attenuations in the designed frequency. In addition, phase jumps are also created at each resonating frequency. Here each resonator acts as a bandstop filter. Then this signal is retransmitted by the transmit (Tx) antenna of the tag. The presence of a resonator creates attenuation in the frequency spectrum and creates phase jumps, whereas the absence of a resonator does not create either attenuation or phase jumps.

As explained in previous chapters, as a response to the combined effects of the presence and absence of the bandstop filter (resonators), the resonant circuit creates a frequency signature in the received signal from the Rx antenna of the tag. Then the signal with frequency signatures is retransmitted with the transmit antenna of the tag. The frequency signature can be controlled by varying the resonant frequencies and using different combinations of resonant frequencies.

If the response of the chipless tag in the frequency domain is $H(f)$, then the amplitude response $A(f)$ and phase response $\phi(f)$ of an $N$-bit tag (if there are $N$ number of resonators) can be expressed respectively as:

$$A(f) = \prod_{n=1}^{N} |H_n(f)| \tag{4.1}$$

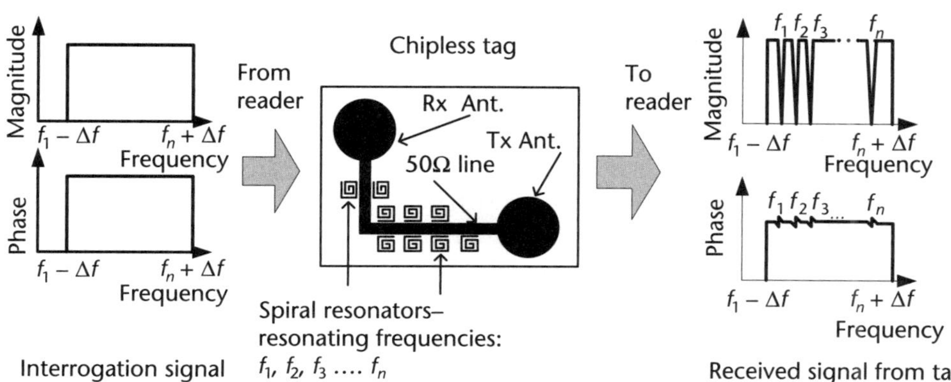


**Figure 4.2**   Structure and operation of a multiresonator based chipless RFID tag.

and

$$\varphi(f) = \sum_{n=1}^{N} \angle H_n(f) \tag{4.2}$$

Using (4.1) and (4.2) it is possible to describe the frequency response (frequency signature) of any multiresonator based chipless RFID tag.

The detection of tags and decoding of data require identification of the attenuations and phase jumps. The two types of readers discussed in this chapter decode data by analyzing the frequency signature in two different ways:

- *Identification of the features of frequency signatures (type-1 readers)*: In this method, the reader's RF section directly compares the signal transmitted toward the tag and the signal received from the tag. A differential microwave gain/phase detector (GPD) processes the raw data in analog gain and phase data obtained from the tag. Based on the comparison of the two signals, attenuations and phase jumps are identified using a simple signal processing technique in the digital section of the reader. In this method, the identification of the attenuations and phase jumps is primarily an RF electronics (hardware) based process. The RF section outputs contain the gain and phase information of the detected tag and the microcontroller merely analyzes the outputs of the RF section.

- *Reconstructing (or recovering) the frequency signature of the chipless RFID tag (type-2 readers)*: In this method, the frequency signature of the chipless RFID tag is reconstructed at the reader using the signal transmitted toward the tag and the signal received from the tag. The output of the RF section is captured and the Hilbert transform (HT) based signal processing technique is applied at the digital section of the reader to recover the frequency signature of the chipless tag being read. In this reader, the output of the RF section does not provide any amplitude or phase information for the tag. It outputs only a specific waveform. This waveform is processed with the HT signal processing technique and the amplitude and phase information of the tag is recovered at the digital section of the reader.

In other words, one type of frequency domain based reader presented in this chapter detects the features of $A(f)$ and $\phi(f)$, while the other reader reconstructs (or recovers) the functions $A(f)$ and $\phi(f)$ using HT based signal processing techniques. The main difference between the two types of the readers is that in type-1 readers, the RF section output contains two separate gain and phase information signals, whereas the RF section output of type-2 readers does not provide such information and those two pieces of information should be recovered with signal processing techniques in the digital section.

The following sections describe the operating principles and the design of the readers in detail.

### 4.2.1   Detecting the Features of the Frequency Signatures of Chipless Tags: Type-1 Readers

The operation of the first type of frequency domain based chipless RFID tag readers is presented in this section. Here, the operating principle of the chipless reader is introduced with a functional block diagram of the process of detecting the features of the frequency signatures of chipless tags. Next the operation of the tag reader is explained by means of a component-level block diagram. The mathematical formulation of the operating principle provides a deeper understanding of this tag reading process.

*Generation of a Wideband Interrogation Signal*
To extract encoded data from a chipless tag, the reader should transmit a suitable interrogation signal and receive the retransmitted signal from the tag. For the interrogation signal, a signal should be generated with a sufficiently wide band to cover all resonating frequencies of the multiresonating circuit. Such a wideband signal can be generated using a VCO, as shown in Figure 4.3. A voltage ramp is applied as the tuning voltage over a certain period of time $(T)$. The frequency output of the VCO varies over the $T$ period of time from $f_1$ to $f_2$, generating a frequency sweep over $f_1$ to $f_2$ with a constant amplitude of wideband microwave power. This signal has a bandwidth  that can be used as the interrogation signal for reading chipless RFID tags. Since the DAC generates discrete voltage steps, the output of the VCO becomes a stepped frequency sweep. However, by applying very small voltage steps (in the millivolt range), the tuning voltage signal can be approximated into a linear voltage ramp. Therefore, the output of the VCO can be approximated into a continuous frequency sweep. Due to the finite settle-in time for generation of each frequency point of the VCOs, it is difficult to achieve very low chip durations $(T)$. A sufficient period of time should be allowed per tuning voltage step to settle down the VCO for providing a stable frequency step. The settle-in time varies with the performance of the VCO. Faster VCOs are typically expensive, and cheaper versions require more time to settle down. Therefore, in the reader design process, we will be required to optimize between the cost and the performance of the VCO.

*Process of Detecting the Features of Frequency Signatures:*
To decode the encoded data on the chipless tags, we must identify the features of the frequency signatures (attenuations and phase jumps introduced by the multiresonating circuit). The identification of the features of the frequency signatures

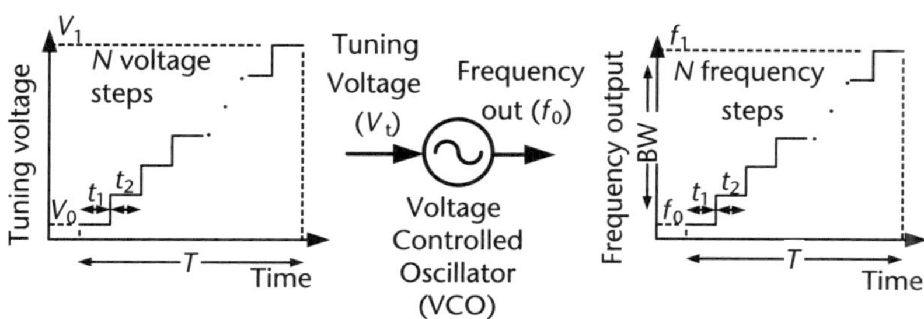


**Figure 4.3**   Generation of a wideband linear stepped frequency signal using a wideband VCO and a DAC.

is done by using a GPD module. The functional block diagram of the process is shown in Figure 4.4. The received signal from the tag and the transmitted signal are compared with each other as shown in Figure 4.4; the differences in amplitude and gain between the transmitted and received signals can be identified in a GPD. The AD8302 GPD is such a module [1], and it is commercially available. At attenuated frequencies, gain and phase information gives peak values for the DC voltage outputs of the GPD. Since the resonance frequencies are known, by observing the peaks, bit 0 can be identified. If there is no peak of gain information or phase information voltage, that can be identified as bit 1. Because this reading process only identifies the attenuations and phase jumps of the frequency signatures, the process can be referred to as "detecting the features of frequency signatures." Decoding of data is done by analyzing the shape of the gain and phase information output signals of the GPD. The output waveforms are sampled with an ADC and stored in the memory of the digital control section. Then the data are processed and converted into digital binary data.

Block diagram of the reader that detects features of frequency signatures: This section provides an in-depth description of the operation of the RFID reader that detects the features of chipless tags with the aid of equations. The component-level block diagram of the reader is shown in Figure 4.5.

The instantaneous frequency of the linear stepped frequency interrogation signal can be expressed as:

$$f(n) = f_0 + Kn \tag{4.3}$$

where, $n = 0, 1$; $K = (f_1 - f_0)/N$, $N$ is the number of steps in the frequency sweep, f0 and $f_1$ are the starting and end frequencies of the sweep, respectively.

Then, the output interrogation signal can be expressed as:

$$X_{TX}(t) = A_0 \cos\left[2\pi(f_0 + Kn)t\right] \tag{4.4}$$

for $t_n < t < t_n +_1$ and A0 is the amplitude of the output signal.

As shown in Figure 4.5, a coupler is used to extract a portion of the transmitted signal for coherent detection as a reference signal. The portion of the transmitted signal extracted from the coupling port of the coupler as a reference signal is:

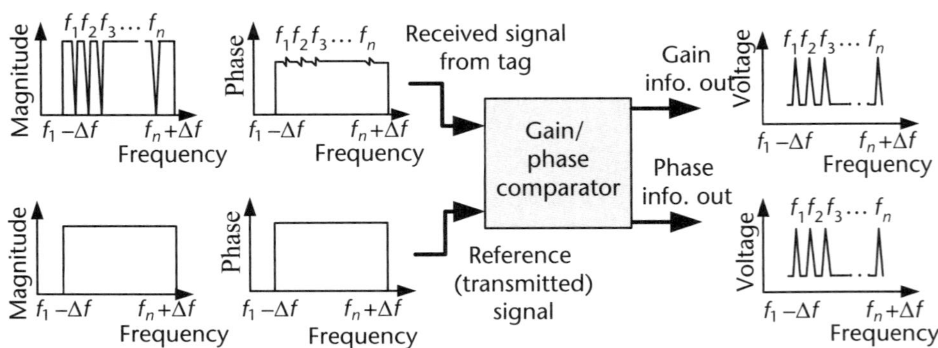


**Figure 4.4** Operation of the reader that identifies the features of frequency signatures.

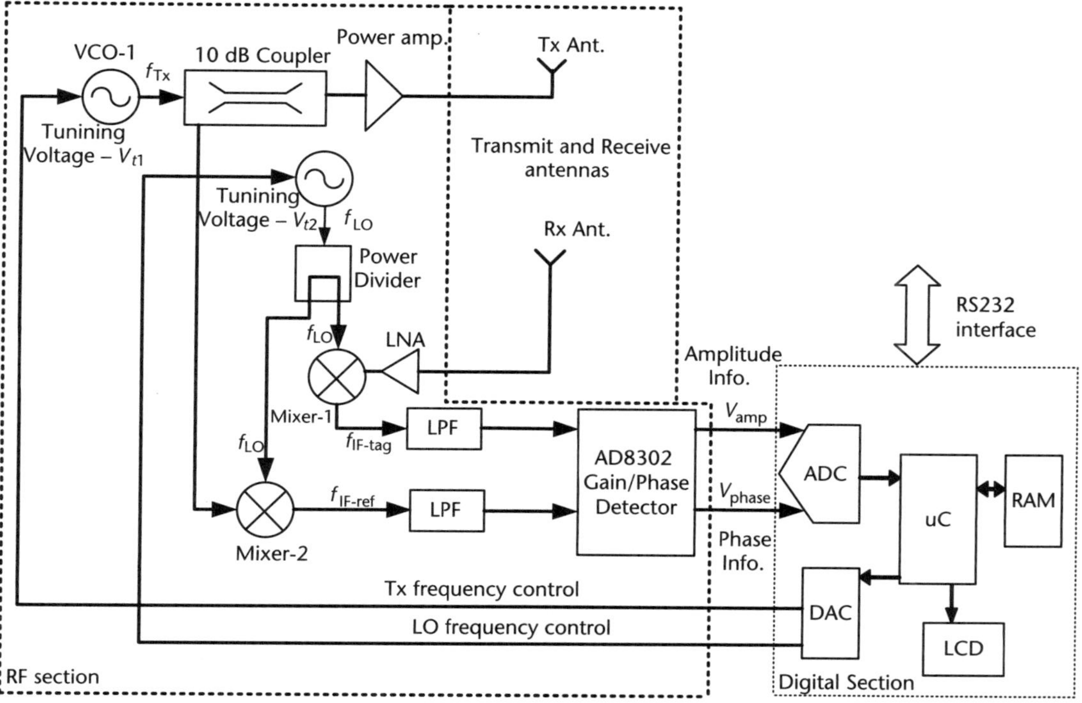


**Figure 4.5**  Block diagram of the reader that identifies the features of frequency signatures.

$$X_{REF}(t) = k_1 X_{TX}(t) \tag{4.5}$$

where $k_1$ is the coupling coefficient of the coupler. The extracted reference signal $X_{REF}(t)$ is directly fed into Mixer-2 as shown in Figure 4.5.

We assume that there are no phase and amplitude distortions at the Tx and Rx antennas of the reader and the tag. A constant path loss over the operating frequency range $L_{total}$ is assumed and the time delay introduced by the tag and the wireless propagation is $\tau_1$.

Then, the instantaneous frequency of the received signal at the receive antenna can be expressed as:

$$Y_{RX}(t) = L_{total} A(f') A_0 \cos\left[2\pi(f_0 + Kn)(t - \tau_1) + \varphi(f')\right] \tag{4.6}$$

where $L_{total}$ is the total loss that occurs between the Tx and Rx antennas over the air interface through the chipless tag, $\tau_1$ is the time taken by the signal to travel from the Tx antenna to the Tx antenna, and $A(f')$ and $\phi(f')$ are the frequency response of the chipless tag. Here, $f' = f_0 + K_n$ where $n = 0, 1, ..., N$.

As shown in Figure 4.5, the output of the variable LO is equally divided into two parts using a two-way power divider. The outputs of the power divider $X_{LO}(t)$ are expressed as:

$$X_{LO}(t) = A_1 \cos 2\pi f_{LO} t \tag{4.7}$$

where $f_{LO}$ can be defined as:

$$f_{LO} = f_0 + \Delta f + Kn \tag{4.8}$$

The VCO-1 and VCO-2 are driven by the outputs of a DAC that is controlled by the microcontroller of the digital control section to keep constant the frequency difference between the two frequency outputs. As shown in (4.3) and (4.8), there is always a constant frequency difference of $\Delta f$ between the two VCO outputs.

The down-converted signals using Mixer-1 with $X_{LO}(t)$ are:

$$\begin{aligned} Y_{IF-RX}(t) &= Y_{RX}(t).X_{LO}(t) \\ &= L_{total}A(f')A_0\cos\left[2\pi(f_0+Kn)(t-\tau_1)+\phi(f')\right].A_1\cos 2\pi f_{LO}t \end{aligned} \tag{4.9}$$

Equation (4.9) can be simplified as:

$$Y_{IF-RX}(t) = k_{IF}\left\{\cos(\theta_1+\theta_2)+\cos(\theta_1-\theta_2)\right\} \tag{4.10}$$

where $\theta_1$, $\theta_2$, and $k_{IF}$ and can be expressed as:

$$\begin{aligned} \theta_1 &= 2\pi(f_0+Kn)(t-\tau_1)+\varphi(f') \\ \theta_2 &= 2\pi f_{LO}t \end{aligned} \tag{4.11}$$

and

$$k_{IF} = L_{total}A_0.A_1$$

where $k_0$ is a constant.

Then the signal $Y_{IF-RX}(t)$ is passed through a LPF to filter out the high-frequency components associated with $\cos(\theta_1+\theta_2)$. The output signal of the LPF then can be approximated to:

$$\begin{aligned} Y_{IF-RX2}(t) &= k_{IF}A(f')\cos\left[2\pi(f_0+Kn-f_{LO})t+\phi(f')-2\pi(f_0+Kn)\tau_1\right] \\ &= k_{IF}A(f')\cos\left[2\pi\Delta ft+\phi(f')-2\pi(f_0+Kn)\tau_1\right] \end{aligned} \tag{4.12}$$

Similarly, the down-converted signal by Mixer-2 can be expressed after passing through the LPF as:

$$\begin{aligned} Y_{IF-REF2}(t) &= k_{REF}\cos\left[2\pi(f_0+Kn-f_{LO})t\right] \\ &= k_{REF}\cos\left[2\pi\Delta ft\right] \end{aligned} \tag{4.13}$$

From (4.12) and (4.13), it can be observed that the frequencies of the two signals are the same. However, there are differences in amplitude and total phase angle. The two signals are fed into the two inputs of GPD to detect the amplitude and phase differences.

*The Operation of the AD8302 Gain Phase Detector*

The functional block diagram of the AD8302 GPD is shown in Figure 4.6. The two inputs to the GPD are amplified with two 60-dB logarithmic amplifiers as shown in the block diagram. Then the outputs of the two amplifiers are passed through a voltage summing circuit to obtain the difference of the two output signals. The output of the summing circuit is amplified further to scale up into the range of 0V to 1.8V. Similarly, to extract the phase difference of the two signals, they are passed through a phase detector as shown in the block diagram. The output of the phase detector is amplified again to match the output range of 0V to 1.8V with an amplifier. The AD8302 GPD allows offset values to be set to the magnitude information output (Vamp) and phase information output (Vphase) signals. However, in this application, setting an offset value to the magnitude and phase information outputs is not required and that feature is not used.

The operation of the GPD can be expressed with two equations as shown in (4.14) and (4.15) according to the block diagram shown in Figure 4.6. The two outputs of the GPD, gain information and phase information voltages are defined as [1]:

$$V_{amp} = V_{SLP} los\left(\frac{V_{IN-A}}{V_{IN-B}}\right) \tag{4.14}$$

and

$$V_{phase} = V_{\phi}\left[\phi(V_{IN-A}) - \phi(V_{IN-B})\right] \tag{4.15}$$

where, $V_{SLP}$ and $V_{\phi}$ are constants [1].

Using the (4.14) and (4.15), the two outputs of GPD $V_{amp}(t)$ and $V_{phase}(t)$ can be expressed as:

$$\begin{aligned} V_{amp}(t) &= V_{SLP} \log\frac{k_{IF}A(f')}{k_{REF}} \\ &= V_{SLP}\left[\log A(f') + \log(k_{AMP})\right] \end{aligned} \tag{4.16}$$

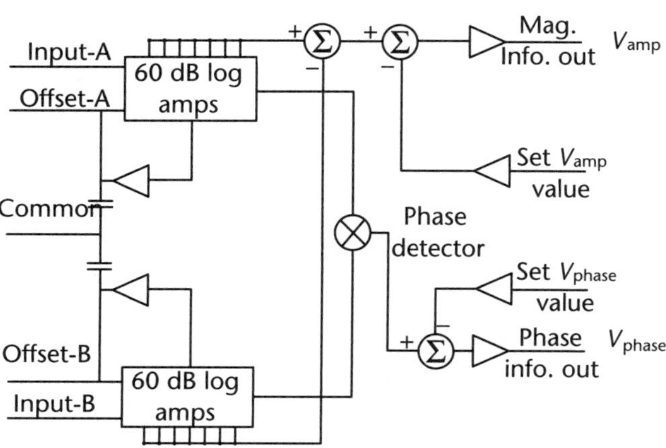


**Figure 4.6**   Functional block diagram of AD8302 gain phase detector [1].

and

$$V_{phase}(t) = V_\phi \phi(V_{IN-A}) - \phi(V_{IN-B})$$
$$= V_\phi \left[ 2\pi\Delta ft + \phi(f') - 2\pi(f_0 + Kn)\tau_1 \right] - \left[ 2\pi\Delta ft \right] \qquad (4.17)$$
$$= V_\phi \left[ \phi(f') - 2\pi(f_0 + Kn)\tau_1 \right]$$

where $k_{AMP} = k_{IF}/k_{REF}$.

Expressions (4.16) and (4.17) shows that the outputs of the GPD vary according to the amplitude and phase differences of the two input signals. The two voltages are sampled and processed in the digital section to identify the resonant frequencies of the tag by comparing them with a calibration (reference) measurement. Because the two output voltages show variations in amplitude and phase information outputs, according to the resonances that occur in the tag, this tag reading technique can be identified as extracting the features of the frequency signatures.

The equations were derived assuming that there are no reflections from the surrounding environment and no leakages of the transmitted signal to the receiver sections. However, in practice, a calibration tag without resonators is used to remove reflections that occur due to the surrounding environment and leakages of transmitted signal to the receiver section. Therefore, a calibration measurement is used to achieve more accurate results and better performance from the reading process.

## 4.2.2 Recovering the Frequency Signature of Chipless Tags: Type-2 Readers

The operation of the second type of the frequency domain based chipless RFID reader presented in this chapter is explained here. The reader reconstructs the frequency signature of the chipless RFID tag with the aid of the received signal from the tag and the transmitted signal from the reader. The detailed description of the tag reading process is explained here.

Figure 4.7 shows the block diagram of the frequency domain based reader that recovers the frequency signature of chipless tags. It wirelessly recovers the amplitude and phase functions of the chipless tag consisting of bandstop filters. The RF transceiver (Tx/Rx) section provides the basic function to recover the frequency signature of the tags being read. A coherent detection is achieved with a bi-static FMCW radar architecture with two Tx and Rx antennas. The digital section assisted with a DAC, drives the VCO, controls the overall operation of the reader and performs signal processing of the re-transmitted signal from the chipless tag. Following is the detailed discussion of the different blocks of the reader. The operation of the reader is explained with the use of mathematical derivations.

RF section: A linear RF chirp signal is used as the interrogation signal in this reader. A sawtooth waveform generator [3, 4] and a VCO generate the linear RF chirp signal. The bandwidth of the swept frequency depends on the specification of the VCO operating bandwidth of the chipless tag. The reader described here is designed with a VCO that operates in the 4- to 8-GHz frequency band.

The instantaneous frequency of the linear chirp interrogation signal can be expressed as [3]:

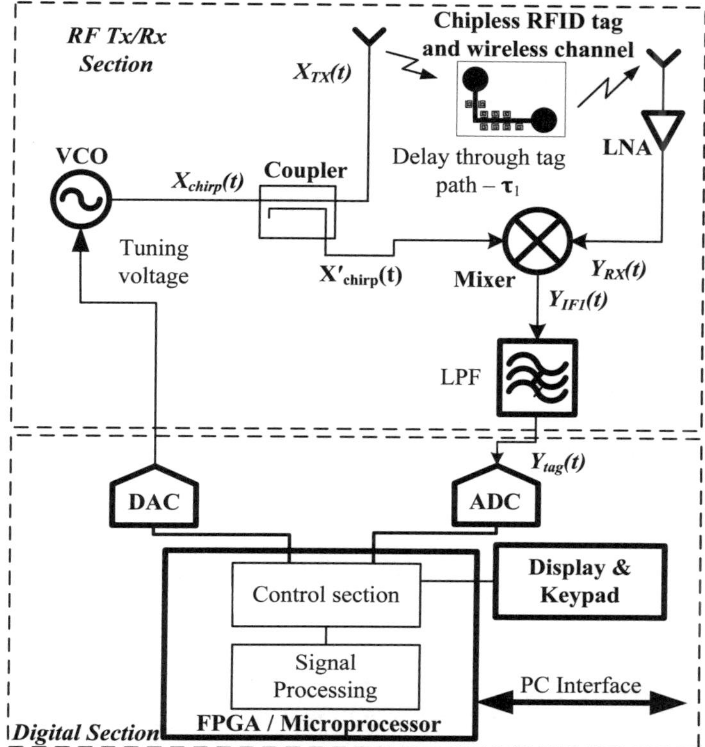


**Figure 4.7**   Block diagram of reader that recovers the frequency signature of chipless RFID tags [2].

$$f(t) = f_0 + Kt \tag{4.18}$$

where $K$ is the chirp rate given by:

$$K = \frac{BW}{T_{chirp}} \tag{4.19}$$

where $f_0$ is the starting frequency of the interrogation signal, BW is the bandwidth of the transmitted interrogation signal, and Tchirp  is the duration of the chirp signal. The output of the interrogation chirp signal generator Xchirp(t) can be expressed as:

$$X_{chirp}(t) = A_1 \cos\left[2\pi\int_0^t (f_0 + Kt')dt'\right]$$
$$= A_1 \cos\left[2\pi\left(f_0 t + K\frac{t^2}{2}\right)\right] \tag{4.20}$$

The output of the chirp signal generator is transmitted toward the tag via the transmit antenna. A coupler is used to extract a portion of the transmitted signal for coherent detection. It is assumed that the antenna responds linearly to the chirped interrogation signal with 100% efficiency (e.g., low-loss microstrip patch antennas). Then the transmitted signal can be approximated as:

$$X_{TX}(t) \approx X_{chirp}(t) \tag{4.21}$$

The portion of the transmitted signal extracted from the coupling port of the coupler is:

$$X'_{chirp}(t) = k_1 X_{chirp}(t) \tag{4.22}$$

where $k_1$ is the coupling coefficient of the coupler. As shown in Figure 4.7, $X'_{chirp}(t)$ is fed directly into the mixer.

As shown in (4.20), the interrogation signal transmitted by the Tx antenna, $X_{TX}(t)$, is:

$$X_{TX}(t) \approx X_{chirp}(t) \approx A_1 \cos\left[2\pi\left(f_0 t + K\frac{t^2}{2}\right)\right] \tag{4.23}$$

The signal propagation and the free-space path loss profile of the tag–reader system is shown in Figure 4.8. As mentioned earlier, it is assumed that the antenna is a linear device and, therefore, there are no phase and amplitude distortions at the Tx and Rx antennas of the reader and the tag. A constant path loss over the operating frequency range [2, 5] is $L_{total}$ and the time delay introduced by the tag and the wireless propagation is  as shown in Figure 4.8.

The total phase $\theta_{RX}(f, t)$ of the received signal $Y_{RX}(t)$ and the amplitude $A_{RX}(f, t)$ at the Rx antenna can be written as:

$$A_{RX}(f,t) = L_{total} A_1 A(f') \tag{4.24}$$

and

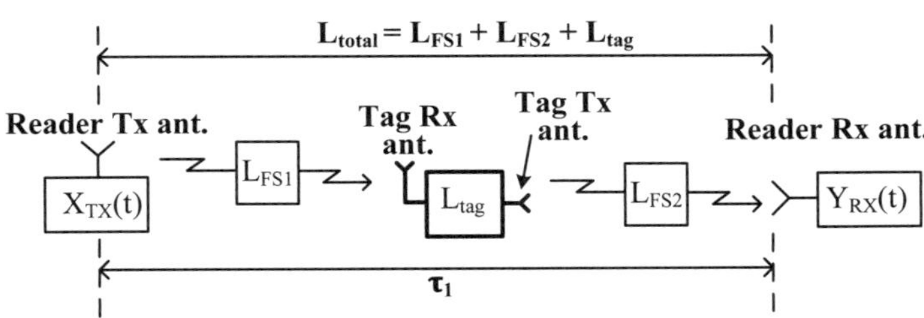


$\tau_1$: Time taken for  the travel from reader Tx ant. to reader Rx ant.

$L_{FS1}$ & $L_{FS2}$ : free space path losses

$L_{tag}$ : insertion loss of the tag

**Figure 4.8** Signal propagation and free-space path loss profile of the reader that recovers frequency signature of tags [2].

$$\theta_{RX}(f,t) = \left[2\pi f_0(t-\tau_1) + \pi K(t-\tau_1)^2 + \phi(f')\right] \tag{4.25}$$

where $f'$ is the normalized time-varying phase of the transmitted interrogation signal, which can be written as:

$$f'(t) = f_0 t + K\frac{t^2}{2} \tag{4.26}$$

The received signal YRX(t) can be written using (4.24) and (4.25) as:

$$\begin{aligned}
Y_{RX}(t) &= A_{RX}(f,t)\cos\left[\theta_{RX}(f,t)\right] \\
&= L_{total}A(f')A_1 \cdot \\
&\quad \cos\left[2\pi f_0(t-\tau_1) + \pi K(t-\tau_1)^2 + \varphi(f')\right]
\end{aligned} \tag{4.27}$$

The IF signal $Y_{IF_1}(t)$ can be written as:

$$\begin{aligned}
Y_{IF_1}(t) &= X'_{chirp}(t).Y_{RX}(t) \\
&= k_1 A_1 \cos\left[2\pi\left(f_0 t + K\frac{t^2}{2}\right)\right].L_{total}A(f')A_1 \\
&\quad \cos\left[2\pi f_0(t-\tau_1) + 2\pi K(t-\tau_1)^2 + \varphi(f')\right]
\end{aligned} \tag{4.28}$$

which can be simplified as:

$$Y_{IF_1}(t) = k_2\left\{\cos(\theta_1 + \theta_2) + \cos(\theta_1 - \theta_2)\right\} \tag{4.29}$$

where $\theta_1$, $\theta_2$, and $k_2$ can be expressed as:

$$\begin{aligned}
\theta_1 &= 2\pi\left(f_0 t + K\frac{t^2}{2}\right) \\
\theta_2 &= 2\pi f_0(t-\tau_1) + 2\pi K(t-\tau_1)^2 + \varphi(f')
\end{aligned} \tag{4.30}$$

and

$$k_2 = \frac{k_1^2 A_1 L_{total}}{2}$$

where $k_2$ is a constant.

The IF signal $Y_{IF_1}(t)$ is passed through a lowpass filter (LPF). Then the high-frequency component that is associated with $\cos(\theta_1 + \theta_2)$ is filtered out. Then the output signal $Y_{tag}(t)$ can be approximated to:

$$Y_{tag}(t) \approx k_2 A(f') \cos\left[2\pi \frac{BW}{T_{chirp}}\tau_1 t + 2\pi f_0 \tau_1 + \phi(f')\right] \tag{4.31}$$

Due to practical limitations such as the high-chirp bandwidth and the required settle-in time for the VCO to stabilize the frequency output between the steps of the frequency sweep, it is difficult to achieve very short chirp durations. Although, it can be seen that the frequency of the output signal varies with the parameters: $T_{chirp}$, $\tau_1$, and BW. The chirp duration Tchirp is the main parameter that determines the frequency of $Y_{tag}(t)$. Therefore, the frequency of the signal $Y_{tag}(t)$ gets very low values in practice and depends more on $T_{chirp}$ than on $\tau_1$ and $BW$.

Since the $Y_{tag}(t)$ contains the information about the frequency signature (phase and amplitude information) of the chipless RFID tag, by analyzing the $Y_{tag}(t)$, the two functions $A(f)$ and $\phi(f)$ can be recovered to identify the data bits encoded in the chipless RFID tag.

Recovering the amplitude and phase responses of the chipless tag using Hilbert transform: Hilbert transform based complex analytical signal representation is used to extract the frequency signatures of chipless tags. Using the HT, it is possible to reconstruct both amplitude and phase responses of the chipless tags.

Any real signal that can be expressed as the sum of a series of sinusoids, a transform can be applied to shift each sinusoidal component in phase by $(\pi/2)$. This transform is called the Hilbert transform [6]. Assume that $x(t)$ is a real signal and the HT output $y(t)$ can be denoted as . The signal $z(t)$ is then called the complex analytical signal of $x(t)$ and it can be expressed as [6, 7]:

$$z(t) = x(t) + i[y(t)] = x(t) + i\mathrm{H}[x(t)] \tag{4.32}$$

The complex analytical signal $z_{tag}(t)$ of the signal represented in (4.31) can be transformed as follows:

$$z_{tag}(t) = k_2 A(f')\{\cos(2\pi K\tau_1 t + 2\pi f_0 \tau_1 + \phi(f')) + i\sin(2\pi K\tau_1 t + 2\pi f_0 \tau_1 + \phi(f'))\}$$
$$= k_2 A(f')\exp i\left(2\pi \frac{BW}{T_{chirp}}\tau_1 t + 2\pi f_0 \tau_1 + \phi(f')\right) \tag{4.33}$$

The envelope of the signal  is given by the magnitude of the complex analytical signal $z_{tag}(t)$, which is expressed by:

$$\left|z_{tag}(t)\right| = \left|k_2 A(f')\exp i\left(2\pi \frac{BW}{T_{chirp}}\tau_1 t + 2\pi f_0 \tau_1 + \phi(f')\right)\right| = k_2 A(f') \tag{4.34}$$

It can be seen that (4.34) represents a scaled amplitude response of the chipless tag. The argument of $x_{tag}(t)$ gives the phase angles of the signal $Y_{tag}(t)$ as follows:

$$\arg\left[z_{tag}(t)\right] = \arg\left[k_2 A(f')\exp i\left(2\pi\frac{BW}{T_{chirp}}\tau_1 t + 2\pi f_0\tau_1 + \phi(f')\right)\right]$$

$$= 2\pi\frac{BW}{T_{chirp}}\tau_1 t + 2\pi f_0\tau_1 + \phi(f') \tag{4.35}$$

The first two terms of (4.35) show a linear variation of phase with time with a constant phase shift of $2\pi f + \tau_1$ due to the time taken by the signal to travel from the Tx antenna of the reader through the tag and be received by the Rx antenna of the reader as shown in Figure 4.8. The third term of (4.35) is easily identifiable as the phase information, $\phi(f')$ of the frequency signature of the chipless tag.

In summary, by applying HT to the output of the RF section, we can reconstruct the frequency signature. Using this information it is possible to identify the bits encoded in a chipless tag.

## 4.3    Design of Frequency Domain Based Chipless RFID Readers

This section covers the design of two major subsections of the readers, the RF section and the digital section. The RF and digital sections consist of a number of active and passive components. A detailed description of each component used in the readers is presented in Chapter 8. In this section, the emphasis is on the system-level design of each subsection of the reader.

### 4.3.1    Design of a Reader That Detects the Features of Frequency Signatures of Chipless RFID Tags

The design of the reader is done in two parts, the RF section and the digital section. The two sections are made on two separate PCB modules that can be plugged into a metallic housing. The fabricated reader is shown in Figure 4.9. The RF section is mounted on top of the digital section using two connectors. The reason for designing two PCBs is the operating frequency ranges of the two sections. The RF section

**Figure 4.9**    Fabricated reader that identifies the features of frequency signatures.

operates over a wide frequency band of 4 GHz (4 to 8 GHz). However, the digital section and other associated components work up to the maximum frequency range of 20 MHz. Because of the wide frequency bandwidth of operation of the RF section, a high-performance PCB substrate is required to achieve the desired performance. Taconic TLX-0, 0.5-mm-thick high-performance RF substrate is used in the RF PCB design. It is possible to use the same PCB substrate for the digital section and power supply unit. However, the high-performance PCB substrate is very expensive. It is not economical to use a single PCB for the whole reader design. Therefore, the digital section is designed on an FR-4 substrate, which provides a more economical solution for the PCB design.

The digital control section controls the entire operation of the reader, and the operation can be divided into three functions: (1) transmitter control, (2) receiver control, and (3) signal processing and data decoding. The digital control section is designed to use a reader for tags with different numbers of bits without changing its hardware configuration. To generate a frequency sweep, a voltage ramp is generated using an MCP4912 digital-to-analog converter (DAC). According to the specifications of the VCO, to operate in the frequency range of 4 to 8 GHz, the voltage ramp applied to the tuning voltage should be between 0V and 20V. The generation of the tuning voltage is achieved by changing the output of the DAC with 1024 steps, which gives 1.96-MHz (2 GHz/1024) resolution per input tuning voltage step, with a DAC.

For sampling the gain and phase voltages, the built-in 10-bit analog-to-digital controller of a PIC 18F452 microcontroller running at 20-MHz clock speed is used. It has an 8-to-1 built-in analog multiplexer, which can be used to sample multiple analog voltages if multiple RF sections are connected to the same digital controller. According to the data sheet of the AD8302 GPD, gain and phase outputs vary between 30 mV/dB and 10 mV/deg, respectively. Since the two analog outputs of the GPD vary between 0V and 1.8V, the reference voltage of the ADC was set to 1.8V, which gives 1.77-mV (1.8V/1024) resolution and provides adequate resolution for sampling the two analog voltages. The sampled data are stored in a 32-kbyte RAM chip and addressing of the RAM is done using a 15-bit binary counter driven by a microcontroller pin. Although the use of a counter to address the RAM increases the access time, it reduces the number of utilized pins of the microcontroller so that a low-cost microcontroller can be used in the design. An RS-232 interface is added to the design for communicating with a PC, as shown earlier in Figure 4.5.

The procedure and sequence of operation explained in the previous section was implemented on a PIC 18F452 microcontroller. Specifications of the PIC 18F452 microcontroller can be found in its data sheet. C language was used to develop the firmware for the microcontroller. The flowchart of the whole controlling algorithm and data decoding algorithm is shown in Figure 4.10.

### 4.3.2  Design of a Reader That Recovers the Frequency Signatures of Chipless RFID Tags

The design of this reader is also done in two parts: as the RF section and the digital section. The RF section of the reader architecture was realized on Taconic TLX-0 ($\varepsilon = 2.45$, tan $\delta = 0.02$, $h = 0.5$ mm) high-performance RF substrate. The digital section was realized on low-cost FR-4 PCB substrate. The two sections are made on

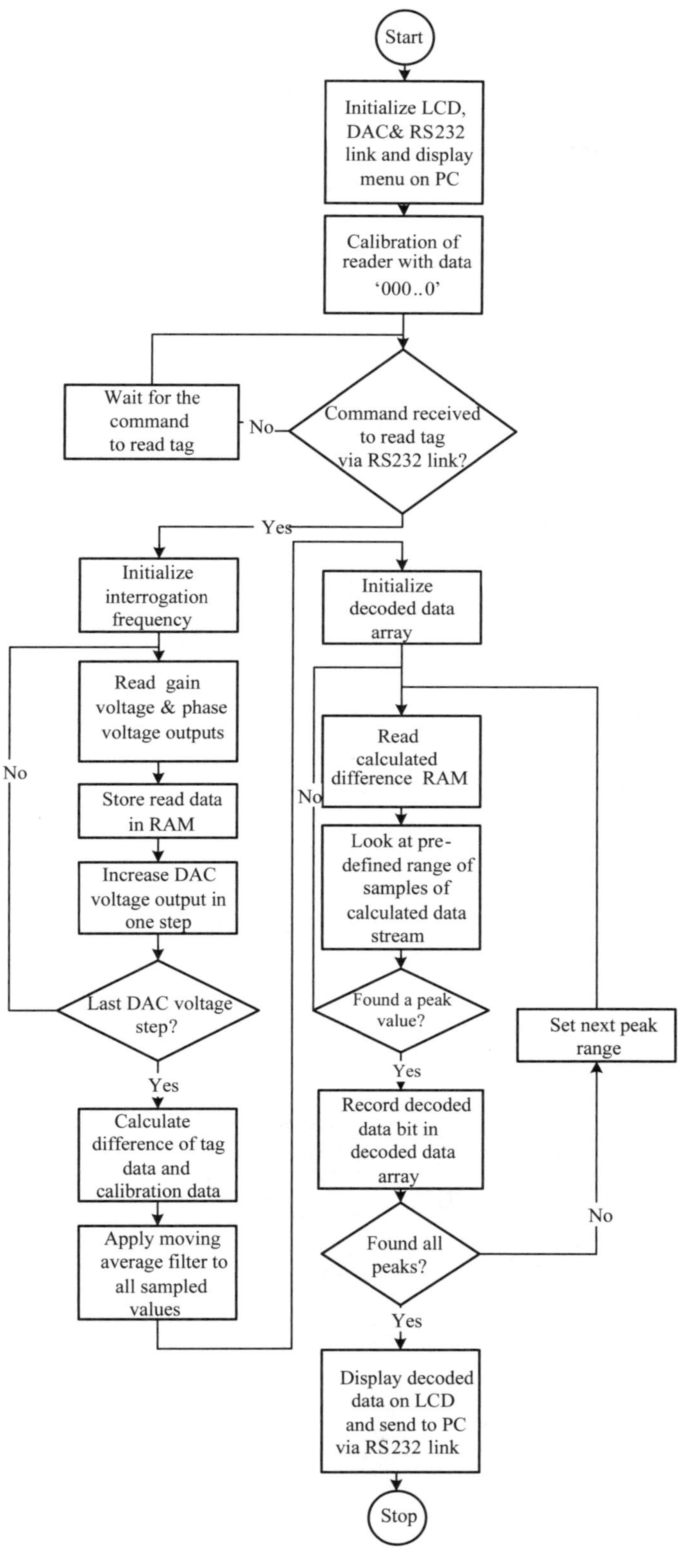



**Figure 4.10**  Flowchart of the signal processing function of the reader that identifies the features of the frequency signatures of chipless tags [2].

two separate PCBs that can be plugged in together, similar to the previous one, to minimize the cost of the expensive PCB substrate. The fabricated reader is shown in Figure 4.11. The reader is designed to operate in the frequency range of 4.0 to 8.0 GHz, and all components were chosen to operate in this frequency range. A Richardson RVCSD6000F wideband VCO [tuning bandwidth = 4 to 8 GHz, tuning sensitivity (fairly constant) = 200 MHz/V, output power = 10 dBm, phase noise at 1 MHz offset = −80 dBc/Hz] is used in the reader design as the signal source. The VCO operates in the frequency band of 4 to 8 GHz. A −15-dB asymmetric coupled line coupler was realized on the same RF PCB section. Surface-mount LPFs with a cut-off frequency of 600 MHz and a Mini-Circuit SIM-14+ double balanced mixer are used to construct the reader along with other parts. A noninverting amplifier is used at the output of the RF section, and the amplified output is interfaced with the digital control section.

The output of the RF section $Y_{tag}(t)$, is sampled with the internal ADC of the PIC 18F452 microcontroller, and the sampled signal is sent to a computer for further processing with Matlab for extracting the frequency signature. In this work, Matlab is used since it is more flexible and easier than using a microprocessor or FPGA based signal processing algorithms with an experimental setup. However, the necessary signal processing can be easily done at the reader by using a microprocessor or an FPGA once the signal processing algorithms have been finalized and fine tuned to suit the application. The different components and units in the RF and digital section require different voltage levels. Therefore, a power supply section is also included in a part of the digital section PCB.

As shown earlier in Figure 4.7, the digital section consists of an ADC, a DAC, a digital control section, and an FPGA or a microcontroller. The frequency of the output signal from the RF section $Y_{tag}(t)$ depends on the parameters $T_{chirp}$, $\tau_1$ and $BW$ of the interrogation signal. Due to the settle-in time required for the VCO, it is difficult to achieve very high speed frequency sweeps. Further, the value of $T_{chirp}$ can be controlled by the digital section. Since, the frequency of $Y_{tag}(t)$ mainly depends on the parameter $T_{chirp}$, it is possible to achieve 2-GHz frequency sweeps with 5-MHz frequency steps in 500 ms using a VCO with average performance.

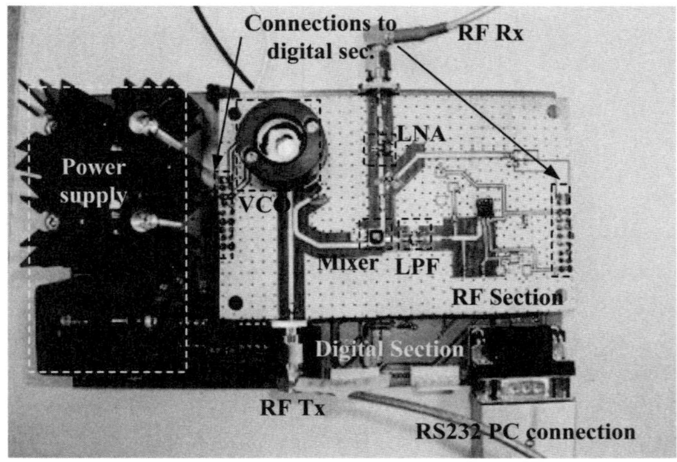


**Figure 4.11**    Fabricated reader that recovers the frequency signatures of chipless RFID tags [2].

Therefore, the output of the RF section generates very low frequency outputs. Therefore, a very low sampling rate is used in the ADC in the digital processing of the reader. A microprocessor or an FPGA can be used to control the ADC and DAC and to carry out the signal processing of the reader [8, 9]. Decoded data can be transferred to a PC using a serial or parallel interface or displayed on the LCD screen of the reader.

The flowchart of the firmware of the digital control section that performs the detection and display of data is shown in Figure 4.12. The number of samples to be acquired over the frequency span of the linear chirp is set at first. The output of the RF section, $Y_{tag}(t)$, is progressively sampled $N$ times and stored after generating the chirp signal. Next, the HT of the stored data is carried out followed by the computation of the analytical signal $z_{tag}(t)$. The computed complex analytical signal represents the amplitude and phase response of the chipless tag. Using the two pieces of information contained in the complex analytical signal, attenuations and phase jumps are identified to decode the encoded data on the tag. The decoded

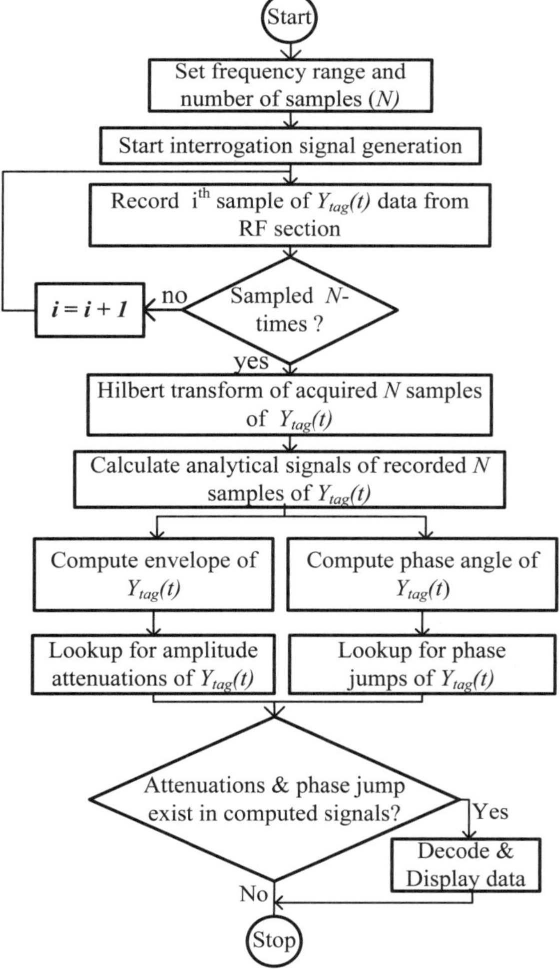



**Figure 4.12**   Flowchart of the signal processing for a reader that reconstructs the frequency signature of chipless tags [2].

data are finally transferred to a PC or displayed on the LCD screen of the reader. The control and signal processing algorithm can be implemented in the form of the firmware of a microprocessor or an FPGA [8, 9].

## 4.4  Results

Here, the results of the two frequency domain based readers are presented. The results presented in this section are obtained using the fabricated prototypes of the two types of readers. The results presented in this section show the functionality of the readers and how the data bits are decoded from the outputs obtained from the RF sections of the readers. To prove the operation of the reader, 9-bit multiresonator based chipless RFID prototype tags with 9 bits encoded were tested with the reader. The tested tags were realized on Taconic TLX-0 substrate ($\varepsilon = 2.45$, tan$\delta$ = 0.0019, $h = 0.7874$ mm). The operating frequency range of the tags is 4 to 6 GHz. The measured resonant frequencies of the multiresonator based chipless RFID tags used for the experiments are listed in Table 4.1. The measurements were recorded using an Agilent E8361A performance network analyzer (PNA).

### 4.4.1  Results: Detecting the Features of Frequency Signatures

Figure 4.13 shows the measured results of reading two 9-bit chipless RFID tags. The sampled gain voltages give a local minimum point with respect to the adjacent areas when a resonance occurs in the multiresonating circuit. Figure 4.13(a) shows nine saddle points in the amplitude information, and Figure 4.13(b) shows nine phase ripples of the phase information for the tag with encoded data "000000000." The difference between the calibration tag (tag with data "111111111" or with no resonators) gives clear peaks in amplitude as shown in Figure 4.13(a), which make easier to identify the resonances. This information can be easily converted into binary data bits. However, the difference of phase information of the two tags does not provide a distinguishable signal that can be used to identify the resonance

**Table 4.1**  Measured Resonance Frequencies of the Multiresonator Circuit of a Chipless RFID Tag and Attenuations and Phase Angle Jumps of the Frequency Signature

| Bit No./ Resonator No | Resonance Frequency/ GHz | Magnitude of Resonance/dB | Variation from Linear Phase |
|---|---|---|---|
| 1 | 4.09 | −6.7 | 29.8 |
| 2 | 4.37 | −7.1 | 32.3 |
| 3 | 4.53 | −5.9 | 20.7 |
| 4 | 4.74 | −6.8 | 26.0 |
| 5 | 4.96 | −6.9 | 26.8 |
| 6 | 5.10 | −7.5 | 32.3 |
| 7 | 5.43 | −8.4 | 36.4 |
| 8 | 5.65 | −9.0 | 34.2 |
| 9 | 5.83 | −9.0 | 37.4 |

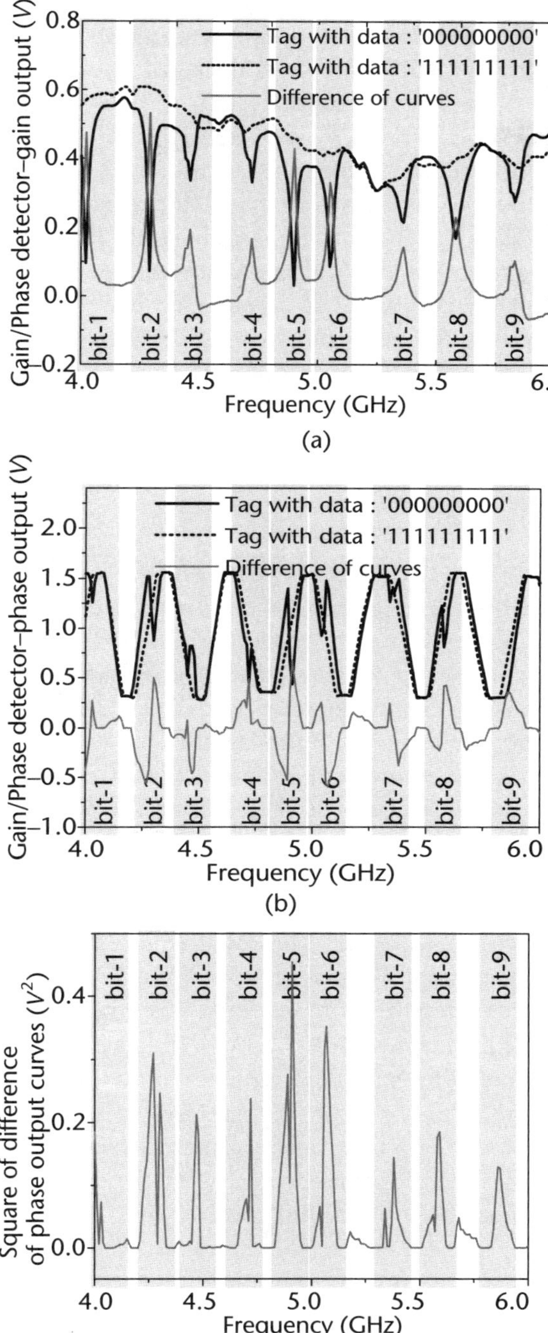



**Figure 4.13**   Measured results of the reader that identifies the features of frequency signatures: (a) measured GPD , amplitude information outputs for two tags with encoded with data "000000000" and "111111111"; (b) measured GPD phase information for two tags with encoded with data "000000000 " and "111111111"; and (c) square of the difference of phase voltage outputs for tag with data "000000000."

frequencies. Therefore, if the signal is squared, it shows very clear peaks with respect to nonresonating frequencies, as shown in Figure 4.13(c). Therefore, it can be seen that these two pieces of information (amplitude and phase) can be extracted using the reader and they can be used to identify the data encoded in any multiresonator based chipless RFID tag. From the results, it is clear that the reader identifies the resonance frequencies by extracting the differences that occur in amplitude and phase information generated by the GPD.

Reading and decoding time takes about 1 second per tag. The reason for taking such a long time per reading is that, when generating the voltage ramp to obtain the frequency sweep, a 1-ms time period should be allowed to obtain a stable frequency output from the VCO. If the bandwidth is increased, it will take more time because more frequency steps will be generated. However, it is possible to reduce the reading time by only sweeping through the frequencies where resonances occur for a chipless tag with known resonance frequencies. Data processing time is very much shorter than the frequency sweep time. Therefore, to increase the tag reading speed, a faster frequency sweep should be used or a different technique is required. A time domain based faster tag reading technique is presented in Chapter 5.

### 4.4.2    Results: Recovering the Frequency Signature of Chipless RFID Tags

The results of the tag readings done with the built reader prototype that reconstructs the frequency signature of chipless RFID tags are presented in this section. The same tags that have 9 encoded bits (operating in the 4- to 6-GHz frequency range) were used with this reader. Since the tags operate only in the 4- to 6-GHz frequency band, the tuning voltage for the VCO was generated using the DAC of the digital section only to sweep the frequency from 3.9 to 6.1 GHz (2.2 GHz BW). A 5-MHz frequency step was used and the chirp duration Tchirp was 500 ms.

The computed envelope of the signal $Y_{tag}(t)$ and the measured output signal of the RF section, $Y_{tag}(t)$, are shown in Figure 4.14. Figure 4.14(a) shows the amplitude response of the tag as expected with proper attenuations according to the resonators of the chipless tag. The calculated phase angle of the tag signal $Y_{tag}(t)$ shows clear phase ripples (varying from 5° to 20°) at their corresponding resonators of the chipless tag. The phase ripples are highlighted in Figure 4.15(b). These two plots of the measured results show that the reader reconstructs both the amplitude and phase responses of the frequency signature of the chipless tags. Using the reconstructed amplitude and phase responses, it is possible to develop an algorithm that detects the saddle points in the amplitude response and discontinuities in the gradient profile of the phase response. Using the amplitude and phase information, the tag is decoded as "000000000" and "010101010."

## 4.5    Conclusion

Two designs for low-cost frequency domain based readers for reading multiresonator based chipless RFID tags were presented. The two readers work on two different operating principles. A type-1 reader detects the features of the frequency signatures of chipless tags, whereas a type-2 reader reconstructs the frequency signatures using the received signals.

The reader that detects the features of frequency signatures uses two VCOs in the RF section. The two VCOs were driven in such a way as to keep a constant

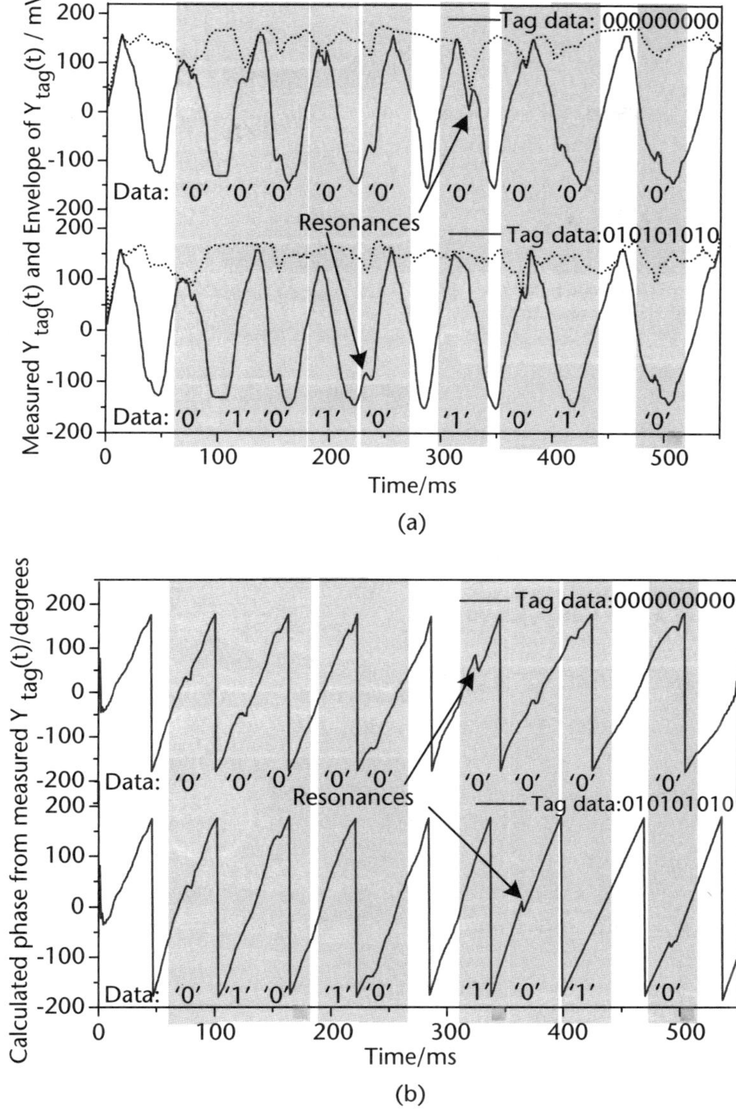



**Figure 4.14**   Measured results of the reader that recovers the frequency signature of chipless tags using Agilent ADS software [2]. (a) Measured outputs of RF section (solid lines) and computed amplitude response of tags (dotted lines), and (b) computed phase response of tags.

frequency difference between the two outputs using the tuning voltages generated by the DAC in the digital section. Using this VCO driving technique, it was possible to achieve a constant IF over the entire operating frequency range of the reader. A GPD was used to compare the transmitted and received signals to identify the attenuations and phase ripples generated by the resonators of the chipless RFID tag. The reader prototype presented in this work operates in the 4- to 8-GHz frequency range.

The type-2 reader, which reconstructs the frequency signature of chipless tags, needs only one VCO, which is the most expensive component in the reader system.

Mixing of the transmitted signal with the received signal and chirped interrogation signal relinquishes the requirement for the second VCO as a variable local oscillator. A compact, less expensive, and simpler RF section is also achieved in type-2 readers compared to type-1 readers, which have two VCOs. Moreover, the type-2 reader does not use a calibration tag to decode data. This is another advantage of this reader. The simple signal processing algorithm based on HT provides for the extraction of the amplitude and phase functions of the frequency signature of the chipless RFID tag.

Both theoretical analysis and experimental validation were carried out to validate the operation of the two readers. The accurate detection of a 9-bit chipless RFID tag using both amplitude and phase information was achieved with both types of readers.

The concept of the reader with a single antenna and a single VCO is proposed for achieving a more compact and less expensive reader device. In that reader, the same antenna is used for the communication and the VCO is used as the signal source. The receiver section contains a detector and filter network followed by a microprocessor-assisted signal processing unit.

Future development will concentrate on reducing the cost and enhancing the compactness of the reader and improving the reading range and reading speed of the reader. The operating bandwidth of the reader is determined only by the operating frequency bandwidths of the components used in the design. By using components that can be operated in wider frequency bandwidths, the operating frequency bandwidth of the reader can be further expanded.

The frequency domain based tag reading techniques consume a considerable amount of time to generate a suitable interrogation signal, and subsequent signal processing methods are needed for the detection process. This limits the usability of the frequency domain based readers in applications that require high-speed tag reading. Therefore, to reduce the tag reading time and enable high-speed tag reading, a time domain based chipless RFID reader is proposed in Chapter 5.

## Questions

1. How would chipless RFID technology load information on the tag?
2. What are the differences between first-generation and second-generation readers?
3. What are the advantages of second-generation readers compared with first-generation readers?
4. What is the disadvantage of using of a counter to address the RAM?
5. Why it is difficult to achieve very high speed frequency sweeps?
6. What is the average time for the second-generation VCO to sweep per step? What is the IF output frequency in this case?
7. Why is the IF of second-generation readers so low?
8. What factors affect the IF of second-generation readers?
9. How long does it take for a current reader to read one tag?
10. How can tag reading time be reduced?

# References


[1] *AD8302: 2.7GHz RF/IF Gain Phase Detector*," June 2002, http://www.analog.com/static/imported-files/data sheets/AD8302.pdf (accessed June 12, 2012).

[2] R. V. Koswatta and N. C. Karmakar, "A Novel Reader Architecture Based on UWB Chirp Signal Interrogation for Multiresonator-Based Chipless RFID Tag Reading," *IEEE Trans. on Microwave Theory and Techniques*, Vol.60, No. 9, pp. 2925–2933, 2012.

[3] B. Edde, *Radar: Principles, Technology, Applications*. Upper Saddle River, NJ: Prentice Hall, 1999.

[4] J. Detlefsen et al., "UWB Millimeter-Wave FMCW Radar Using Hilbert Transform Methods," in *2006 IEEE Ninth International Symposium on Spread Spectrum Techniques and Applications*, Manaus-Amazon, Brazil, 2006, pp. 46–48.

[5] S. Mukherjee, "Chipless Radio Frequency Identification by Remote Measurement of Complex Impedance," in *European Conference on Wireless Technologies*, Munich, Germany, 2007, pp. 1007–1010.

[6] S. L. Hahn, *Hilbert Transforms in Signal Processing*. Norwood, MA: Artech House, 1996.

[7] I. Kollar, R. Pintelon, and J. Schoukens, "Optimal FIR and IIR Hilbert Transformer Design via LS and Minimax Fitting," in *7th IEEE Instrumentation and Measurement Technology Conference (IMTC-90)—Conference Record*, San Jose, CA, 1990, pp. 240–243.

[8] M. A. Hassan, A. M. Youssef, and Y. M. Kadah, "Embedded Digital Signal Processing for Digital Ultrasound Imaging," in *28th National Radio Science Conference (NRSC 2011)*, Cairo, Egypt, 2011, pp. 1–10.

[9] J. H. Chang, J. T. Yen, and K. K. Shung, "A Novel Envelope Detector for High-Frame Rate, High-Frequency Ultrasound Imaging," *IEEE Trans. on Ultrasonics, Ferroelectrics and Frequency Control*, Vol. 54, No. 9, pp. 1792–1801, September 2007.

## CHAPTER 5

# Time Domain Based Chipless RFID Reader

## 5.1 Introduction

This chapter presents the operating principle for a time domain (TD) based chipless RFID reader. The frequency signature of a chipless tag is identified using a short ultra-wideband (UWB) RF pulse as the interrogation signal and a TD based detection technique. The received TD signal from the tag can be directly processed with a fast Fourier transform (FFT) based signal processing technique to obtain the frequency signature of the tag. However, direct processing of the received signal requires high-performance analog and digital hardware. This greatly increases the cost of the reader. In this chapter, a simple and low-cost TD based reader receiver architecture, which comprises a low noise amplifier (LNA), a 1-to-n-way-power divider, narrow bandpass filters (BPFs), envelope detectors, and comparators, is proposed. The reader yields high-speed reading capability. The proposed reader architecture is augmented with theory, design, simulation, and performance benchmark results.

The previous chapter dealt with the design of two types of chipless RFID readers using frequency domain (FD) reading techniques. The two FD readers operate over the UWB frequency band. The readers extract the frequency signatures of a chipless RFID tag fully within the FD. In the FD reading method, a few issues hinder high-speed reading: (1) the settling time and linearity of the VCO output, (2) the number of frequency points required to determine the frequency signatures, (3) the sampling rate, (4) the synchronization of VCO driving signals with sampling of the received signals, (5) the nonlinearity of other active and passive components, (6) the transmitted power, and (7) the number of data bits to be extracted.

An alternate approach to improve reading performance and speed is a TD based impulse radio (IR) technique. In a TD-IR based reader, a short RF pulse that spreads the spectral contents (real power) over the UWB frequency band is transmitted. The result is that a very fast reading speed is achieved via the UWB-IR based reader. In traditional TD based analytical techniques, the waveforms of time-varying signals at various stages of a system are captured to examine the performance of the system. In the proposed reading method, the time-varying waveforms and their representative frequency domain spectral contents will be examined at each important stage of the reader. This anatomical approach to the processed signal of a frequency signature based chipless tag interrogated with a time domain impulse will create a unique discipline.

As stated above, the tag reading speed of a FD based reader is limited by various factors. The UWB FD chipless tag may operate in a very wide frequency band from 3.1 to 10.6 GHz[1]. It is difficult to generate a very high speed interrogation signal with a sufficient number of frequency steps to interrogate a chipless tag with a wideband VCO. With the operating frequency bandwidth and the number of frequency points generated by the VCO, the required settling time for stable frequency outputs increases the tag reading time. As an example, our current reading time for a 9-bit chipless tag is approximately a few seconds. To increase the tag reading speed, use of a high-performance VCO with a low settling time is envisaged. However, this solution would increase the cost of the reader. Therefore, it is imperative to investigate alternative options that generate interrogation signals that have very high reading speeds and cover the required operational bandwidth.

In UWB communication systems, short RF pulses are used as transmitting signals. These short RF pulses have a several gigahertz operational bandwidth. The typical width of a single UWB RF pulse is less than 10 ns. Therefore, the use of a short RF pulse as the interrogation signal opens up a new discipline to achieve high-speed reading of chipless RFID tags. Figure 5.1 shows a comparative study of frequency and time domain reading techniques and the advantages of TD based reading techniques over FD based reading techniques. As shown in the figure, the TD based reader enjoys the unique advantages of short transmission times and a simple architecture.

Chapter 3 gives a detailed description of the design and operation of a commercial TD based SAW chipless RFID tag reader. In addition to the SAW RFID tag reader, there are a few TD based chipless tag reading techniques reported in the scientific literature. However, they are still in the inception phase with only laboratory prototypes reported. A comprehensive review of recently reported works on UWB-IR based chipless RFID tag reading systems is given below. This review exposes the significance of the proposed reading technique described in this chapter. As mentioned in Chapter 3, to the authors' best knowledge there is no reported UWB-IR based commercial chipless RFID system except for the RF-SAW. Therefore, this chapter presents a fundamental concept in the field of chipless RFID tag readers.

TD based chipless RFID systems [1–6] use UWB short pulses as the interrogation signals. The transmitter of the reader comprises a UWB impulse generator that generates a short RF pulse. The width of the RF pulse is in the range of a few hundred picoseconds to a few nanoseconds [4, 5] based on required specifications. Most of the TD-based chipless RFID tags [2, 4, 5] are based on the principle of time domain reflectometry (TDR). The reader's receiver decodes the encoded data bits in a chipless RFID tag in the TD. The receiver block contains a high-speed sampling device. The reported experimental work has been carried out using either a PC based high-speed sampler [4] or a high-speed digital storage oscilloscope (DSO) [1, 5]. The sampled signal is processed with different signal processing techniques, described as follows.

Lazaro et al. [4] used a continuous wavelet based signal processing technique to detect a chipless RFID tag based on TDR. A time coded tag with a terminated

---

1.    The authors' group designed a UWB FD chipless tag for the 4- to 8-GHz frequency band [27].

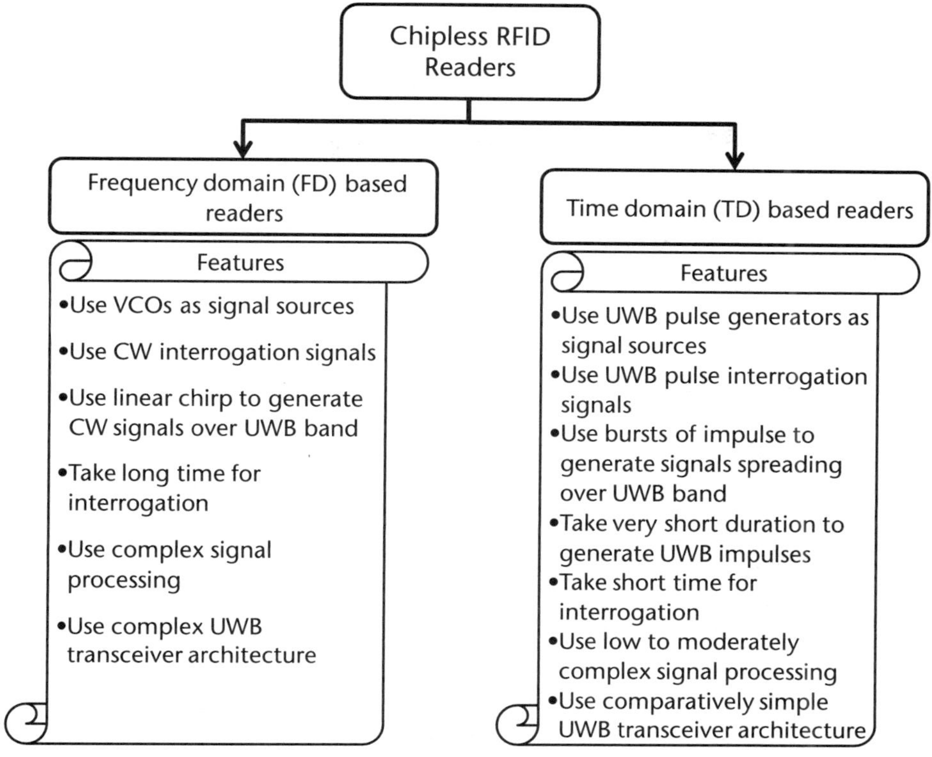

**Figure 5.1**  Comparison of the salient features of frequency domain based and time domain based chipless RFID readers.

transmission line is used in this experiment. Hu et al. [5] analyzed the antenna mode scattering for a UWB pulse of the tag. The tag has a simple passive structure with one antenna and a terminated transmission line. Kalansuriya and Karmakar [2] proposed a different tag design with a multiple spiral resonator loaded transmission line and a UWB monopole antenna. They use a technique similar to that proposed in [5] to identify the signature of the tag. Another TDR based chipless tag is reported by Zheng et al. [3]. A transmission line with capacitive discontinuities is used to encode the data. By analyzing the reflected signal recorded with high-speed DSO, the bits are identified [3]. A pulse position modulation (PPM) based UWB chipless RFID tag has been proposed by Gupta et al. [1]. The chipless tag contains two cross-polarized antennas as the receiving (Rx) and transmitting (Tx) antennas. A dispersive delay structure (DDS) is used to obtain different group delays for different frequencies. The retransmitted pulses are recorded with a high-speed DSO and analyzed to obtain the encoded data [1]. The use of these types of expensive laboratory equipment does not suit commercial applications. A commercial-grade reader needs to have a low-cost design if it is to meet market demands. So far no integrated chipless RFID reader architecture has been proposed in the open literature. Therefore, there is room for further development of a TD based reader in an integrated package. This chapter highlights some of the practical issues of a TD-based chipless tag reader.

The FD chipless RFID tag uses a frequency signature to store data bits in a fully passive microwave structure. The chapter has adopted a TD based technique to interrogate the FD chipless tag with a UWB-IR signal. The TD based receiver decodes the frequency signature of the tag into meaningful ID information.

The chapter is organized as follows. First, various techniques for generating UWB short RF pulses are described. Then the operating principle of the TD based chipless RFID system is presented. The whole tag and reader system is designed on an Agilent ADS 2009 platform. A simulation is performed to examine the performance of the reader. Practical implementation issues, including the availability of components to develop a commercial-grade reader, are presented. Finally, results are discussed and conclusions drawn.

## 5.2 Theory of Operation of a Time Domain Based Chipless RFID Reader

The operation of a TD based chipless RFID reader is described in this section. The reader can be divided into two main functional units. The first functional unit is the UWB short RF pulse generator. Because the short RF pulse is used as the interrogation signal, it is important to understand the operation of the impulse signal generation. The second functional unit comprises the receiver and a data decoding section. The operation of the reader is described in this section based on these two main functional units.

### 5.2.1 Generation of UWB Short Pulses

Figure 5.2 shows the two common types of UWB short pulse generators in UWB communication systems [7]. The two commonly used techniques are:

- Use of a pulse generator and an up-converter ;
- Use of a short pulse generator and a pulse shaping circuit.

The two pulse generation techniques are described here in detail.

### 5.2.1.1 UWB Short RF Pulse Generation Using a Pulse Generator and an Up-Converter

In this method, a UWB short pulse is generated by mixing a baseband pulse with a microwave frequency signal. A local oscillator (LO) generates the microwave

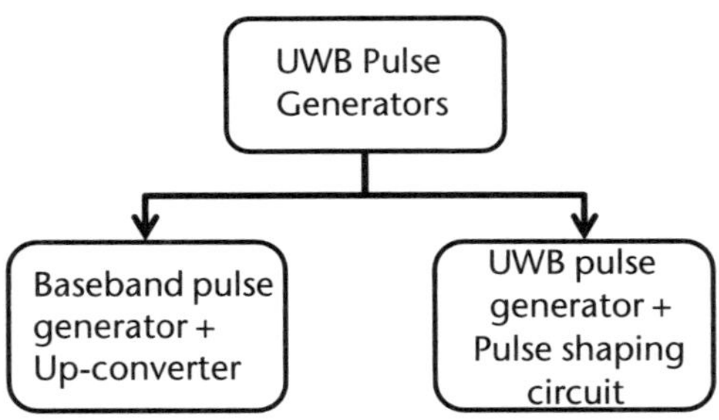


**Figure 5.2** Types of UWB pulse generators.

frequency signal that lies in the UWB frequency band [7]. A mixer up-converts the pulse with the LO signal. The up-converted pulse passes through a bandpass filter (BPF). The block diagram of the UWB pulse generation circuit is shown in Figure 5.3. As shown in the figure, control signals are used to control the pulse generator as well as the LO. Usually a digitally controllable pulse generator is used to allow binary phase-shift keying (BPSK) and other types of modulation on the pulse.

Note that the bandwidth of the generated pulse is around 550 MHz. To cover the entire UWB frequency band from 3.1 to 10.6 GHz, the output of the pulse generator is up-converted with a signal generated by a tunable LO in a mixer. The LO provides a continuous RF signal centered at discrete frequencies within the UWB frequency band. The output of the mixer is an up-converted version of a short RF pulse covering the entire UWB frequency band. The frequency spectrum of the RF pulse is centered at the LO frequency, which lies inside the UWB frequency band [7]. Figure 5.4 shows the schematic diagram of the impulse generating circuit. As can be seen, this method of UWB short pulse generation is complex due to the number of modules. In addition to the complexity of the circuitry, this method consumes significant power for the operation. However, the main advantage of this method is that the frequency spectrum of the output UWB short pulse is more easily controllable compared to the other technique, which is described in next section.

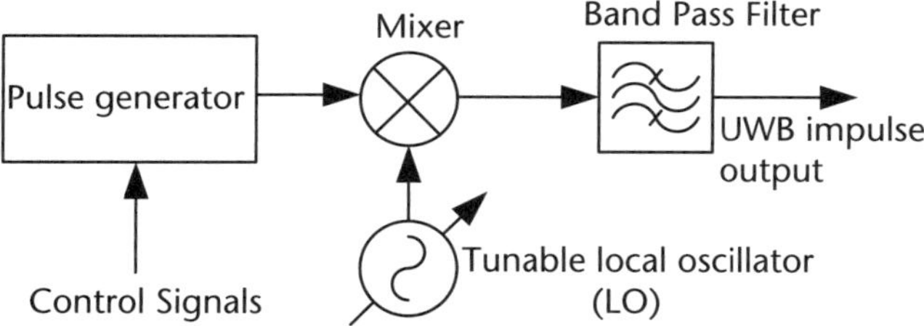


**Figure 5.3**   Generation of UWB short RF pulses by using a pulse generator and mixing a baseband pulse with a signal generated by a local oscillator [7].

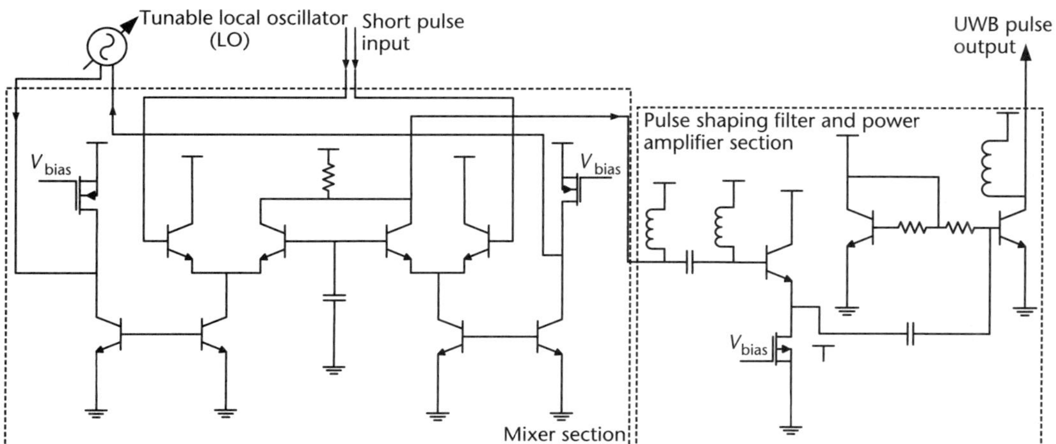


**Figure 5.4**   Schematic diagram of a UWB short RF pulse generator that uses a pulse generator and mixes a baseband pulse with a signal generated by a local oscillator [7].

### 5.2.1.2  UWB Short RF Pulse Generation with a Short Pulse Generator and Pulse Shaping Circuit

A short pulse generator is used to directly generate RF pulses that have frequency spectra lying in the UWB frequency band. A block diagram of this UWB short RF pulse generation is shown in Figure 5.5. As shown in the figure, the output of the pulse generation circuit is sent through a delay line directly to an AND or XOR gate. Due to the time delay caused by the pulse transmitted through the delay line, the output of the gate becomes a narrow pulse as shown in Figure 5.5. Then this narrow pulse is sent through pulse shaping circuitry to meet the spectral requirements of the UWB frequency band [8, 9].

The short pulse generator with a pulse shaping circuit is less complex than the previous technique. The use of typical digital logic gate building blocks makes this technique very popular when designing chips for UWB applications [8–11]. The use of standard CMOS designs and manufacturing processes makes this type of UWB short RF pulse generation technique more economical as well. Figure 5.6 shows the schematic diagram of the impulse generator with a pulse shaping circuit.

UWB pulse generation circuits that are made using some lumped components have been reported [12–14]. The operation of the pulse generation circuit is still the same as that shown in Figure 5.6. The only difference is that they do not use digital logic blocks for generating narrow pulses. Instead of digital building blocks, they use typical diodes or RF transistors that operate over microwave frequency bands to generate narrow pulses. In addition to the UWB pulse generators, the circuitry for pulse shaping and signal subtraction is required to generate a very narrow pulse. This is done using a reflective network consisting of shorted stubs. The

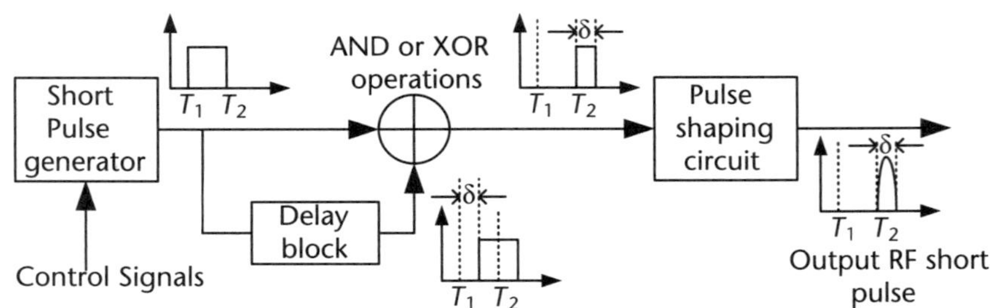

**Figure 5.5**  Generation of UWB short RF pulses by using a short pulse generator and a pulse shaping circuit [8].

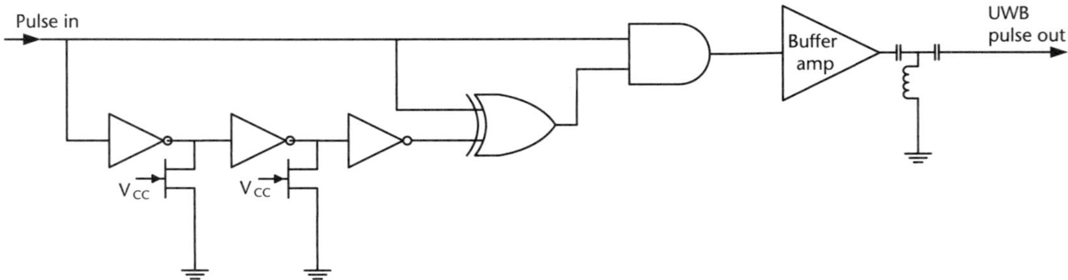

**Figure 5.6**  Schematic diagram of a UWB short RF pulse generator that uses a short pulse generator and a pulse shaping circuit [8].

length of the shorted stub is designed to provide enough delay to generate a very narrow output pulse as described in [12, 13].

Some reports have been made of pulse generators that can be tuned to obtain different pulse widths [15–17]. They are considered when tunable pulse generation circuits are needed in some special applications. However, obtaining the required center frequency of the UWB pulse spectrum is more difficult than in the previous type of pulse generators. The fact that the process for setting the center frequency is less flexible. This is the major disadvantage of this class of pulse generators. Therefore, these types of impulse generators are not considered here.

### 5.2.1.3    Laboratory Equipment Used to Generate UWB Short Pulses

Arbitrary signal generators with various pulse shapes in a tunable bandwidth are used for various time domain experiments in laboratory settings. For such applications there are ranges of laboratory equipment that can be used according to project requirements. In such applications the pulse generators presented above may not offer enough flexibility in project development and experimentation. However, both pulse and arbitrary waveform generators are used in laboratory settings. Some commercially available pulse and arbitrary waveform generators that are used for laboratory testing of time domain project development and experimentation are described next.

*Pulse Generators*
There are a range of commercially available pulse generators. All of them provide flexibility in terms of setting the pulse width, rise and fall times, amplitudes, and the shapes of pulses. The Picosecond pulse generator [18] is popular equipment that is used to generate short RF pulses for UWB applications. Agilent [19] and Quantum Composers [20] also provide a range of pulse generators with flexible pulse generation capability.

*Arbitrary Waveform Generators*
An arbitrary waveform generator is another piece of equipment that can be used to generate UWB pulses. It is possible to feed in data points from a text file into an arbitrary waveform generator. One possible use is to generate the shape of a UWB pulse that has the required width and bandwidth using a software tool such as Matlab and then record the data points on a text file and feed it into the waveform generator. Some versions of the arbitrary waveform generators provide predefined shapes of waveforms such as Gaussian pulses. By changing the associated parameters of the waveform, it is possible to generate UWB short RF pulses. LeCroy's ArbStudio [21], Agilent's series of arbitrary waveform generators [22], and Tektronix's AWG7000 series [23] are popular and are commonly used in laboratory experiments. Although these pieces of laboratory equipment provide much flexibility in setting the properties of a pulse such as pulse width, amplitude, and bandwidth, they are bulky and expensive. Therefore, they are not suitable for low-cost commercial applications.

Several specific mathematical functions such as Gaussian [24] , truncated sine, and shifted sinc [24, 25] have wide frequency spectra that can be used to generate wideband signals. Gaussian pulses are easier to generate using electronics than truncated sine and shifted sinc waveforms [24]. Therefore, a Gaussian function and its derivatives are widely used waveforms in short pulse UWB applications [26]. The expression of the waveform is given in (5.1) [25]:

$$f(t) = e^{-\frac{1}{2}\left(\frac{t-\mu}{\sigma}\right)^2} \cos\left(2\pi f_c t\right) \tag{5.1}$$

where $\sigma$ is the standard deviation of the Gaussian pulse in seconds, $\mu$ is the position of the pulse in time, and fc is the frequency of the modulated sinusoidal signal. The time domain waveform and frequency spectrum of a Gaussian modulated sine wave are shown in Figure 5.7.

The waveform shown in Figure 5.7 is a 4-GHz sinusoidal signal modulated as a Gaussian pulse with $\sigma = 0.6$ and $\mu = 0.5$ ns. As shown in Figure 5.7(b), it has a 40% 3-dB and 60% 10-dB bandwidth. The frequency spectrum is centered at 4 GHz as shown in the figure. It shows the possibility of generating wideband signals with derivatives of Gaussian pulses. The next section presents the architecture and operation of the receiver and data decoding unit.

### 5.2.2 UWB Short Pulse Based Interrogation

As mentioned in previous chapters, the multiresonator based chipless RFID tag can be modeled as a block of cascaded narrowband bandstop filters attached to two orthogonally polarized Tx and Rx antennas [27]. The Tx and Rx antennas are oriented in the orthogonal polarization to minimize the coupling between the two antennas. The reader antennas are also oriented in orthogonal polarization to maximize the transmitted and received signals to and from the tag, respectively.

The UWB short RF pulse is transmitted via the reader's Tx antenna toward the chipless RFID tag. The Rx antenna of the chipless tag captures the transmitted signal, and it passes through the multiresonator circuit. Next, the signal reaches the Tx antenna of the chipless tag and reradiates. The reradiated pulse is captured by the reader's Rx antenna. This whole process of signal flow from the transmitter to the receiver via the multiresonator based chipless RFID tag is shown in Figure 5.8.

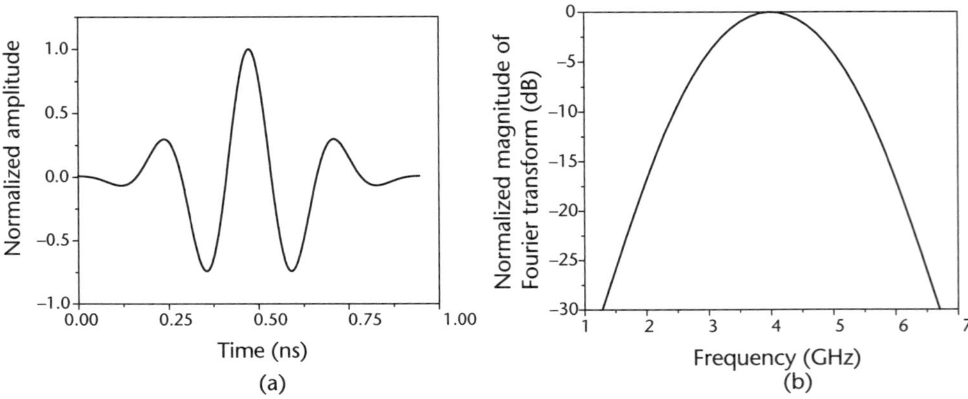

**Figure 5.7** (a) Trace of a Gaussian modulated sinusoidal signal and (b) frequency spectrum of the Gaussian modulated sinusoidal signal.

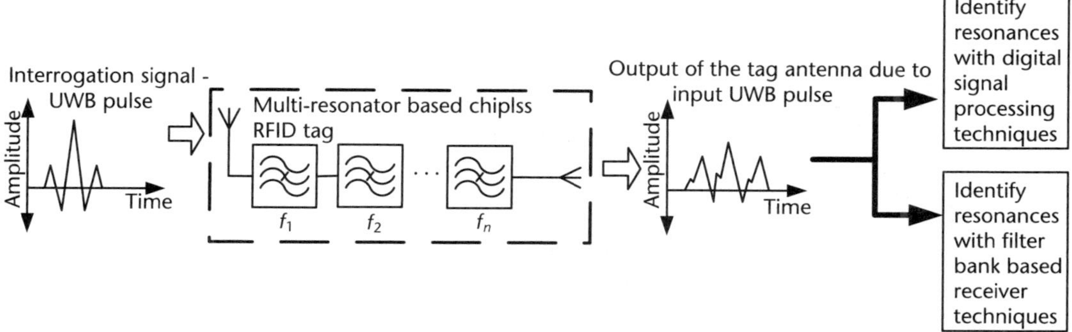


**Figure 5.8**   Reading process of the chipless tag using the UWB short RF pulse based interrogation technique.

The signal captured by the Rx antenna of the reader is processed in the reader's receiver to identify the frequency signature of the chipless RFID tags. Two processing methods can be used to process the captured signal from the chipless RFID tag. The first method is based on a *digital signal processing technique*. The second method is based on *microwave channelized receiver architecture*. The operating principles and the design of the two methods are described in following sections.

### 5.2.2.1   DSP Technique

Figure 5.9 shows the signal processing technique in a UWB-IR reader architecture. In this method, the received pulse captured from the chipless RFID tag is digitized with an ADC in the receiver section, as shown in Figure 5.9. Because the operational bandwidth of the chipless RFID tag is several gigahertz, the ADC should be selected carefully. When choosing the ADC, there are two options. According to the Nyquist sampling theorem, the sampling frequency of the signal should be at least twice the bandwidth. However, in practice at least five times the bandwidth of the signal is used as the sampling frequency for faithful reconstruction of the processing signal. Therefore, the required sampling rate should become several gigasamples per second (GS/S). These types of high-speed ADCs are extremely expensive. In addition, very high performance hardware platforms are also required to control the high-speed ADCs.

To alleviate this problem, the *equivalent time sampling technique* [28, 29] can be used to sample the repetitive signals with lower sampling rates than are required using the conventional method. The following section explains the theory and operation of the *equivalent time sampling technique.*

*Equivalent Time Sampling (ETS) Technique*
The ETS technique uses a much lower sampling rate than real-time sampling rate ADC to sample the signals that have very large bandwidth. Although the maximum sampling rate of the ADC is low, it is essential to have an analog front-end with enough bandwidth that is able to handle the sampled wideband input signal. However, ETS can be applied only to repetitive signals or signals that can be repeated with precise repetition time throughout the entire sampling cycle. The ETS process is illustrated in Figure 5.10. As shown in the figure, the input signal is repeated with

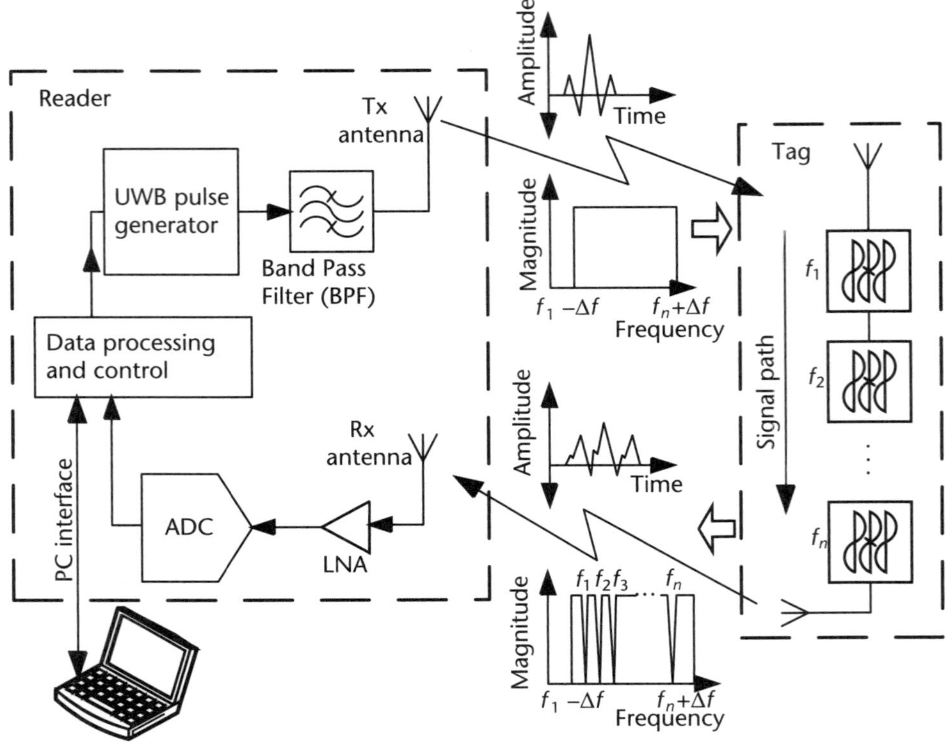



**Figure 5.9** Proposed time domain based signal processing architecture for UWB short RF pulse interrogation based reader.

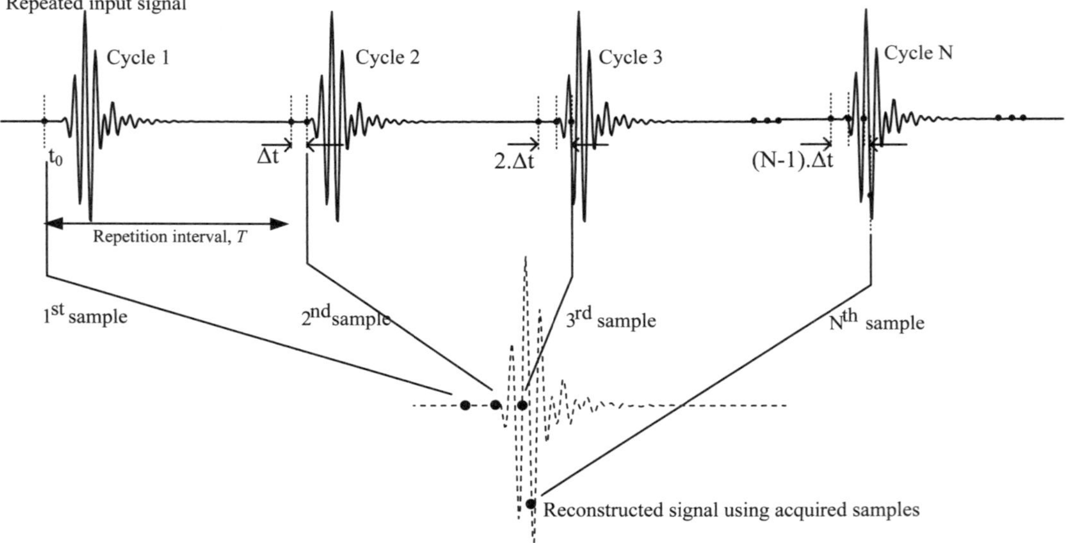



**Figure 5.10** Illustration of the process of reconstructing an input signal using a large number of acquired samples with the equivalent time sampling technique.

a time interval T that has a lower repetition rate than the maximum sampling rate of the ADC. The sampling interval T is obtained with a triggering mechanism or using a counter. As shown in the figure, in the first cycle of the repetitive input signal, only one sample is acquired by the ADC. Then in the next cycle, a small time delay of $\Delta t$ is added before sampling the input signal. This time delay is progressively incremented over the entire sampling cycle as the sampling continues. Therefore, to sample the Nth cycle, a time delay of $(N-1) \cdot \Delta t$ is required after the place where the first trigger point is applied. Since only one sample is acquired per cycle, the acquisition time is no longer limited by the sampling rate of the ADC. It is possible to acquire N number of samples from the input signal by repeating the input signal N number of times. Then the signal is reconstructed with the acquired samples with correct timing information. However, the use of ETS increases the complexity of the signal processing techniques. As explained in [30, 31], the reconstruction of the original signal by using the acquired samples requires a considerable amount of signal processing in order to achieve accurate results. In conclusion, the high-end ADC is traded off with the advanced signal processing algorithm.

In the chipless RFID application, the interrogation signal generator is controlled by the reader's control section. The interrogation signal can be repeated as required as long as the tag produces the same response for the repeated interrogation signal. Therefore, the signal at the reader's receiver is a repetitive signal. The ETS technique repetitively samples the received signals from the tag that have several gigahertz bandwidth without using very high speed expensive ADCs.

After digitizing the received signal, the fast Fourier transform (FFT) can be applied to obtain the frequency spectrum of the received signal. The FFT of the received signal shows the frequency signature of the chipless RFID tag. Therefore, by analyzing the computed frequency spectrum, it is possible to decode the identification data of the chipless tag.

By implementing a low sampling rate ADC and ETS technique, it is possible to design a low-cost reader architecture for a chipless RFID system. The reader architecture can read a variety of chipless RFID tags with different operational bandwidths. The next section discusses a low-cost receiver architecture that can be used with this time domain reading technique.

### 5.2.2.2   Microwave Channelized Receiver Architecture

Figure 5.11 shows the block diagram for a chipless RFID reader that identifies the frequency signature of the chipless tag with a filter bank based RF receiver architecture. As shown in the figure, after the receiver antenna, an LNA is used to amplify the received signal. To keep the received UWB pulse undistorted after amplification, the LNA must meet two important requirements: good phase linearity (or fewer variations in group delay performance of the LNA) [32–34] and a flat gain response over the operational bandwidth [33, 35]. The nonlinear phase response of the amplifier distorts the shape of the amplified pulse [34]. As a consequence, the LNA may distort the frequency spectrum of the received signal. In addition to the group delay and gain requirements, a flat noise figure (NF) performance over the operational bandwidth and low power consumption are also desirable [33]. Followed by the LNA, a power divider is used with a number of branches. The output branches of the power divider are terminated to a set of bandpass filters. This configuration

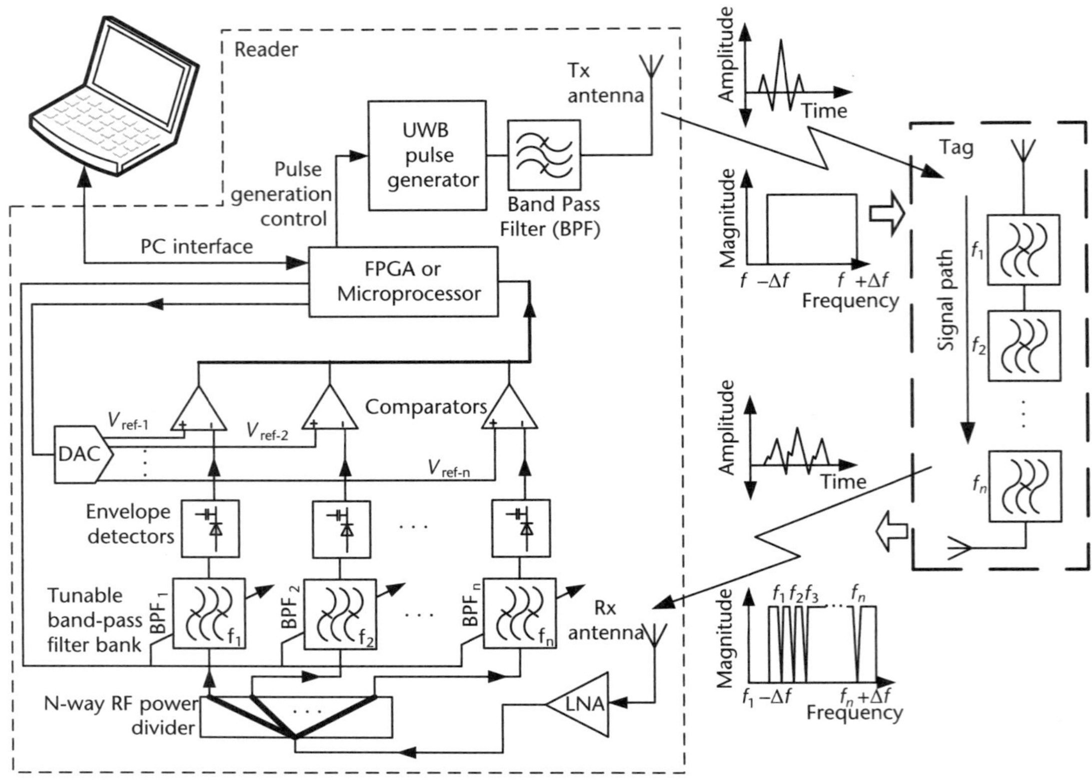


**Figure 5.11**  Proposed UWB short RF pulse interrogation based reader with channelized receiver for data decoding.

of a receiver is called the channelized receiver architecture (or filter bank based receiver) [36]. This receiver architecture has been widely used in wideband radar receiver systems for a long time [37, 38].

As shown in Figure 5.11, a channelized receiver typically consists of a 1-to-$n$-way power divider and a set of bandpass filters connected at each output port of the power divider. This configuration divides a wideband signal into a set of narrowband channels. Then each channel is processed separately. This simple technique greatly reduces the complexity of wideband receiver architectures [36]. The concept of channelization is being used in digital receiver architectures as well. By using a digitized signal with digital down-conversion and digital filtering, division of the wideband signal into a number of narrowband channels has been reported [39, 40]. However, due to the limitations in signal processing power in ADC and DSP hardware, the bandwidth of the processed signal and the operating frequencies of the digital channelization techniques are not enough for the UWB chipless RFID application. Solutions are discussed next.

The output of the power divider produces the copies of the original signal received by the Rx antenna of the reader. At each output of the power divider, a narrow bandpass filter is used. Each bandpass filter has the center frequencies of the resonators of the chipless tag. The output of each bandpass filter passes only the energy contained in the passband of the input pulse. Then the envelope detectors connected to the output of the bandpass filters produce a pulse output that can be detected with typical digital hardware systems. If there is a resonance in the chipless RFID tag, the output pulse from the chipless tag does not contain that particular frequency band. Then the corresponding channel corresponding to that particular resonator does not output a signal. On the other hand if no resonator resonates at a certain frequency, the output pulse from the chipless tag contains that frequency band. Then the corresponding channel of the receiver outputs a pulse indicating the absence of the resonator and the presence of the particular frequency band in the received signal. Based on this operating principle, by observing the output channels of the receiver section of the reader, it is possible to identify the resonances of the chipless RFID tag.

The next section presents the design and the simulation setup of the proposed chipless RFID reader using Agilent's ADS2009. ADS2009 is a very powerful tool for designing and evaluating a complex system. The system-level simulation of the TD based reader architecture validates the concept.

## 5.3  Design of UWB TD Based Chipless RFID Reader

The important design steps and the simulation setup in Agilent ADS2009 are described in this section. Out of the two reader designs presented in Figures 5.9 and 5.11, only the reader with the channelized receiver architecture is simulated. The operation of the reader that uses signal processing techniques is self-explanatory and not simulated.

As shown in Figure 5.11, each output branch of the power divider has an RF power divider network and a bandpass filter. The power divider and the filters can be implemented using microstrip lines on the RF PCB board. It is preferable to use a power divider type with sufficient isolation between output ports. A 1-to-$n$-way fork-type power divider with high isolation on the order of 30 to 40 dB is reported in [41]. Similar multi-stage power dividers can be implemented in the current receiver architecture. Even for a flexible receiver configuration, a reconfigurable power divider can be used.

The bandpass filters are tuned to the center frequency of each resonant frequency of the multiresonator based chipless RFID tag. The bandpass filters also can be implemented on the same RF board of the power divider. Implementation of the bandpass filters can be done either by using lumped components or by using microstrip technology without a lumped component. However, the use of microstrip technology to implement the filters may occupy more space on the RF PCB board. Therefore, proper optimization of the cost and the size of the RF PCB should be done when choosing the type of bandpass filter. The pass bandwidth of the bandpass filter must be kept in the range of the bandwidth of the corresponding bandstop resonator of the chipless tag. If the passband is not synchronized with the bandwidth of the corresponding bandstop filter of the tag, the receiver will give erroneous results at the outputs of each channel of the RF section. A sufficient guardband is needed to isolate adjacent resonances.

The outputs of each bandpass filter are then connected to an envelope detector. For the envelope detector a rectifying diode and a capacitor are used to keep the cost of the reader to a minimum. When choosing the components for the reader, it is important to choose the ones that cover the whole operating frequency bandwidth of the reader.

The output of the envelope detector is then connected to an FPGA or a microcontroller to identify which branch gives a voltage output. If a certain frequency is attenuated at the tag, then at the receiver the corresponding branch of the envelope detector will give a lower voltage output. If a certain frequency is not attenuated at the tag, it will give a higher voltage output than the earlier case.

However, this reader architecture is not as flexible as the one that uses signal processing techniques for identifying the resonances and data decoding. For reading a tag with different resonant frequencies, a different RF section is required. An RF section with tunable bandpass filters can accommodate the variations of frequency responses in the chipless tag due to variations in the printing quality, material properties, and addition of environmental noises and interferences. Such a reconfigurable RF section will bring new breakthroughs in chipless RFID tag reader development. A list of components that can be used in the design of the chipless RFID readers described above is shown in Table 5.1.

### 5.3.1 Simulation Setup in Agilent ADS2009

This section describes the simulation setup of the reader that is based on the channelized receiver architecture. The modeling of the UWB short RF pulse generator, chipless RFID tag, and receiver section is presented in detail. The important steps and methods used in the reader configuration to obtain the results at each section of the design are also described along with illustrations of waveforms. Note: The

**Table 5.1**  Specifications of Selected Components for Chipless RFID Readers

| Component | Specifications |
|---|---|
| Picosecond 10050A Programmable Pulse generator | Pulse width: 100 ps–10 ns |
| | Maximum amplitude: 10V |
| | Pulse repetition rate: 100 kHz |
| LNA: Mini Circuits™ ERA-1+ | Bandwidth: 0.1–8 GHz |
| | Gain: 10.9 dB |
| | Noise figure: 4.3 dB |
| BPF: @output of pulse generator | Passband bandwidth : 3–8 GHz |
| Microstrip-line BPFs or tunable BPFs | Passband: as required by tags—custom designs on PCB |
| | Passband bandwidth: 100 MHz |
| Fork power divider | Operating bandwidth: 3–9 GHz |
| | Made on PCB |
| Comparators: TI TLV3501 | Propagation delay: 4.5 ns |
| Envelope detectors | Made on PCB with discrete components |
| DAC: Analog Devices™ AD7845 | Resolution: 12 bits |
| | Conversion rate: 200 KSPS |
| ADC: TI™ AD5523 | Resolution: 12 bits |
| | Maximum sampling rate: 210 MSPS |
| Reader antennas | Operating bandwidth : 3–8 GHz |
| | Gain: >5 dBi |

*Note:* ™ represents commercially available components.

Agilent ADS2009 software package and hereafter the term ADS are used as the nomenclature of ADS the software development platform.

Figure 5.12 shows the schematics of the block diagram of the simulation setup of the reader system used in ADS. The three main units of the reader system are marked on the block diagram. When creating the simulation project, the "Digital Signal Processing Network" option is used. For the simulation controller, the "Data Flow Controller" should be used in simulations. A 20-ps time step is used for the simulation.

In developing the modular equivalent circuit for a chipless RFID tag reader system, a few unique models for antennas, chipless tags, and readers have been derived. For example, for a UWB antenna, the model comprises (1) a gain block of $G_n$ to represent the gain of the $n$th antenna, (2) a bandpass filter to represent the finite bandwidth of the antenna, and (3) an antenna that is used as a spatial transformer, the coupling interface between free spaces of intrinsic impedance of 377 $\Omega$ to a 50-$\Omega$ microwave system, a transformer with a turns ratio of $377\Omega : 50\Omega$ completes the antenna model. The model is used in the transmitter and receiver of the reader and in the tag. Figure 5.12 show the model for the UWB antenna, the tag, and the reader as used in ADS. All modules are connected via wires. The free-space path loss is represented with an attenuator with an attenuation factor of $A \propto 1/R^2$, where $R$ is the free-space path length. Following are detailed descriptions of the models for the antennas, UWB-IR transmitter, and the receiver.

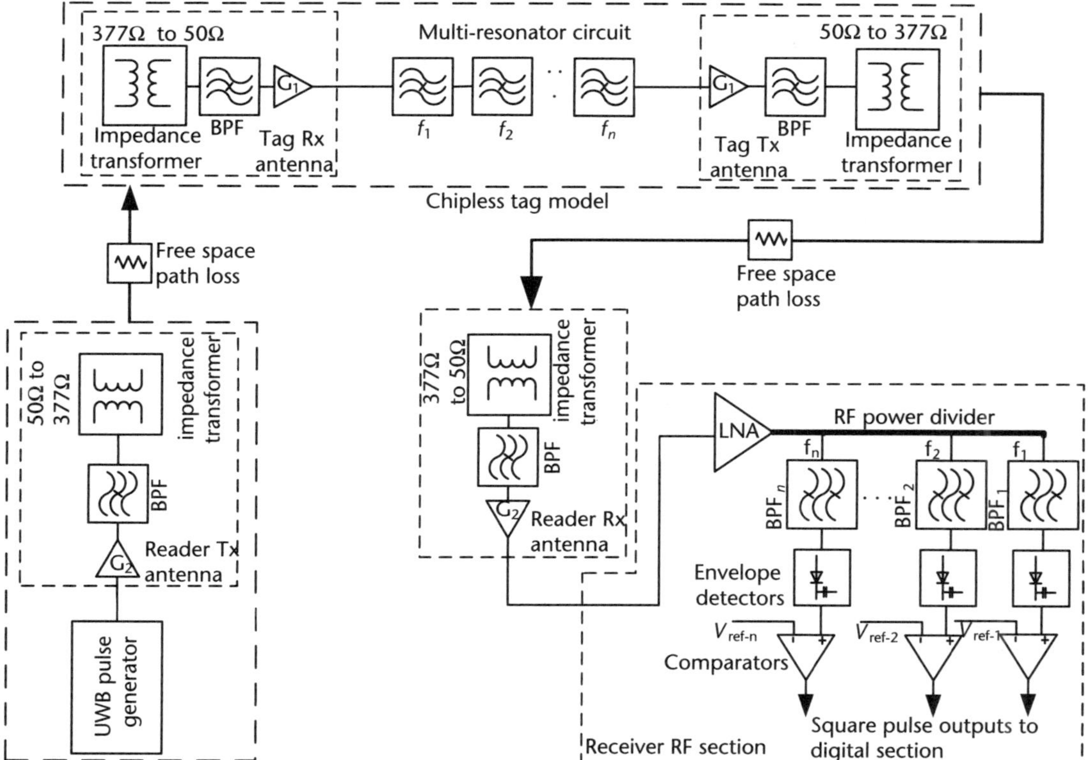


**Figure 5.12**   Block diagram of the simulation model for an UWB short RF pulsed based reader simulated in ADS2009.

*Modeling Chipless Tag*

A series of bandstop filters was used to model the tag with the UWB Tx/Rx antennas. For the model a 4-bit chipless tag is used in the investigation. Hence, only four resonant frequencies were used: 4.5, 4.8, 5.1, and 5.4 GHz. Data conventions were as follows: the presence of a resonator (attenuation at a designated frequency)—logic 0; the absence of a resonator (no attenuation at a designated frequency)—logic 1. Figure 5.13(a) shows the schematic model of the tag.

The system-level simulation model of the tag created in ADS is shown in Figure 5.13(b). The bandstop filters were created using the "SBlock" module available in the "Timed Linear" library. The characteristics of the filter are specified with a text file created in the Touchstone™ format. The four filters represent a multiresonator based 4-bit chipless tag. As shown in Figure 5.13(b), parallel unity gain blocks are used to remove the bandstop filter from the simulation setup easily without deleting components from the simulation setup. Just disabling an "SBlock" and enabling the gain block connected in parallel to it removes the bandstop filter from the simulation without interrupting the signal flow through the deactivated resonator block.

*Modeling the UWB Pulse Transmitter*

The UWB pulse transmitter consists of a UWB pulse generator and a transmitting antenna of gain $G2$ as shown in Figure 5.14(a). The built-in Gaussian second-order derivative pulse generator is used to model the UWB short RF pulse generator as shown in Figure 5.14(b). A 100-ps pulse width is used to generate a wideband short RF pulse. The UWB antenna model with a $G2$ gain block, a bandpass filter with a 3- to 7-GHz passband bandwidth, and a wire-to-air interface transformer is used after the impulse generator.

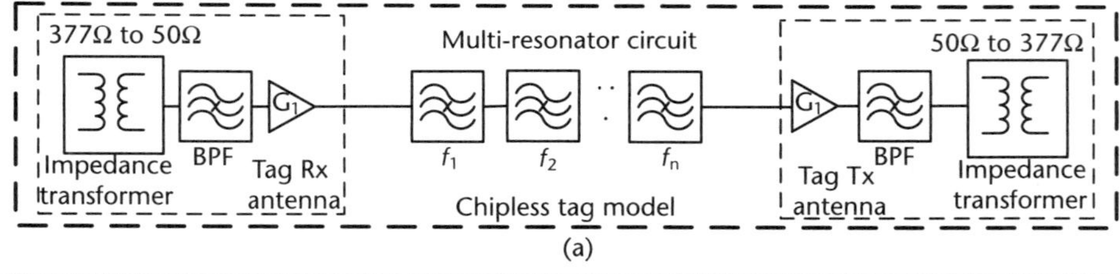


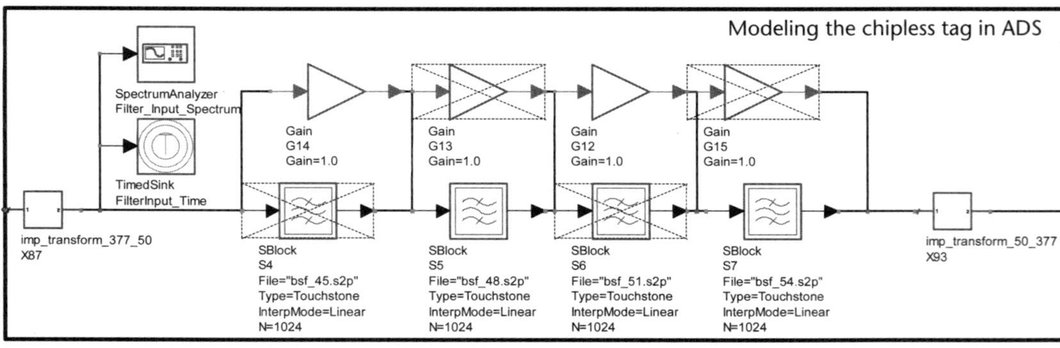


**Figure 5.13**  System-level model of multiresonator based chipless RFID tag used in the simulation setup in ADS2009.

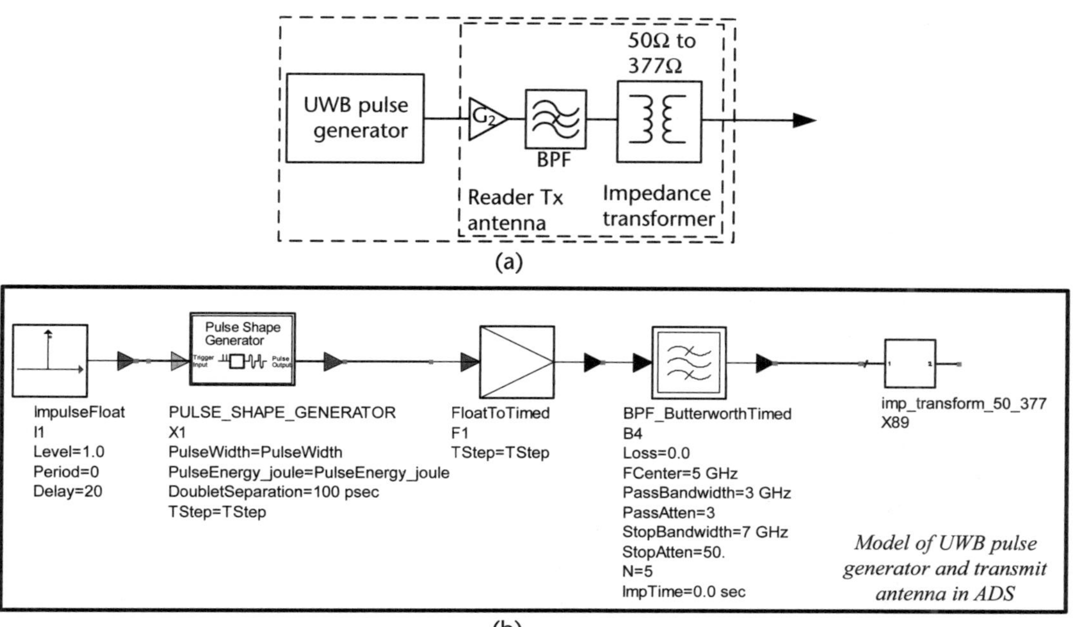



**Figure 5.14**    (a) System level model of UWB short RF pulse generator and antenna section of the reader. (b) Simulation setup in ADS2009 (Antenna and LNA are not shown in ADS system level model).

*Modeling Receiver Section*

The receiver section consists of (1) a receiving antenna, (2) an LNA, (3) a 1-to-n-way power divider, (4) bandpass filters, (5) envelope detectors, and (6) comparators. A built-in, four-way fork power divider is used as the power divider simulation model. The bandpass filters were modeled with SBlock, which has characteristics defined by a text file in the Touchstone format similar to the bandstop filter models used in the chipless tag model. The envelope detectors were modeled with the in-built models of a rectifier and a lowpass filter. To obtain the time domain waveforms at different points of the setup, "TimedSink" data monitors available in the "Sinks" library in ADS are used.

The outputs of each envelop detector are connected to a voltage comparator. The comparator converts the analog pulse output into a square pulse that can be easily interfaced into a digital electronics device such as a microcontroller or a FPGA. For the simulation, same reference voltage is used for all output channels. However, it is possible to provide different reference voltage levels for each comparator using a DAC as shown in Figure 5.11 to provide more flexibility to determine suitable reference values for each channel.

The complete system-level simulation model of a chipless RFID tag-reader system is used in Agilent ADS2009. Three main blocks—the pulse transmitter, the tag, and the receiver—are hardwired. The air interfaces between the three blocks are represented by the 377 Ω : 50 Ω transformer blocks incorporated in the antennas. The air interfaces are modeled with attenuators. The next section presents the system-level simulation results obtained using this simulation model of the chipless tag-reader system.

## 5.4  Results

Here, the simulation results for a UWB TD based chipless RFID reader are presented. The results presented in this section show the functionality of the reader, how the data bits are encoded in the tag and decoded from the outputs obtained from the RF sections of the reader, and the time domain waveforms and their frequency spectrum at each important observation point of the system. Places where important changes of the input signal take place are connected with "TimedSinks" (equivalent oscilloscopes) and "Spectrum Analyzers" on the block diagram of the simulation model.

The system is modeled with two 4-bit frequency signature based chipless tags with tag IDs of "0101" and "1111." As can be seen in the figure, the transmitter generates a Gaussian pulse and transmits through the antenna to the tag. The tag encodes the pulse as it passes through the resonators. The output pulse contains the frequency signature of the tag embedded in the waveform. A detailed analysis of the pulse can be found in [2]. The FFT of the signal yields the encoded data bits in the frequency domain. The tag sends the encoded signal to the receiver. The receiver's LNA amplifies the signal and sends it to a 1-to-4-way power divider. The BPFs channelize the UWB signal into their prescribed narrowband channels. The outputs of the filter provide refined frequency domain data bits. The filter outputs pass through the envelope detectors followed by comparators. A square pulse output is generated by the comparator that can be easily interfaced with digital electronic devices such as microprocessors or FPGAs.

As shown in Figure 5.15, the UWB pulse only exists for 0.5 ns and this waveform has a 3-dB bandwidth of 3 GHz and 10-dB bandwidth of 6 GHz. The analysis of the time domain impulse when it passes through a bandpass filter is shown in Figure 5.15. Figures 5.16(a) and (b) show the signal when it passes through the tag. The time domain waveform and the frequency spectrum depict the encoded data bits "0101" and "1111" as shown in the figure. For ID "0101" two obvious resonances in the frequency spectrum are observed whereas for ID "1111" the spectrum flattens. Similar phenomena are also observed in the time domain signals.

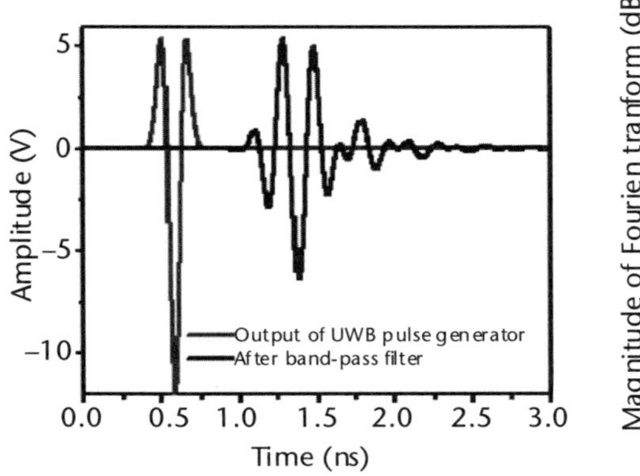


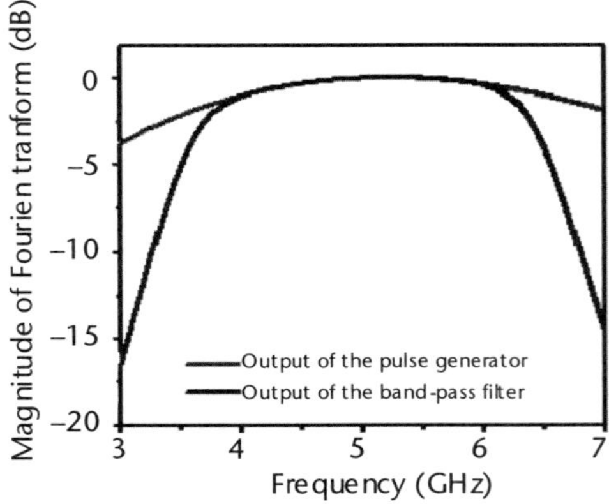


**Figure 5.15**  Output of the UWB pulse generator and bandpass filer (a) waveform in time domain (b) normalized frequency spectrum.

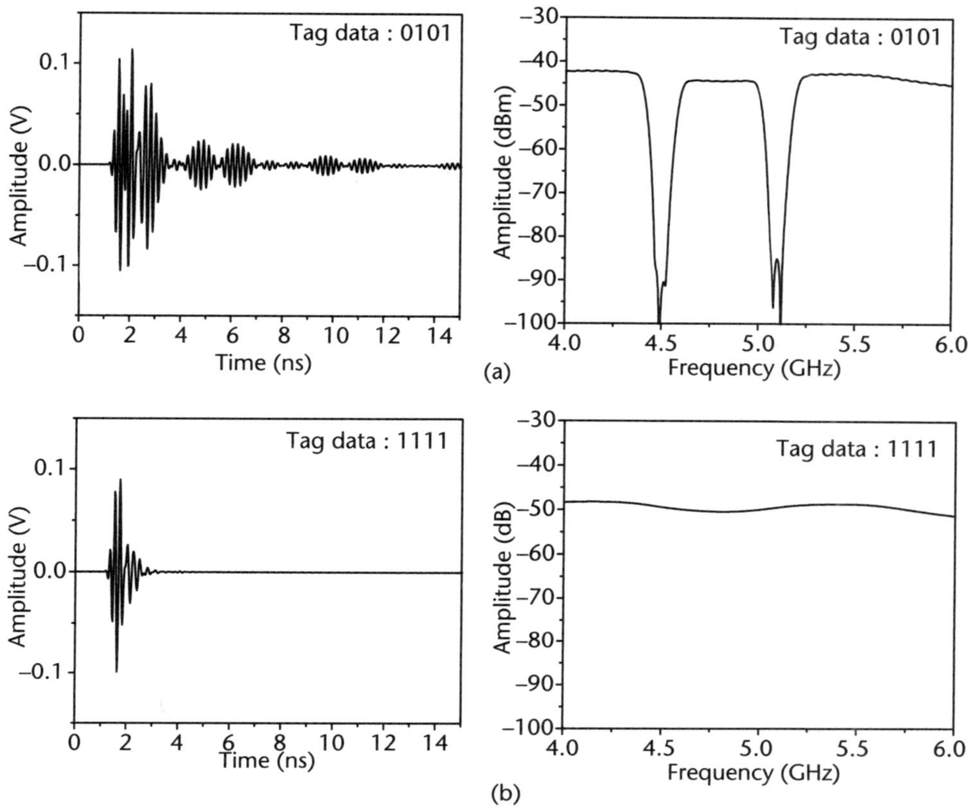

**Figure 5.16**   Time domain waveform and frequency spectrum of the impulse signal after passing through the multiresonator based tag models with (a) encoded data "0101" and (b) encoded data "1111."

Figures 5.17(a) and (b) show the outputs of the bandpass filters of the receiver. The attenuation at the selected frequency bands can be seen clearly. This attenuation of the resonance frequency does not provide enough power to create an output at the envelope detector, hence the outputs of the envelope detectors remain near zero. The outputs of envelope detectors and comparators for tags encoded as "0101" and "1111" are shown in Figure 5.18. The comparator outputs can be interfaced directly with the FPGA/microcontroller pins. The FPGA/microcontroller identifies the data as "0101" or "1111." The maximum time taken for this decoding process is 9 ns according to the simulation. This shows the improvement of the data decoding speed and demonstrates the suitability of this technique for use with high-speed tag reading. For frequency domain reading of the same 4-bit tag, the reading time is 500 ms [42]. This shows the improvement in the data decoding speed and the suitability of this technique for use in high-speed tag reading and when reading tags on the move.

## 5.4.1   Discussion of Results

The design and simulation results were performed on the Agilent ADS2009 platform. The results have proved the concept of the proposed UWB-IR based reader

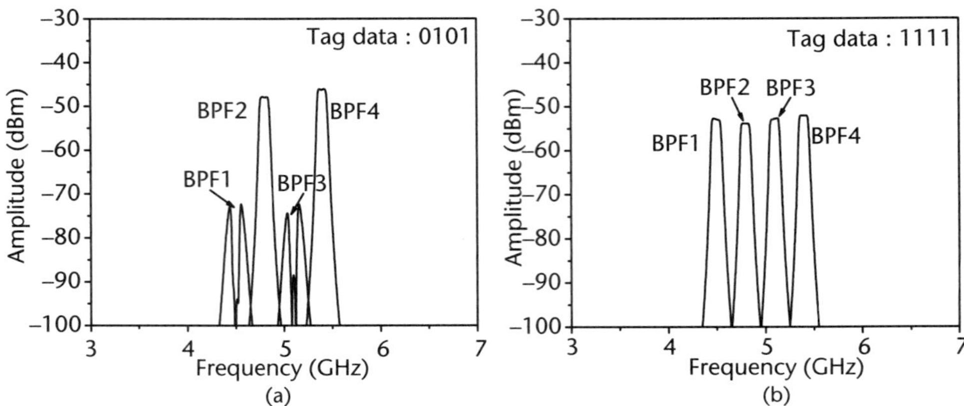


**Figure 5.17**  Frequency spectrum of outputs of four bandpass filters for encoded data "0101" and "1111."

and shown promise for practical implementation. However, many practical issues need to be addressed when practical design and system-level integration are considered for the proposed reader system. The contrasting issues of the simulations are as follows:

- ADS considers ideal components and work in their linear operating regions. The nonlinear behavior and transient analysis of the components may be incorporated in future simulation studies of the reader system.

- We have hardwired components assuming the antenna to be the spatial filter with an impedance transformer of ratio of 377 $\Omega$ : 50 $\Omega$ for reception and 50$\Omega$ : 377$\Omega$ for transmission. The dynamic free-space path loss and practical channel model are missing in this simulation.

- We also did not consider the distance between the tag and the reader as we hardwired the system blocks. Therefore, the propagation path loss and surrounding interference and the attenuation of power levels with distances are missing in the simulation.

- We also did not consider the phase noise, delay spread, antenna temperature, system thermal noise, and nonlinearity of devices in the simulation. What will be the effect of group delay in the antenna and filter with respect to the time domain impulse response? Some antennas are good at UWB impulse response compared to other types of antennas. The reader antenna should be planar, low weight, and low profile in design and implementation. How will the antenna be responsive to the UWB-IR based interrogation signals?

- In the receiver the UWB channels are subdivided into narrowband channels with a 1-to-n-way power divider followed by the narrow passband filters. Is adequate channel equalization considered in the simulations?

The results of the individual blocks have been discussed in a sequential order as presented in Section 5.4. Here some critical issues are addressed for practical implementations of system blocks: impulse transmitter, chipless RFID tag, and receiver.

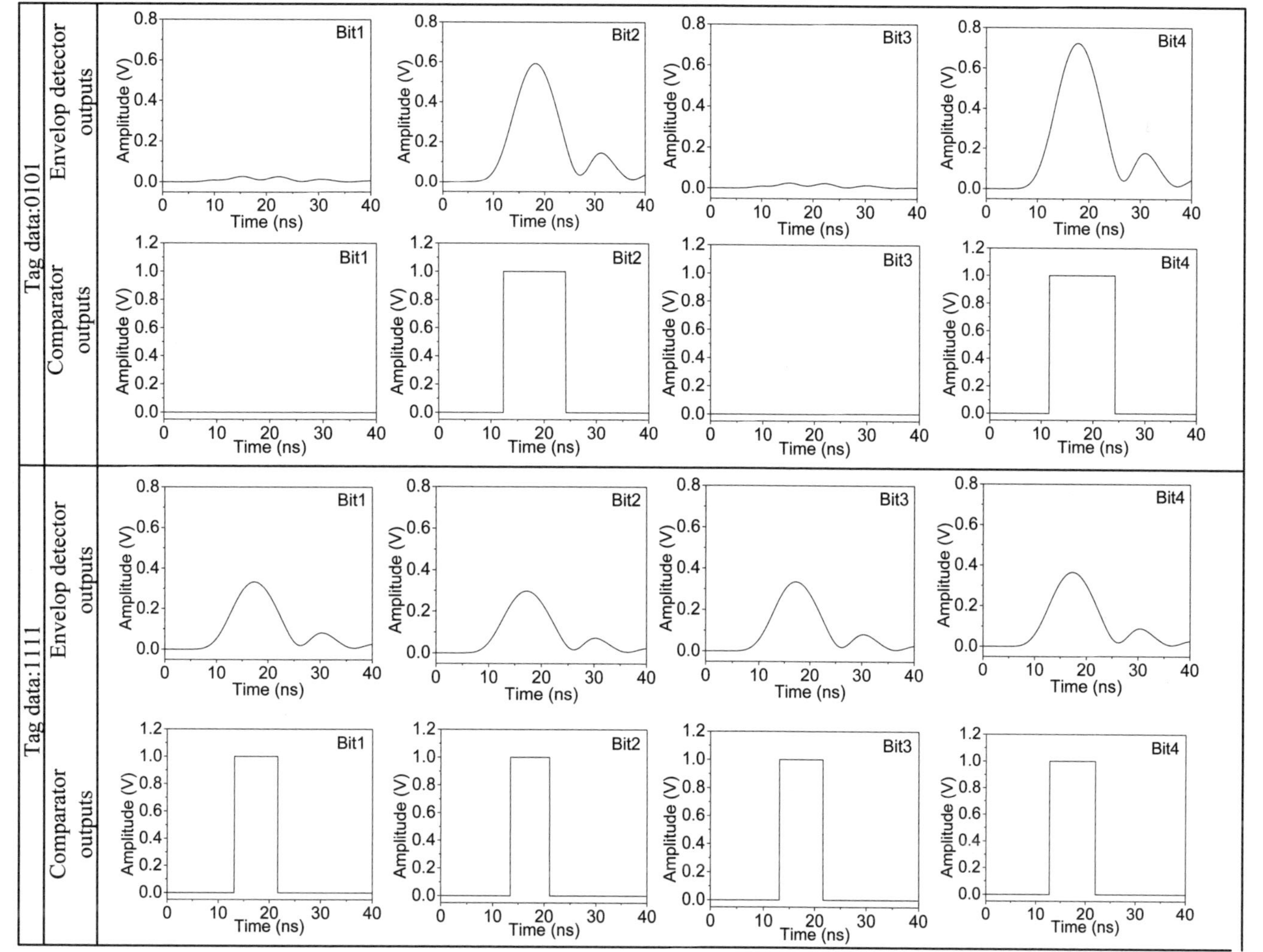

**Figure 5.18**  Outputs of envelope detectors and outputs of comparators for encoded data "0101" and "1111."

*Impulse Transmitter*

The impulse transmitter is modeled with an impulse generator followed by a band-pass filter. The impulse distorts when the filter is applied as shown in Figure 5.18. A few issues are considered for UWB-IR based communication systems as follows: The impulse responses of passive components such as antennas could be major issues in UWB communications systems. A dipole-like antenna is less susceptible to the time domain signal compared to a planar antenna such as a microstrip patch antenna. Will the delay spread be the optimum choice for a low-profile printed antenna such as a microstrip patch antenna? Microstrip patch antennas are conformal to handheld and gantry-based readers. They are readily available in the commercial market. Therefore, the proposed reader may consider microstrip antennas as suitable candidates.

*Chipless RFID Tag*

The chipless RFID tag comprises two orthogonally polarized UWB disk-loaded monopole antennas and a series of multiresonator circuits [27]. A few critical issues may need to be considered as follows:

- Planar bandpass and bandstop filters suffer from group delays of time domain transmitted signals. Will there be acceptable group delays of the incoming UWB impulse signals in both transmission and reception modes via the chipless RFID tag?

- The isolation level between the frequency responses of a logic 0 (with resonator) and logic 1 (without resonator) is on the order of 25 dB. For very weak signal reception, this constitutes a very minimal difference between the two logic levels. What will be the receiver sensitivity to discriminate these minute differences between the two logic levels and to extract the information carrying data bits in the reader?

- Printing errors of chipless RFID tags, variations of base materials on which the tags will be printed, and ambient noise may contribute to the signal distortion and variation in frequency responses (both amplitude and phase) when the tags are in the reading zone of the reader. What types of adaptive threshold technique and reconfigurable filters will be utilized to overcome the reading errors in the system?

*Receiver*

The receiver comprises the receiving antenna, a LNA, a power divider, narrow bandpass filters for channelization of the UWB signal, envelope detectors, and comparators. The receiver is the most susceptible to interference and noise. Antenna temperature and its impulse response, the noise figures, the gain and linearity of the LNA, the delay spread (group delays) of passive components such as power dividers and filters and their interconnects (such as SMA connectors and interconnecting microstrip transmission lines), and the intermodulation products of the envelope detectors and the comparators may contribute to significant distortions and impose challenges to recovering the encoded data bits from the chipless RFID tag.

*System-Level Considerations*
During the integration of the system using the modular approach as shown in Figure 5.16, a few system-level considerations need to be addressed as follows:

- Signal-to-noise ratio of the overall system;
- Aggregated antenna noise temperature of the system;
- System noise temperature;
- Noise due to inappropriate shielding, grounding, and filters (all electromagnetic compatibility and electromagnetic interference issues and their remedies);
- Noise introduced by the tag due to printing errors and variations in substrate materials and design inaccuracies due to these consequences.

Yield analysis and Monte Carlo simulation may provide some degrees of flexibility in the system-level design considerations. All of these practical issues must be considered in the practical design of the system.

## 5.5   Conclusion

This chapter has presented a UWB reader architecture based on a time domain UWB-IR interrogation signal. UWB is a license-free frequency band technology that has gained popularity for short-distance, high-data-speed communications between household devices. In UWB communications, devices are designed in two formats: UWB microwave circuit design and impulse radio based system design. In the first method, all active and passive components are designed targeting a bandwidth of more than 500 MHz. However, in the time domain the requirements of UWB microwave circuits are not relinquished. Here devices that provide a linear response to the UWB impulse with minimum distortion are envisaged. The main advantage of UWB-IR based devices is the minimal time required to generate the transmitted signal. A burst of wideband signals can be sent in a very short duration via a very narrow pulse. This advantage of very short time transmission, which is contrary to a VCO swept frequency based source, has been exploited in the proposed time domain based reader architecture. However, since most microwave communications circuits are traditional frequency domain circuits, the time domain devices are maturing. There are also many challenges for time domain devices. They are basically transient responses and nonlinearity may disrupt the system performance. The chapter has taken a challenging stride to propose a new TD based reader architecture for the chipless RFID system.

First, the chapter reviewed time domain based chipless RFID systems. The review revealed only a few reported works [1–6] based on a TD reading technique. These works are mainly based on time domain reflectometry (TDR) based chipless RFID tags. Only the authors' group has reported multiresonator frequency signature based chipless tag detection using time domain signals [2]. In this regard, the chapter has contributed some fundamental knowledge in the discipline. The proposed reader architecture comprises an impulse transmitter and a channelized receiver. The transmitter is comprised of a modified Gaussian pulse generator followed by a UWB antenna. The receiver is comprised of an LNA block, a 1-to-$n$-way power divider followed by narrow bandpass filters, envelope detectors, and comparators. The system is capable of reading any frequency signature based chipless RFID tag such as a retransmission chipless tag [27] and a backscatter chipless tag [43] with minor modifications if required. To implement the TD-based reader architecture the following tasks have been performed:

- Various impulse generators and pulse shaping circuits have been studied. The study has led to three categories of impulse generators: (1) a pulse generator with a reconfigurable up-converter, (2) a pulse generator containing a pulse shaping circuit, and (3) commercially available equipment. As stated earlier, most reported works used laboratory equipment to prove their concepts. Therefore, there is a big gap in the research and development efforts. An impulse generator has been proposed that can be directly implemented in the prototype time domain based chipless RFID tag reader. In this regard, among the three pulse generation circuits, the pulse generator with an up-converter circuit has been selected as the candidate signal source. This is mainly because these types of pulse generators yield the most controlled impulse for a chipless RFID tag reader.

- After selecting the source, we have proposed two types of detection methods in the receiver of the reader: (1) a signal processing method and (2) a hardware method. Both techniques have pros and cons in processing the returned echoes from the tag.

- The chipless RFID is a UWB system. Processing the UWB signal in state-of-the-art processors requires very expensive DSP chip sets, high sampling ADCs, FPGAs, and a powerful algorithm. In this chapter, we have concentrated on the hardware based receiver.

- The Agilent ADS2009 platform with "Digital Signal Processing Network," "Data Flow Controller," and 20-ps time step has been used to implement the proposed TD reader architecture. The three modules the impulse transmitter, the multiresonator chipless RFID tag, and the hardware based receiver were fully designed in the schematics of Touchstone of ADS2009 and simulated for performance validation. For the proof-of-concept ADS implementation, a 4-bit chipless RFID tag operating over a 4- to 6-GHz frequency band was designed. The reader comprises the transmitter and the receiver.

- In modeling the three modules in ADS2009, a few unique contributions are made. Since the system is a modular design, the transmitter, the tag, and the receiver are hardwired for simulation.

- The antennas for the reader and the tag are modeled as a spatial filter with a gain block representing the gain of the antenna, a bandpass filter, and a transformer with an impedance ratio of 377 $\Omega$: 50$\Omega$ for transmission and 50$\Omega$ : 377$\Omega$ for reception. A variable attenuator with attenuation A $\quad$ 1/R2 can be used to represent the free-space path loss. Here R is the reading distance of the tag from the reader.

- The universal model of the frequency signature based chipless RFID tag is comprised of a tandem of parallel gain blocks and stopband filters. Based on the presence and absence of the multiresonator (stopband filter) circuits, the gain block and the filter were activated. Both transmitting and receiving antennas with appropriate gain blocks, bandpass filters, and impedance transformers were connected with the resonator blocks. This design is a unique equivalent circuit model representing any frequency signature based chipless RFID tag

- The transmitter has been designed with an impulse generator and an antenna. A 100-ps pulse width is used to generate a wideband short RF pulse. The pulse bandwidth is constricted with the UWB antenna, namely, the bandpass filter. However, the pulse shape generates appropriate spectral content spanning from 4 to 7 GHz with approximately uniform amplitude over the frequency band.

- The receiver is the most complex system of all modules in the proposed reader architecture. The receiver uses a UWB receiving antenna, an LNA gain block, a power divider, narrow bandpass filters, envelope detectors, and comparators. The unique feature of the receiver is the method of channelization using the power divider and the narrow bandpass filters. The down-conversion to a baseband signal and filtering are done with envelope detectors. The final baseband signal is sent for further processing in the voltage comparators assisted decision circuit such as FPGAs and microcontrollers and their computer interfaces.

- The designed receiver was able to successfully decode two 4-bit chipless tags having the IDs "1010" and "1111." A comprehensive system-level design and signal flow diagrams helped depict the performance of the proposed reader architecture in every stage of signal transitions of the tag–reader system. This has developed huge confidence in the practical implementation of such reader architecture in real-world scenarios.

## Questions

1. Which signal processing scheme can be applied directly on the received time domain signal to obtain the frequency signature of the tag?
2. What issues associated with frequency domain reading hinder high-speed operation?
3. How does an impulse radio based time domain reader operate?
4. What is the typical width of a UWB RF pulse used in communication systems?
5. What are the main advantages of a time domain based reader?
6. What impulse radio based UWB chipless RFID system is available commercially?
7. What is the working principle for most time domain based chipless RFID tags?

8.  What are the commonly used techniques for generating short UWB pulses?

9.  What laboratory equipment can be used to generate UWB pulses?

10.  Explain the operation of a channelized receiver architecture.

# References


[1]  S. Gupta, B. Nikfal, and C. Caloz, "Chipless RFID System Based on Group Delay Engineered Dispersive Delay Structures," *IEEE Antennas and Wireless Propagation Letters*, Vol. 10, pp. 1366–1368, 2011.

[2]  P. Kalansuriya and N. Karmakar, "Time Domain Analysis Of A Backscattering Frequency Signature Based Chipless RFID Tag," in *Asia-Pacific Microwave Conference Proceedings (APMC2011)*, December 2011, pp. 183–186.

[3]  L. Zheng et al., "Design and Implementation of a Fully Reconfigurable Chipless RFID Tag Using Inkjet Printing Technology," in *IEEE International Symposium on Circuits and Systems, ISCAS 2008*, May 2008, pp. 1524–1527.

[4]  A. Lazaro et al., "Chipless UWB RFID Tag Detection Using Continuous Wavelet Transform," *IEEE Antennas and Wireless Propagation Letters*, Vol. 10, pp. 520–523, 2011.

[5]  S. Hu et al., "Study of a Uniplanar Monopole Antenna For Passive Chipless UWB-RFID Localization System," *IEEE Trans. on Antennas and Propagation*, Vol. 58, No. 2, pp. 271–278, February 2010.

[6]  S. Shrestha et al., "A Chipless RFID Sensor System for Cyber Centric Monitoring Applications," *IEEE Trans. on Microwave Theory and Techniques*, Vol. 57, No. 5, pp. 1303–1309, May 2009.

[7]  D. Wentzloff and A. Chandrakasan, "Gaussian pulse Generators for Subbanded Ultra-Wideband Transmitters," *IEEE Trans. on Microwave Theory and Techniques*, Vol. 54, No. 4, pp. 1647–1655, June 2006.

[8]  M. Yuce, H. C. Keong, and M. S. Chae, "Wideband Communication For Implantable And Wearable Systems," *IEEE Trans. on Microwave Theory and Techniques*, Vol. 57, No. 10, pp. 2597–2604, October 2009.

[9]  J. Ryckaert et al., "Ultra-Wide-Band Transmitter for Low-Power Wireless Body Area Networks: Design and Evaluation," *IEEE Trans. on Circuits and Systems I: Regular Papers*, Vol. 52, No. 12, pp. 2515–2525, December 2005.

[10]  J. Xia et al., "3-5 GHz UWB Impulse Radio Transmitter and Receiver MMIC Optimized for Long Range Precision Wireless Sensor Networks," *IEEE Trans. on Microwave Theory and Techniques*, Vol. 58, No. 12, pp. 4040–4051, 2010.

[11]  S. Sim, D. W. Kim, and S. Hong, "A CMOS UWB Pulse Generator for 6–10 GHz Applications," *IEEE Microwave and Wireless Components Letters*, Vol. 19, No. 2, pp. 83–85, 2009.

[12]  Z. Low, J. Cheong, and C. Law, "Novel Low Cost Higher Order Derivative Gaussian Pulse Generator Circuit," in *The Ninth International Conference on Communications Systems (ICCS 2004)*, pp. 30–34.

[13]  J. Lee and C. Nguyen, "Uniplanar Picosecond Pulse Generator Using Step-Recovery Diode," *Electronics Letters*, Vol. 37, No. 8, pp. 504–506, 2001.

[14]  P. Protiva, J. Mrkvica, and J. Machac, "Sub-Nanosecond Pulse Generator for Through-the Wall Radar Application," in *2009 European Microwave Conference (EuMC 2009)*, pp. 1904–1907.

[15]  M. Miao and C. Nguyen, "On the Development of an Integrated CMOS-Based UWB Tunable-Pulse Transmit Module," *IEEE Trans. on Microwave Theory and Techniques*, Vol. 54, No. 10, pp. 3681–3687, October 2006.

[16]  H. Kim, Y. Joo, and S. Jung, "A Tunable CMOS UWB Pulse Generator," in *2006 IEEE International Conference on Ultra-Wideband*, September 2006, pp. 109–112.

[17]  R. Jin et al., "Tunable Pulse Generator for Ultra-Wideband Applications," in *Asia Pacific Microwave Conference (APMC 2009)*, December 2009, pp. 2272–2275.

[18]  *Picosecond Pulse Labs—Pulse Generators*, August 2010, http://www.picosecond.com/product/category.asp?pd id=10 (accessed September 12, 2012).

[19]  *Agilent Technologies—Pulse Generator Products*, January 2012, http://www.home.agilent.com/agilent/product.jspx?cc=AU&lc= eng&ckey=1000003131:epsg:pgr&nid=-536902258.0.00&id=1000003131:epsg:pgr (accessed September 12, 2012).

[20]  *Quantum Composers—Pulse Generators*, January 2011, http://www.quantumcomposers.com/products/pulse-generator (accessed September 12, 2012).

[21]  *Lecroy—ArbStudio Arbitrary Waveform Generator*, January 2012, http://www.lecroy.com/arbstudio (accessed September 12, 2012).

[22]  *Agilent Technologies—Function/Arbitrary Waveform Generators*, January 2012, http://www.home.agilent.com/agilent/product. jspx?nid=-536902257.0.00&lc=eng&cc=AU (accessed September 12, 2012).

[23]  *Tektronix Inc.—AWG7000 Arbitrary Waveform Generator*, January 2012, http://www.tek.com/signal-generator/ awg7000-arbitrary-waveform-generator (accessed September 12, 2012).

[24]  M. Ghavami, L. Michael, and R. Kohno, *Ultra Wideband Signals and Systems in Communication Engineering*, New York: Wiley, 2004.

[25]  M.-G. D. Benedetto et al., *UWB Communication Systems: A Comprehensive Overview*, EURASIP Book Series on Signal Processing and Communications), Hindawi Publishing Corporation, May 2006.

[26]  R. Fontana, "Recent System Applications of Short-Pulse Ultra-Wideband (UWB) Technology," *IEEE Trans. on Microwave Theory and Techniques*, Vol. 52, No. 9, pp. 2087–2104, 2004.

[27]  S. Preradovic et al., "Multiresonator-Based Chipless RFID Systemfor Low-Cost Item Tracking," *IEEE Trans. on Microwave Theory and Techniques*, Vol. 57, No. 5, pp. 1411–1419, May 2009.

[28]  *Application Note: Real-Time Versus Equivalent-Time Sampling*, January 2001, http://www.tek.com/application-note/ real-time-versus-equivalent-time-sampling (accessed September 12, 2012).

[29]  A. Toya et al., "32 GS/s Ultra-High-Speed UWB Sampling Circuit for Portable Imaging System," *Electronics Letters*, Vol. 47, No. 3, pp. 165–167, 2011.

[30]  E. Moreno-Garcia, J. de la Rosa-Vazquez, and O. Alonzo-Larraga, "An Approach to the Equivalent-Time Sampling Technique for Pulse Transient Measurements," in *16th International Conference on Electronics, Communications and Computers (CONIELECOMP 2006)*, February 2006, p. 34.

[31]  Y.-C. Jenq, "Perfect Reconstruction of Digital Spectrum from Nonuniformly Sampled Signals," *IEEE Trans. on Instrumentation and Measurement*, Vol. 46, No. 3, pp. 649–652, June 1997.

[32]  J.-H. Lee, C.-C. Chen, and Y.-S. Lin, "0.18 mm 3.1–10.6 GHz CMOS UWB LNA with 11.4 –0.4 dB Gain and 100.7–17.4 ps Group-Delay," *Electronics Letters*, Vol. 43, No. 24, pp. 1359–1360, 2007.

[33]  H.-Y. Yang, Y.-S. Lin, and C.-C. Chen, "2.5 dB NF 3.1-10.6 GHz CMOS UWB LNA with Small Group-Delay Variation," *Electronics Letters*, Vol. 44, No. 8, pp. 528–529, 2008.

[34]  Y. Park et al., "The Analysis of UWB SiGe HBT LNA for its Noise, Linearity, and Minimum Group Delay Variation," *IEEE Trans. on Microwave Theory and Techniques*, Vol. 54, No. 4, pp. 1687–1697, June 2006.

[35]  K.-P. Ahn, R. Ishikawa, and K. Honjo, "Low Noise Group Delay Equalization Technique for UWB InGaP/GaAs HBT LNA," *IEEE Microwave and Wireless Components Letters*, Vol. 20, No. 7, pp. 405–407, July 2010.

[36]  J. B.-Y. Tsui, *Microwave Receivers with Electronic Warfare Applications*, Herndon, VA: SciTech Publishing, 2003.

[37]    G. Anderson et al., "Advanced Channelization for RF, Microwave, and Millimeterwave Applications," *Proc. IEEE*, Vol. 79, No. 3, pp. 355–388, March 1991.

[38]    C. Rauscher, J. Pond, and G. Tait, "Cryogenic Microwave Channelized Receiver," *IEEE Trans. on Microwave Theory and Techniques*, Vol. 44, No. 7, pp. 1240–1247, July 1996.

[39]    W. Namgoong, "A Channelized Digital Ultrawideband Receiver," *IEEE Trans. on Wireless Communications*, Vol. 2, No. 3, pp. 502–510, May 2003.

[40]    D. Gupta et al., "Digital Channelizing Radio Frequency Receiver," *IEEE Trans. on Applied Superconductivity*, Vol. 17, No. 2, pp. 430–437, June 2007.

[41]    N. C. Karmakar and M. E. Bialkowski, "Design of Multistage Multiway Microstrip Fork Power Dividers," *Microwave and Optical Technology Letters*, Vol. 23, No. 3, pp. 141–147, 1999.

[42]    R. V. Koswatta and N. C. Karmakar, "A Novel Reader Architecture Based on UWB Chirp Signal Interrogation For Multiresonator-Based Chipless RFID Tag Reading," *IEEE Trans. on Microwave Theory and Techniques*, Vol. 60, No. 9, pp. 2925–2933, September2012.

[43]    I. Balbin and N. C. Karmakar, "Multi-Antenna Backscattered Chipless RFID Design," in *Handbook of Smart Antennas for RFID Systems*, N. C. Karmakar (Ed.), New York: John Wiley & Sons, pp. 413–443, 2010.

## CHAPTER 6

# Hybrid Chipless RFID Reader

## 6.1 Introduction

In previous chapters chipless RFID systems that generally use a single dimension, such as the frequency or time domain, for encoding information were discussed. In such systems, an elemental encoding structure is used to transform the properties of the dimension used, frequency or time, in order to encode information where the presence of the elemental structure encodes a single bit of information. Also, in the process of reading a chipless tag, RFID readers are required only to analyze the characteristics of the single dimension used for encoding information.

In [1–3] a spiral resonator is used as the elemental encoding structure to transform the amplitude spectrum for encoding information in the frequency domain (refer to Chapter 2). The presence of a spiral resonator causes a sharp attenuation in the amplitude spectrum and also simultaneously causes an abrupt phase jump in the phase spectrum at the resonant frequency of the resonator in the frequency domain. Therefore, one can use either the amplitude spectrum or the phase spectrum as a dimension for encoding information where the representation of a single bit of information requires a unique spiral resonator resonating at a unique frequency. In the time domain reflectometry based chipless RFID systems described in [4–6] the elemental encoding structure is a capacitive stub that introduces a discontinuity in the transmission line forming the chipless RFID tag. Each discontinuity generates a single echo that encodes a single information bit. Hence, each capacitive element can only encode a single bit of information in the chipless RFID tag.

It is clear that the use of a single dimension limits the amount of information that can be encoded into an elemental encoding structure. Also, encoding more information using a single dimension requires the use of multiple elemental structures. This demands more space, requiring the overall size of the chipless RFID tag to be very large, which causes the chipless RFID system to be impractical in real-world applications. To increase the spectral and spatial efficiency in encoding information in chipless RFID systems, each elemental structure needs to encode more than one bit of information. By designing an elemental structure capable of manipulating multiple amplitude levels as opposed to causing an abrupt change of a single level, multiple bits can be encoded into a given dimension such as time or frequency; that is, multiple amplitude level changes at a given resonant frequency for a frequency domain chipless RFID tag or multiple amplitude levels of an echo in a time domain reflectometry based chipless RFID tag. However, using multiple levels for encoding information in a single dimension, particularly in a passive context without any active digital modulation or forward error correction mechanisms together with noise and other detriments introduced by the wireless environment is unreliable and will introduce many errors. Therefore, a better solution to improve spectral and spatial efficiency in chipless RFID systems is to design elemental structures that are capable of encoding information by simultaneously transforming properties in multiple dimensions. Figure 6.1 contrasts the use of a single dimension and multiple dimensions for encoding information in chipless RFID systems. It is evident that the use of extra dimensions enables the efficient use of frequency spectrum and space.

Several works have reported on the use of hybrid encoding approaches to design compact and efficient chipless RFID systems. The earliest reported hybrid RFID systems have elemental encoding structures that make use of frequency, phase, and time for encoding information in a mixed nature [7, 8]. Although the authors of these works do not specifically define the term hybrid chipless RFID, they make use of more than one dimension for encoding information in each elemental encoding structure.

In [8] the elemental encoding structure is a stub-loaded microstrip patch antenna, where the antenna resonates at a unique resonant frequency and also produces a phase shift due to the loading condition. The phase shift depends on the length of the stub attached to the patch antenna. Therefore, each patch antenna can be used to encode more than a single bit of information by using multiple levels of phase shifts.

In [7] the dimensions of phase and time are used for encoding information in a chipless RFID. The structure of the chipless RFID tag discussed in [7] is similar to a time domain reflectometry based system where discontinuities in a transmission

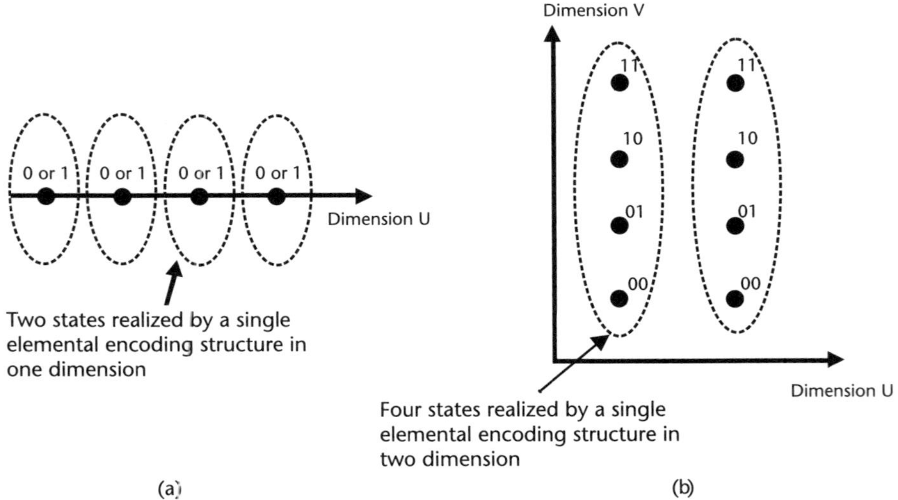


**Figure 6.1**  Realization of a 4-bit chipless RFID system using (a) four distinct 1-bit elemental encoding structures in one dimension and (b) two distinct 2-bit elemental encoding structures in two dimensions.

line result in echoes, which in turn encode information bits. Here in addition to encoding information bits using the time of arrival of the echoes, each discontinuity also introduces unique phase information to each echo, which further enhances the data capacity of the system. The authors of [9] use the two dimensions, time and frequency, for encoding information in a chipless RFID tag. The elemental encoding structure is a spiral resonator. Even though each resonator encodes a single bit, the introduction of a meandered transmission line section in the chipless tag causes a time delay that enables the reuse of the same resonance frequencies in a different time period, as in a time domain multiplexing system. This enables better spectrum utilization; however, the size of the chipless tag becomes larger since the number of bits per elemental encoding structure is not improved.

In [10] the chipless tag uses a pulse position encoding technique where it introduces a frequency-dependent unique time delay to interrogation pulses modulated with different frequencies. The system makes use of the dimensions of group delay and frequency for encoding information. The elemental encoding structure used is a C-section dispersive delay structure. However, the system demonstrated can only encode a single bit per C-section. The authors of [11, 12] also make use of a similar C-section based chipless RFID design to make use of the dimensions of group delay and frequency for encoding information into chipless RFID tags. In a recent work [13], a chipless RFID system that uses the dimensions of frequency and polarization for encoding information is presented. Here, the elemental encoding structure is a set of co-centric split-ring resonators (SRR). In [14] a chipless RFID tag based on a set of C-sections is designed that makes use of the two dimensions of phase deviation bandwidth and frequency for encoding information.

In this chapter the aforementioned hybrid chipless RFID systems that make use of more than one dimension for encoding information are examined in detail. The discussions presented are classified into three groups: (1) frequency-phase, (2) frequency-polarization, and (3) frequency-time as shown in Figure 6.2. The theory involved in encoding information in an elemental encoding structure of the chipless RFID tag and the organization of the RFID reader for each group is discussed in detail. The rest of the chapter is organized as illustrated in Figure 6.3. In Section 6.2 the research reported in [14] is discussed in detail where the two dimensions of frequency and phase are used for encoding information bits in chipless RFID tags. The characteristics and operation of the encoding element are discussed and

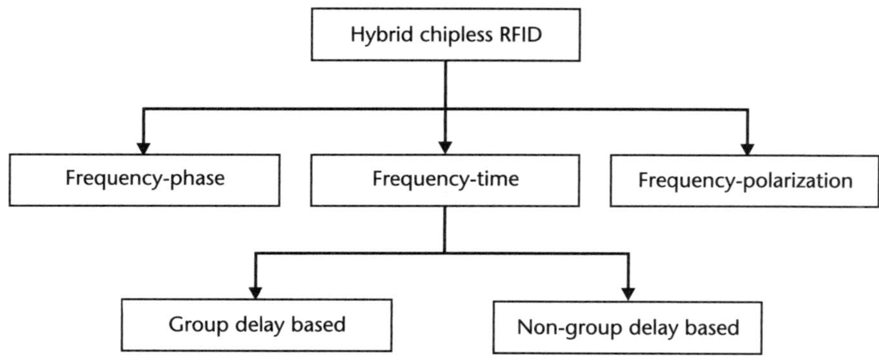


**Figure 6.2**  Classification of hybrid chipless RFID discussed in the chapter.

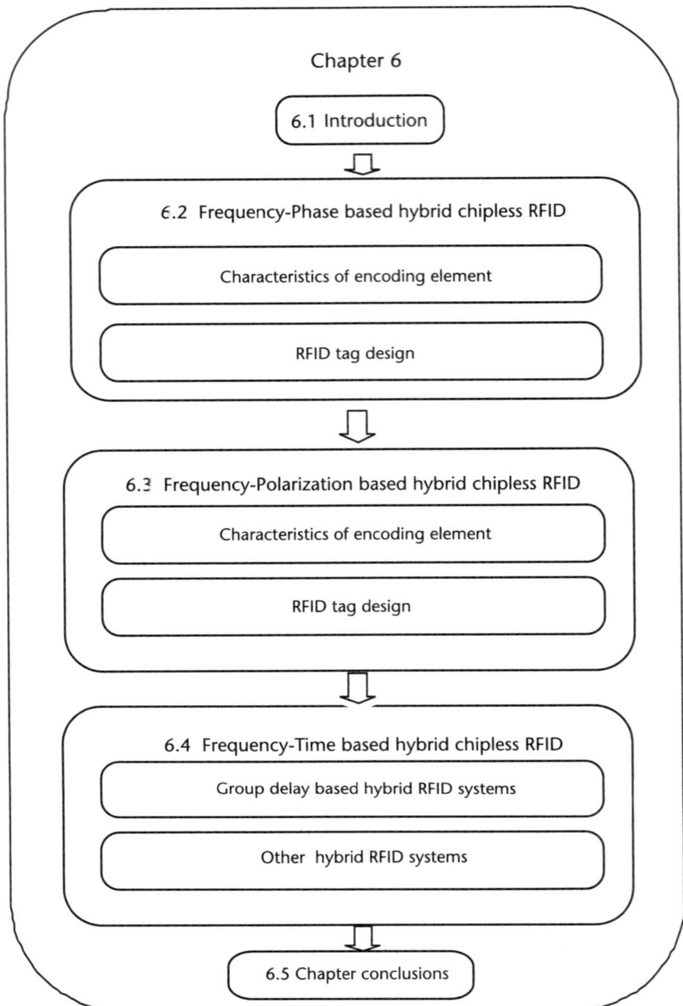



**Figure 6.3**   Chapter overview.

examined in detail. The encoding element's application in the chipless RFID tag design is presented and the corresponding RFID reader design is discussed. Next, the novel technique reported in [13], which makes use of polarization as an additional dimension for encoding information, is examined. The frequency and polarization characteristics of the elemental encoding structure are discussed in detail and its application for data encoding in the hybrid chipless RFID tag design is illustrated. An RFID reader design capable of reading hybrid frequency-polarization based chipless RFID tags is also discussed. Section 6.4 discusses several works [10–12, 15, 16] that make use of frequency and time as dimensions for encoding information in chipless RFID technology. An RFID reader design that analyzes characteristics of both time and frequency is also discussed. Finally, conclusions are drawn in Section 6.5.

## 6.2 Frequency-Phase Based System

In this section the work of Vena et al. [14] on the simultaneous use of the two dimensions of frequency shift and phase deviation bandwidth for data encoding in chipless RFID is discussed. Here, the basic element that encodes information in the chipless RFID tag is a planar C-section.

### 6.2.1 Working Principle of the Encoding Element

Figure 6.4 depicts the elemental encoding structure used in the chipless RFID tag design of [14]. It is a planar C-section consisting of only a single layer of conductive material placed on a substrate material. The dimensions $L$ and $g$ of the C-section determine the resonance frequency and the phase deviation bandwidth.

When the structure shown in Figure 6.4 is illuminated with an electromagnetic wave that is vertically polarized, a highly resonant mode is observed at a specific frequency in the RCS. The resonant frequency is a function of the physical dimensions of the C-section. Figure 6.5 shows the behavior observed in the radar cross section near the resonance frequency of the C-section when $L$ is fixed and $g$ is varied. Each C-section gives rise to a resonant peak and a resonant dip, and the frequency of the resonant peak is determined by the length $L + g/2$. The gap between the peak and dip is related to the ratio between $g$ and $L$. The C-section also transforms the phase near the resonance. From Figure 6.5 we can observe that the phase transformation bandwidth is closely linked with the frequency of the resonant peak and resonant dip. Therefore, we can see that the phase transformation or deviation bandwidth $B$ is also a function of $g/L$. Figure 6.6 shows the behavior of the radar cross section near the resonance when $g$ is fixed and $L$ is varied slightly. The lengths $g$ and $L$ are chosen such that $g$ is always much smaller than $L$. Therefore, a slight change in $L$ does not greatly affect the ratio $g/L$. Hence, it is observed that when $L$ is slightly changed the resonance peak shifts, whereas the phase deviation band-

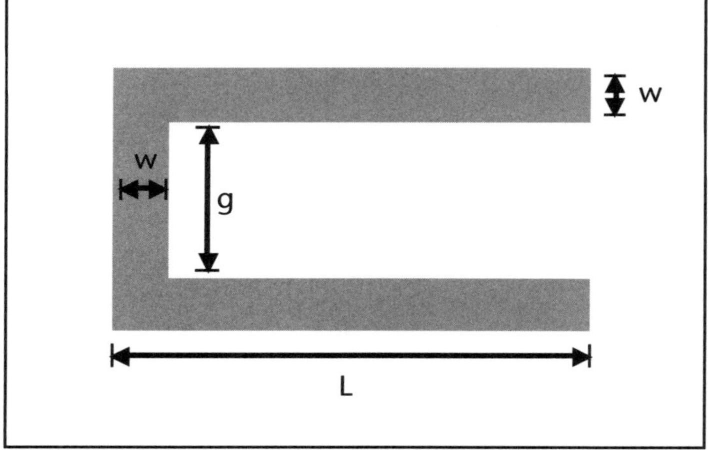


**Figure 6.4** Basic resonant element (C-section) that encodes information simultaneously as a frequency shift and a phase deviation. Resonant properties are determined by the parameters L and g.

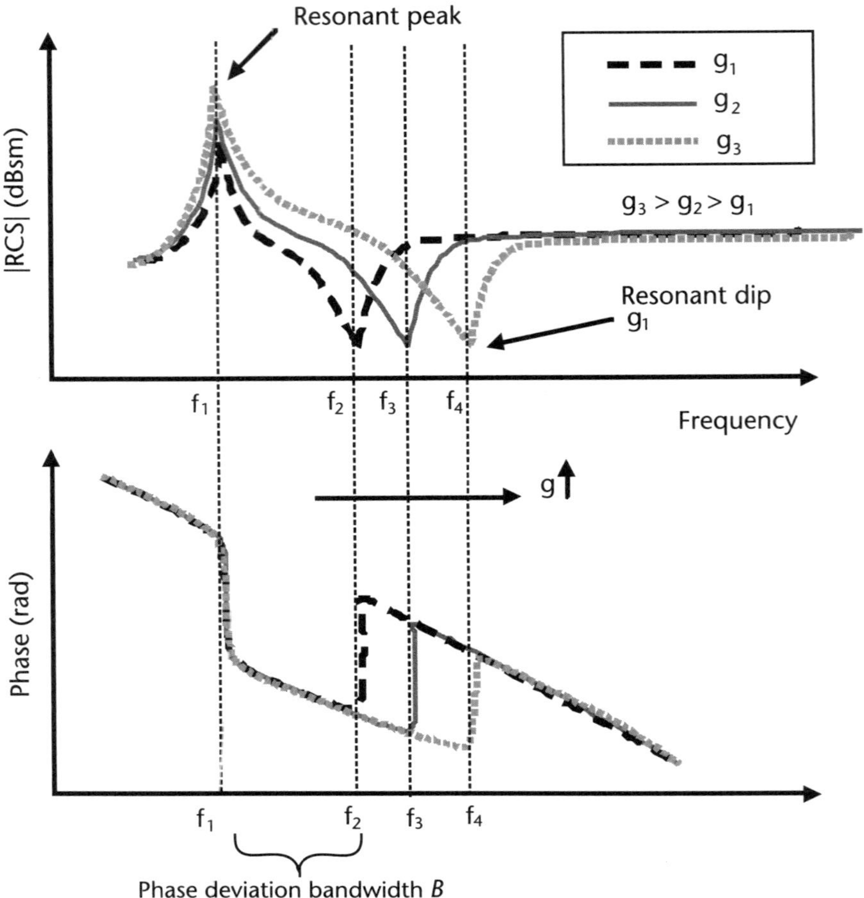



**Figure 6.5**    Resonant behavior of a C-section, when L is fixed and g is varied. (Adapted from [14].)

width stays the same since the ratio is not affected. However, the start frequency of the phase deviation bandwidth shifts together with resonance peak.

The resonant nature of the C-section observed in Figures 6.5 and 6.6 can be analytically expressed using the following transfer function $T(\omega)$ [14],

$$T(\omega) = G\left[\frac{1 + \dfrac{2j\, m_z \omega}{\omega_z} + \left(\dfrac{j\omega}{\omega_z}\right)^2}{1 + \dfrac{2j\, m_p \omega}{\omega_p} + \left(\dfrac{j\omega}{\omega_p}\right)^2}\right] \tag{6.1}$$

where $\omega$ is the angular frequency, and G is the gain characterizing the level of backscattered signals. Here the resonance peak is characterized by the pole at $\omega_p$ with the damping factor $m_p$ where the resonant dip is characterized by the zero at $\omega_z$ with the damping factor $m_z$.

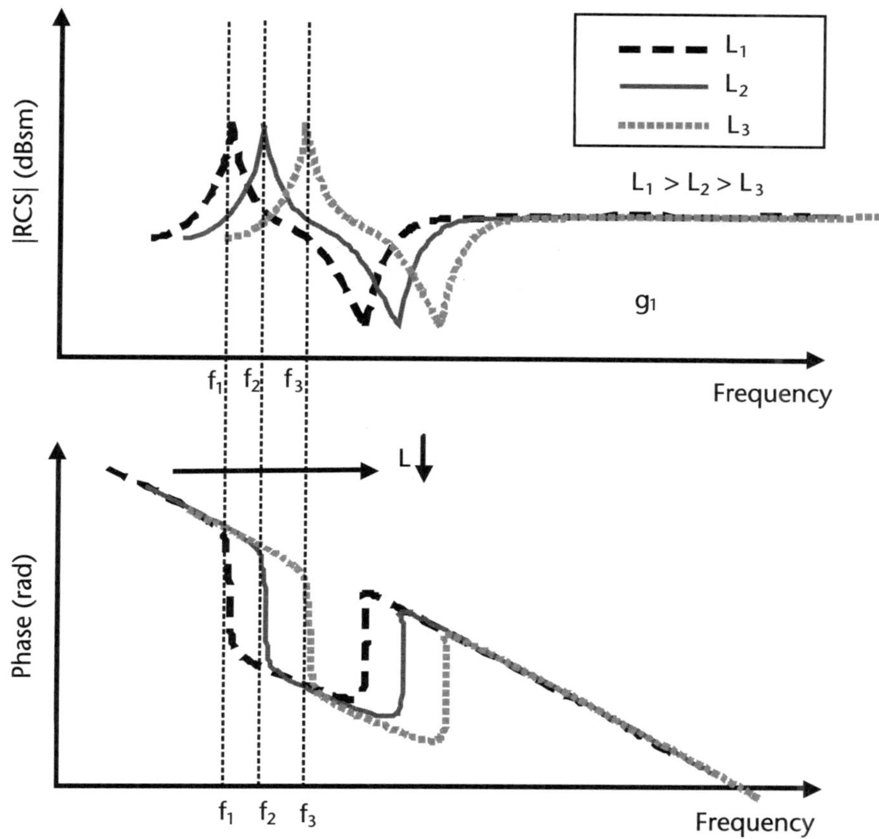



**Figure 6.6**   Resonant behavior of a C-section, when $g$ is fixed and $L$ is varied. (Adapted from [14].)

### 6.2.2   Hybrid Frequency-Phase Chipless RFID Based on C-Sections

By carefully selecting the values of $L$ and $g$, it is possible to assign a unique resonance peak and a phase deviation bandwidth to a C-section so that information
can be encoded using both frequency and phase. Figure 6.7 demonstrates a possible
case where 2 bits can be encoded in a single C-section using a combination of phase
deviation bandwidth and frequency shifts. From Figure 6.7 it is clear that the most
significant bit (MSB) of the 2 bits is determined by the resonant peak frequency;
that is, the MSB is "0" when the resonant peak frequency is $f_1$, and it is "1" when
the C-section has a resonant peak at $f_1 + \Delta f$. The resonant peak frequency is pre-
dominantly controlled by the length $L_1$.

The least significant bit (LSB) of the 2 bits is determined by the phase deviation
bandwidth. When the C-section takes the smaller phase deviation bandwidth, $B_1$,
the LSB is "0" and when it has the larger bandwidth $B_2$ the LSB is "1." Therefore,
through the unique combination of a resonant peak frequency and a phase deviation bandwidth, the C-section will encode one of four 2-bit data combinations,
"00," "01," "10," or "11." Figure 6.8 shows a constellation diagram for a different set of frequency-phase combinations that a C-section can be configured to have
for improved encoding efficiency. It is possible to encode 1 of 10 states to a single
elemental encoding structure. All 16 states shown in the constellation diagram can-

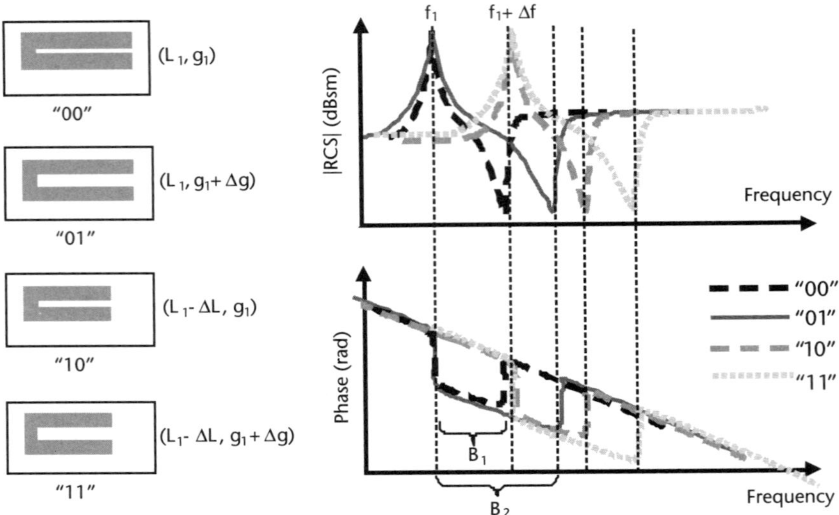



**Figure 6.7**  Implementation of a C-section encoding 2 bits of information. (Adapted from [14].)

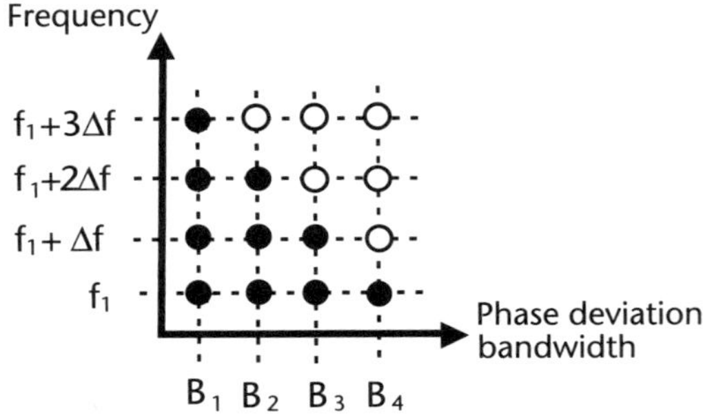



**Figure 6.8**  Constellation diagram showing the combinations of frequency and phase that can be used for encoding information in a C-section. (Adapted from [14].)

not be physically achieved due to overlapping of resonance frequencies; only the 10 states that are shown as dark filled circles can be realized [14].

The chipless tag demonstrated in [14] consists of multiple C-sections, as shown in Figure 6.9, each encoding more than 1 bit of information. The total tag signature will have multiple resonant peaks and phase deviation bandwidths corresponding to each C-section in the chipless tag.

### 6.2.3  Chipless RFID Reader for Hybrid Frequency-Phase Chipless RFID

The RFID reader required for reading a hybrid frequency-phase chipless RFID should be able to simultaneously extract both phase and amplitude spectrums of the backscattered signal. Either of the reader designs proposed in Chapter 4 (frequency domain) and Chapter 5 (time domain) can be employed for implementing an RFID reader for reading hybrid frequency-phase chipless RFID tags. Figure 6.10 illustrates the design of a frequency domain based RFID reader that uses a gain/phase detector (GPD) for reading hybrid chipless RFID.

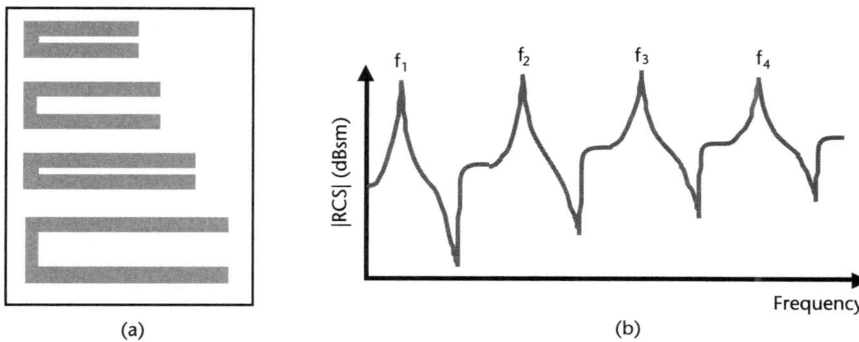

**Figure 6.9**   (a) Chipless RFID tag designed using multiple C-sections and (b) frequency signature of the chipless RFID tag.

The reader design shown in Figure 6.10 is discussed and explained in detail in Chapter 4, Section 4.2.1. It consists of mainly two sections, an RF front-end and a digital section that performs control functions of the RF front-end and processes digitized data. Usually the digital section is implemented using a field programmable gate array (FPGA) or an industrial microcontroller.

The interrogation signal is generated using a voltage-controlled oscillator (VCO), which is controlled using analog control signals produced by a digital-to-analog converter (DAC). The DAC is controlled by the digital section. Here the transmitted interrogation signal is a frequency sweep where the frequency linearly increases from low to high, which corresponds to a control signal having the shape of a sawtooth waveform. The interrogation signal is amplified using a power amplifier before being transmitted through a transmitting antenna.

The backscatter coming from a chipless tag is picked up by the receiving antenna of the reader and amplified using a low noise amplifier (LNA). The reader design makes use of a GPD to extract the phase and amplitude information from the received backscattered signal. The operation of the GPD requires a reference signal that serves as the reference with respect to which the gain and phase of the received signal is calculated. Here, a coupled signal obtained from the transmitted interrogation signal serves as the reference. Because the frequencies considered in a typical chipless RFID application lie beyond the S-band, both the reference and the received signal are first converted into an intermediate frequency (IF) that falls within the operational range of frequencies of the GPD. For this purpose a second VCO is used which is also controlled using a sawtooth voltage waveform. The two VCOs operate at slightly different frequencies (with a constant frequency gap) in order to produce the IF signal. The generation of the IF signals also require mixers and low pass filters (LPF). The GPD outputs voltages proportional to the amplitude difference and phase difference between the reference (IF Ref.) and received (IF Rx) signals. These output signals are digitized using an analog-to-digital converter and are presented to the digital section of the RFID reader. The digital section uses DSP algorithms to filter out any noise present in the data and constructs amplitude and phase spectrums using the gain phase information. These amplitude and phase spectrums are then simultaneously analyzed and examined to extracting the information encoded in them.

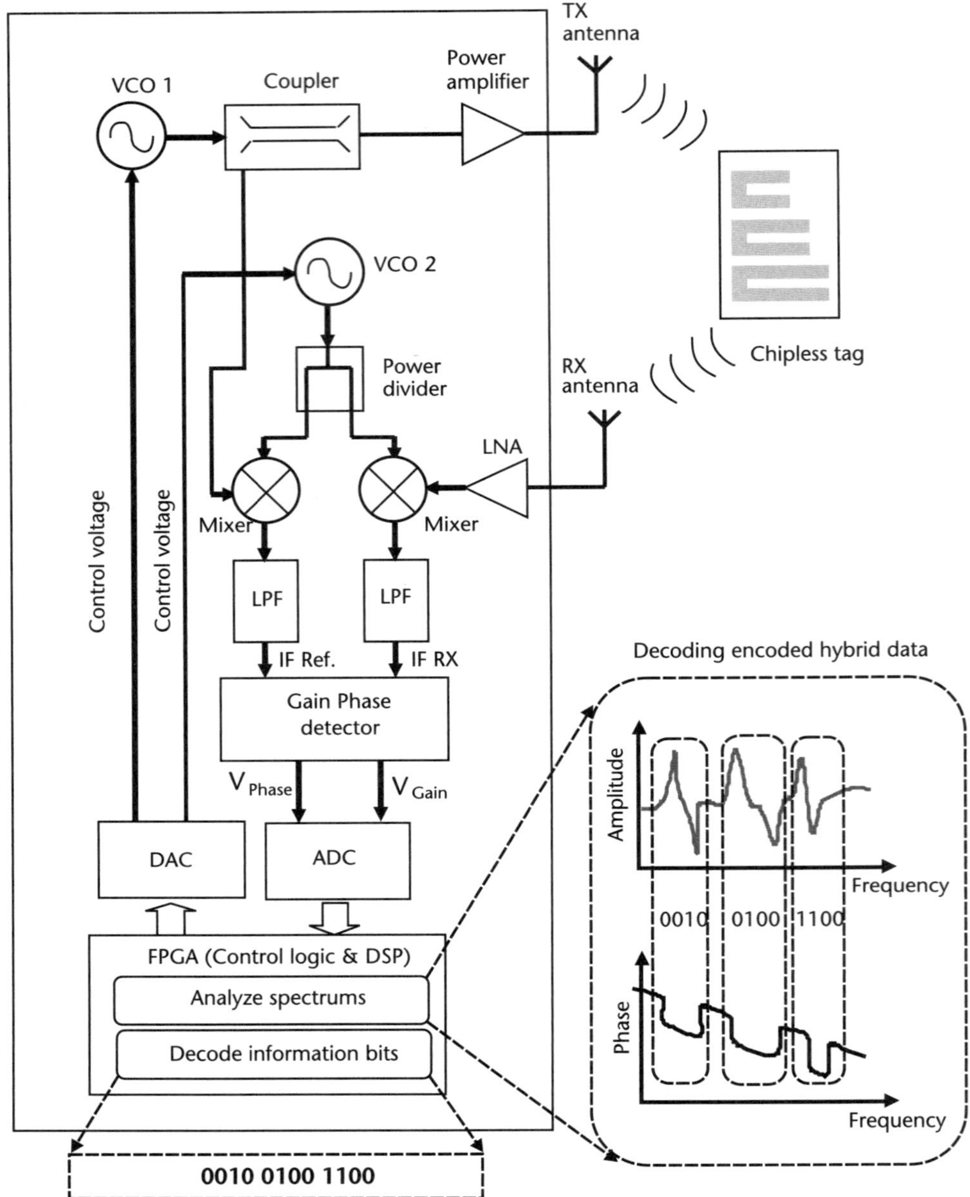



**Figure 6.10**  RFID reader for hybrid frequency-phase chipless RFID.

In the illustration shown in Figure 6.10, a chipless RFID tag having three C-sections is used. Here each C-section is considered to be capable of encoding 16 different frequency-shift and phase deviation bandwidth combinations at the vicinity of a particular resonance frequency. Therefore, when algorithms in the digital section simultaneously analyze the amplitude and phase spectrums, they will generate a 4-bit code for each frequency-phase combination, resulting in decoding of the 12-bit data contained in the chipless tag.

## 6.3  Frequency-Polarization Based System

This section describes chipless RFID systems that make use of the dimensions of frequency and polarization for data encoding. Polarization has been investigated as a means of enhancing data capacity in [13, 17, 18]. In [17, 18] a compact high-data-capacity chipless RFID tag design is introduced that consists of patches loaded with horizontally and vertically polarized slot resonators. The elemental encoding structures in this tag design are these slot resonators. In the design, the information is encoded as sharp amplitude or phase changes in the frequency domain and the polarization (vertical/horizontal) is simply used as a means of frequency reuse for doubling the data capacity for a given frequency spectrum. Therefore it uses the additional dimension of polarization in a limited fashion to achieve better spectrum utilization where the elemental encoding structures in either polarization (vertical or horizontal) can only each encode a single bit. However, because each patch can hold multiple slot resonators in a compact nature, the overall chipless tag design is both spatially and spectrally efficient in encoding data. On the other hand, the research presented in [13] fully utilizes polarization as an extra dimension together with frequency for encoding information. The elemental encoding structure in the chipless RFID tag is a split-ring resonator, which has a unique resonance frequency and also exhibits a characteristic behavior with the polarization angle of the incident electromagnetic wave. Therefore, each elemental encoding structure can encode information using both polarization angle and resonance frequency.

### 6.3.1  Resonant and Polarization Characteristics of the Split-Ring Resonator

Figure 6.11 shows the microstrip split-ring resonator (SRR) used as the elemental encoding structure in [13]. The resonator gives rise to two resonant modes depending on the polarization of the incident electromagnetic wave. When the gap is oriented at the top as in Figure 6.11(a) and the wave is vertically polarized, the half-wavelength resonant mode of the SRR is aligned with half the perimeter of the ring, which gives rise to a lower frequency $(f_0)$ resonant mode. When the gap is rotated $90°$ as shown in Figure 6.11(b) so that the incident wave is horizontally polarized

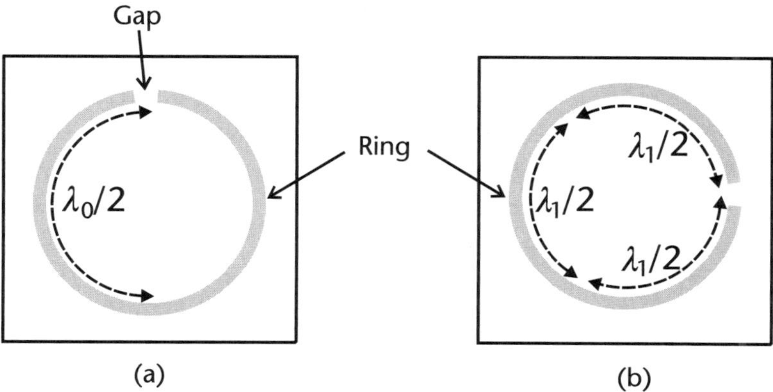


**Figure 6.11**  SRR having radius $R$ gives rise to two resonant modes (on substrate RO4003): (a) lower frequency resonant mode having wavelength $\lambda_0$ and (b) higher frequency resonant mode $\lambda_1$.

with respect to the gap orientation, the half-wavelength resonance spans the perimeter C such that $C = 3\lambda_1/2$. This gives rise to a higher frequency $(f_1)$ resonant mode having a shorter wavelength $\lambda_1$.

The two resonant modes of a SRR having a ring radius of 9.3 mm and a ring width of 1 mm (on substrate RO4003) are shown in Figure 6.12(a). When the incident wave is vertically polarized (polarization angle of 0°), only the lower resonant mode, 3.1 GHz, is visible; and when it is horizontally polarized (polarization angle of 90°), only the higher resonant mode, 4.6 GHz, is visible. With a 45° polarization angle, both resonant modes can be observed together. The behavior of the SRR with the polarization angle at these two resonant modes is shown in Figure 6.12(b). When the polarization angle of the incident wave is gradually incremented, it is observed that both resonant modes show characteristic drops in the magnitude of the radar cross section. For the lower resonant mode, this characteristic drop occurs when the polarization angle is 0°, whereas for the higher resonant mode the drop occurs when the polarization angle is either −90° or 90°. Figure 6.13 shows the behavior of the lower resonant mode with the changing polarization angle and how the orientation of the gap in the SRR affects the response. This characteristic behavior of the two resonant modes of the SRRs can be readily used for information encoding.

### 6.3.2 Encoding Information in a SRR Based Frequency-Polarization Tag

From the characteristic nature of the SRR described earlier, it is clear that the orientation of the split/gap of the SRR is pivotal to its behavior with respect to the polarization angle. That is, when the polarization angle of the incident wave matches the angle of the gap as shown in Figure 6.13, a significant drop in the RCS is observed at the lower resonant mode $(f_0)$. This behavior can be exploited for encoding information bits as shown in Figure 6.14. Here, the gap is oriented in one of four different angles (0°, 45°, 90°, or 135°) for encoding one of four 2-bit data combinations: "00," "01," "10," or "11." In this approach a resonant SRR will encode 2 bits of information using polarization in a given resonant frequency. Hence, the complete chipless RFID tag can encompass multiple SRRs resonating at different frequencies

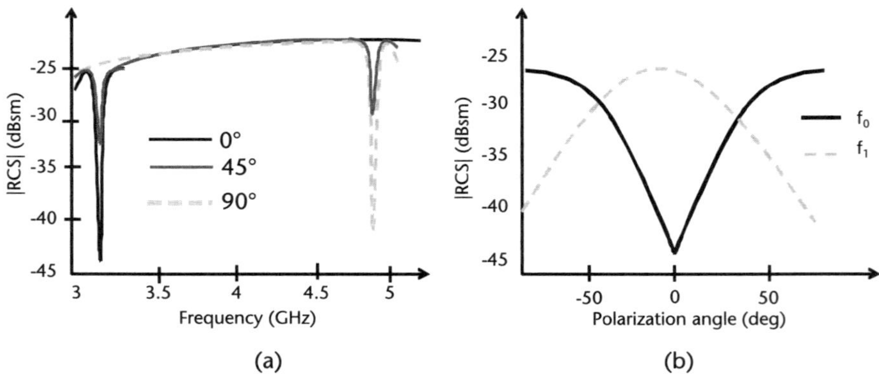


**Figure 6.12** Behavior of the SRR having a ring radius of 9.3 mm and ring width of 1 mm (on substrate RO4003). (a) Two resonant modes of the SRR observed in the frequency domain for different polarization angles of the incident wave. (b) Characteristic behavior of the two resonant modes with the changing polarization angle of the incident wave. (Adapted from [13].)

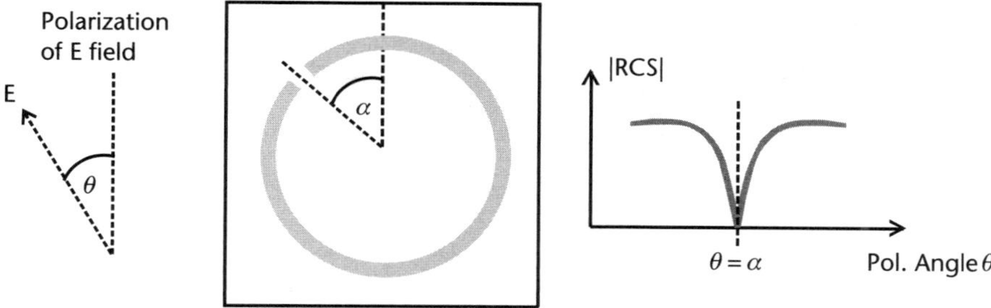

**Figure 6.13**    Characteristic of lower resonant mode of SRR with polarization angle.

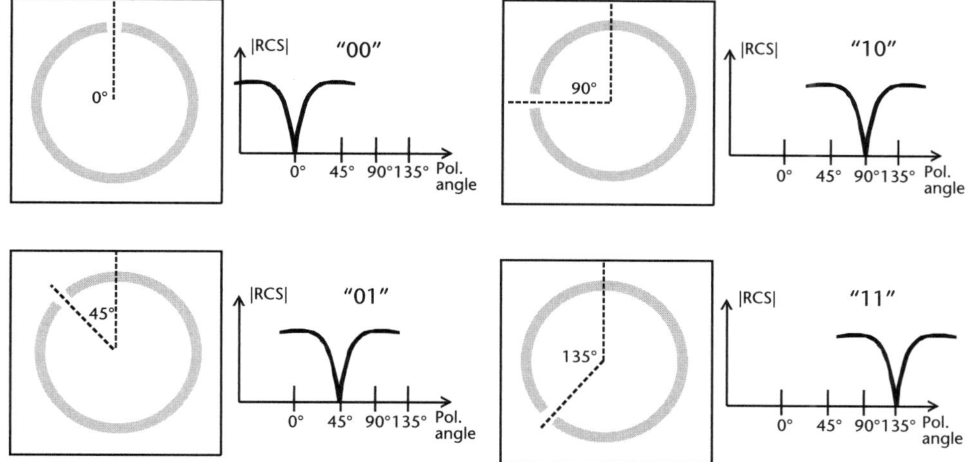

**Figure 6.14**    SRR configured for encoding 2 bits of information using the polarization angle at a given resonance frequency.

for encoding more information. Therefore, the encoding process utilizes both dimensions of frequency and polarization.

The encoding element shown in Figure 6.14 uses the absolute angle of the gap orientation for encoding information. One drawback of this method is that if the chipless RFID tag is physically rotated, the reading obtained would include both the angle of orientation of the SRR gap and the physically rotated angle, causing an error in the detection process. The solution to this problem is to include a dedicated reference SRR and use relative angles with respect to the reference SRR for encoding information. Figure 6.15 illustrates a 4-bit chipless RFID tag that uses three co-centric SRRs having radii $R_1$, $R_2$, and $R_3$. The outermost SRR is used as the reference and the two inner SRRs are used to encode 2 bits each. In the scenario presented in Figure 6.15, where the tag is rotated, the critical polarization angles (where the RCS drops significantly) of the SRRs $R_2$ and $R_3$ are meaningless and do not provide any useful information for decoding by themselves without the knowledge of the angle of tag rotation $\phi$. Using the polarization characteristics of the outermost SRR, which serves as the reference, this rotation angle can be computed. Once the tag rotation angle is found, it is clear that $R_3$ has a critical polarization angle of 45° relative to $R_1$, and $R_2$ has a critical polarization angle of 90° relative to $R_1$. Therefore, $R_2$ encodes "10" and $R_3$ encodes "01." If the MSB is assigned to

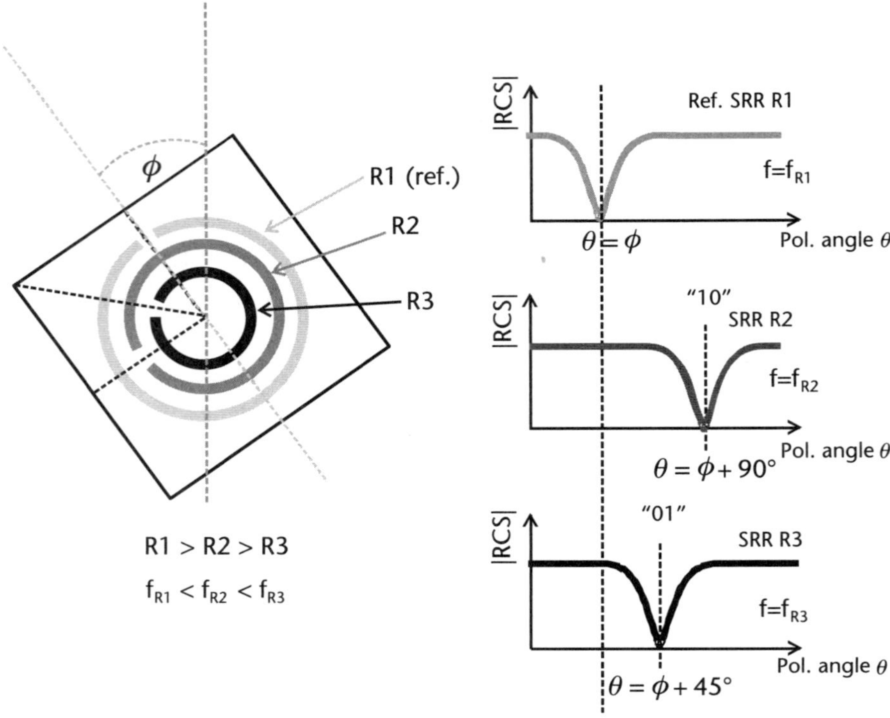


**Figure 6.15**　Four-bit chipless RFID tag consisting of three co-centric SRRs having radii R1, R2, and R3, which resonate at frequencies $f_{R1}$, $f_{R2}$, and $f_{R3}$, respectively. Measurement ambiguity caused by rotation is countered by using the outer reference SRR.

the lowest frequency and the LSB is assigned to the highest frequency, then the tag shown in Figure 6.15 encodes the data "1001."

The SSR based frequency-polarization hybrid chipless RFID tag design discussed here uses the frequency spectrum efficiently since the data encoding is performed using the polarization angle. The encoding is ultimately determined by the orientation of the gaps of the SRRs. The number of bits encoded per SRR is determined by the number of polarization angles considered in the encoding process. In the example presented in Figure 6.15, only four polarization angles (0°, 45°, 90°, and 135°) were considered. By increasing the number of polarization angles, the number of bits encoded per elemental structure can be further increased, making the chipless tag spatially efficient. However, this would increase the receiver complexity since the reader is required to detect subtle changes in the polarization characteristic in fine resolution.

## 6.3.3　Chipless RFID Reader for Hybrid Frequency-Polarization Based Chipless RFID

To read the hybrid frequency-polarization chipless RFID tag, the reader needs to have the ability to transmit and receive signals with different polarizations. This can be achieved by using polarization agile reader antennas or through mechanical rotation of the reader antenna setup. However, mechanical rotation of reader antennas using stepper motors would render the RFID reader bulky and make the interrogation process very slow.

Figure 6.16 shows reader systems for hybrid frequency-polarization chipless RFID. A reader system that utilizes dually polarized antennas (vertical and horizontal) in the transmit section and the receive section of the reader is shown in Figure 6.16(a). This type of reader is suitable for reading the tag design proposed in [17, 18] where the tag consists of horizontally and vertically polarized resonating slots. The RF section of the reader is very similar to that described in Figure 6.10. Here, two IF received signals are produced for each polarization. These signals are then compared against the reference signal in two separate gain/phase detectors to calculate the respective amplitude and phase spectrums of each signal. These spectrums are then analyzed in the digital section of the reader to determine the data encoded in the chipless RFID tag.

A reader system that employs polarization agile antennas is shown in Figure 6.16(b). Here, the antennas used for transmitting the interrogation signal and receiving the backscatter have the ability to dynamically change their polarization. One method of achieving polarization agility is by using a dually fed antenna

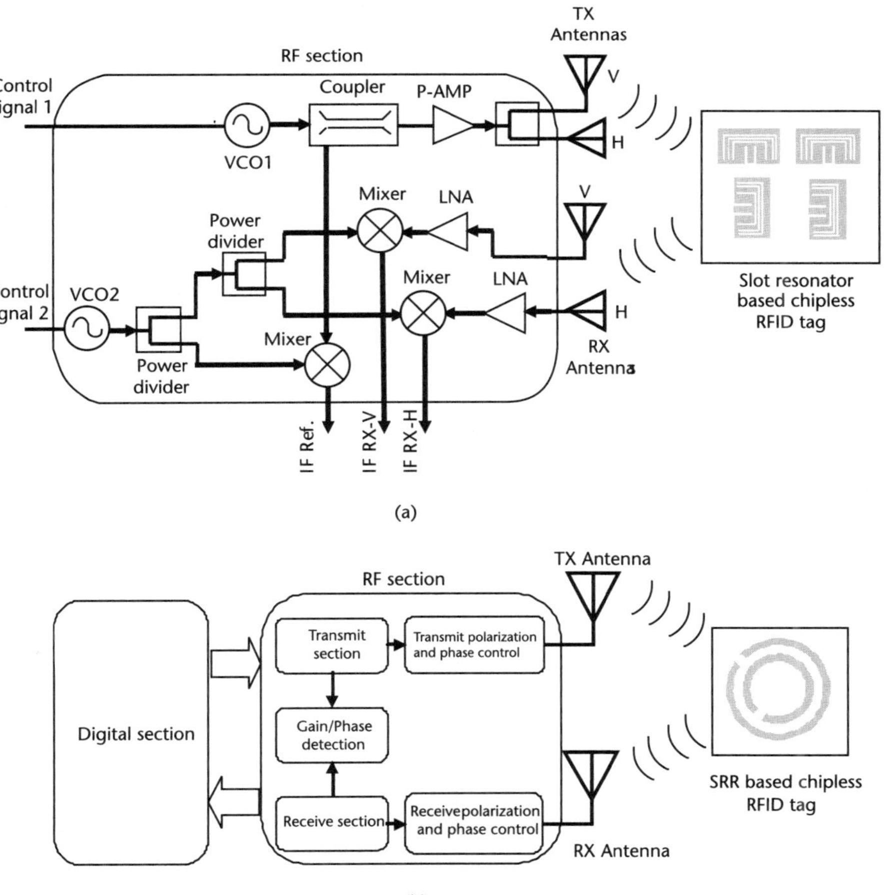


**Figure 6.16**  Reader designs for hybrid frequency-polarization based chipless RFID: (a) reader system utilizing vertical (V) and horizontal (H) antennas for transmission and reception and (b) reader system with polarization agile antennas.

(vertically and horizontally) and controlling the phase difference of the signals fed to the antenna [19]. This technique enables the transmission of vertical, horizontal, left-hand circular, and right-hand circular polarized signals. This type of reader system is capable of reading the tag designs proposed in [13, 20]. For reading the SRR based chipless RFID tags discussed earlier, the reader needs to interrogate the tag with a linear polarization and also vary the polarization angle in order to calculate the profile of the amplitude versus polarization angle for decoding the information contained in the tag. The amplitude of the received signal can be obtained as in the previous reader design through the use of a GPD. In [20] a reader design is proposed where the transmitter and receiver antennas choose between different combinations of polarization and record the response from a chipless RFID tag in order to uniquely identify it.

## 6.4 Frequency-Time Based System

In this section we explore hybrid chipless RFID systems that make use of the dimensions of time and frequency for encoding information. The research on time-frequency based chipless RFID systems available in the literature is mostly based on the group-delay characteristics of microwave structures [10–12, 15, 16]. In these designs the encoding mechanism is a frequency-dependent time delay defined by the group-delay characteristics of the elemental encoding structure used in the tag.

Let $H(j\omega)$ be the frequency domain transfer function of the elemental encoding structure, where $H(j\omega)$ can be defined as:

$$H(j\omega) = A(j\omega)e^{j\phi(\omega)} \tag{6.2}$$

The group delay, $\tau(\omega)$, of this system is defined as [21]:

$$\tau(\omega) = -\frac{d}{d\omega}\phi(\omega) \tag{6.3}$$

The group delay $\tau(\omega)$ is a measure of how long it takes for a signal having angular frequency to transit through the system $H(j\omega)$. Therefore, by utilizing a system having a unique group-delay profile, unique tag ID information can be encoded.

### 6.4.1 Group Delay Based Frequency-Time Hybrid Chipless RFID

The elemental encoding structure used in the group delay based chipless RFID systems reported in the literature [10–12, 15, 16] is a group of microstrip C-sections. The C-section described here is different from that of Section 6.2.1, which is single layer nonmicrostrip design. Each C-section is essentially formed by the end-to-end connection of two coupled microstrip transmission lines. In the chipless tag design, the microstrip C-sections are grouped and cascaded together and a dispersive delay structure (DDS) is constructed. This DDS introduces a frequency-dependent time delay [22] to the signals that travel through it. Figure 6.17 illustrates the DDS used in [16]. For clarity only one C-section is shown per group in Figure 6.17(a), whereas

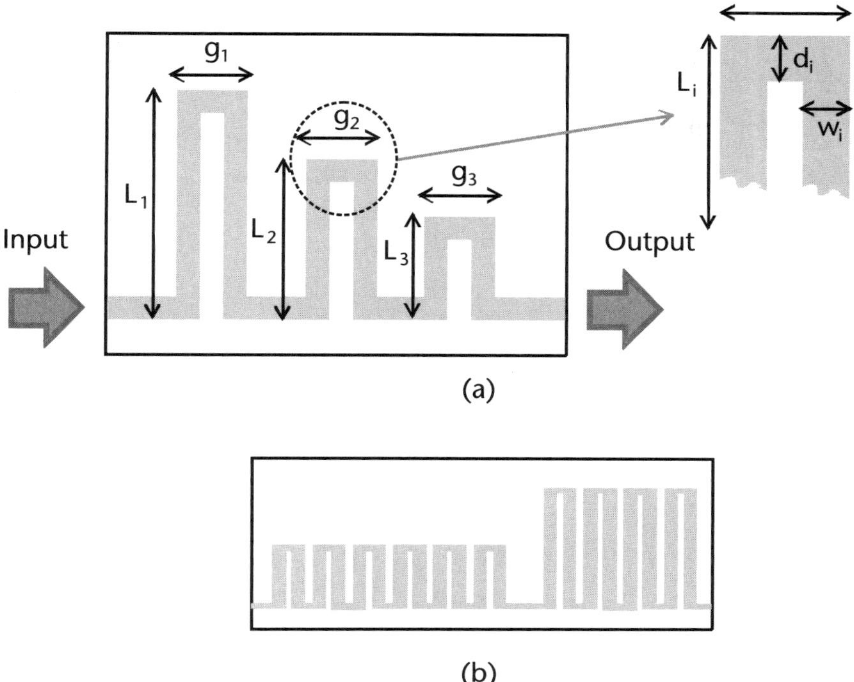

(a)

(b)

**Figure 6.17**  Dispersive delay structure constructed using a set of C-sections: (a) dimensions of the C-sections and (b) DDS consisting of closely grouped C-sections.

in the actual DDS, multiple C-sections are grouped and closely coupled to enhance the group-delay profile as shown in Figure 6.17(b).

The group delay introduced at a particular frequency can be controlled by varying the physical dimensions $L_i$, $g_i$, $d_i$, and $w_i$ of the C-section. The frequency of interest where the group delay peaks can be controlled by varying $L_i$. The group-delay profile of a C-section is periodic with frequency where peaks occur at odd multiples of the frequency of interest [16].

Let Figure 6.18(a) denote the group-delay profile of the structure shown in Figure 6.17(a). The group-delay profile shows three peaks at three distinct frequencies, $f_{L1}$, $f_{L2}$, and $f_{L3}$, which are caused, respectively, by the three C-sections having lengths $L_1$, $L_2$, and $L_3$. By varying the parameters of each C-section, the amount of group delay at each frequency can be controlled. By increasing the length $L_i$, the group delay can be increased; at the same time, however, this causes the corresponding frequency $f_{Li}$ to shift to the left or to decrease. On the other hand, the group delay can be increased without affecting the frequency by reducing $g_i$ and $w_i$. Given this group delay profile, let us consider that a modulated pulse is input into the structure. The pulse is modulated with a frequency $f_m$. When $f_m = f_0$, where $f_0$ is a frequency where the group is a minimum as shown in Figure 6.18(a), the pulse traveling through the structure experiences the lowest propagation delay. Whereas when the pulse is modulated using the frequencies $f_{L1}$, $f_{L2}$, and $f_{L3}$, the output pulse experiences a longer delay since the group-delay profile maximizes at those frequencies. The most delay is experienced when $f_m = f_{L1}$ because the highest group delay is at the frequency $f_{L1}$. Figure 6.18(b) illustrates the time of arrival of the output pulse, using the envelope of the modulated pulse, for different $f_m$.

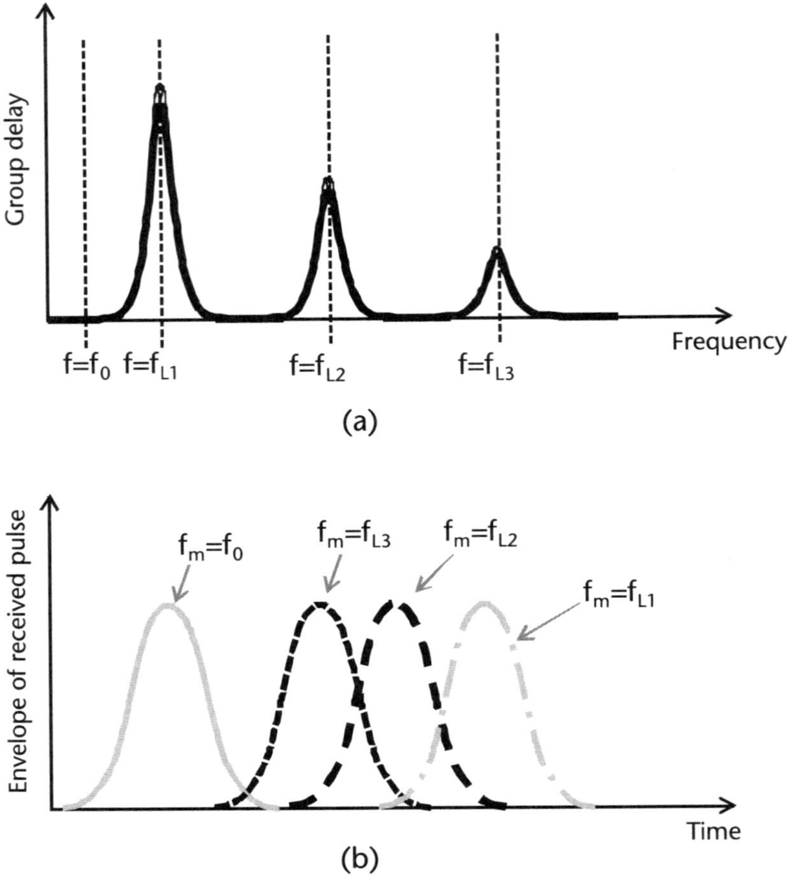


**Figure 6.18**   (a) Group-delay characteristics of the DDS and (b) time of arrival of pulses modulated with different frequencies that are output from the DDS.

It is clear from Figure 6.18 that the DDS constructed using C-sections can be used to introduce a frequency-dependent time lag for a signal traveling through it. This behavior can be readily exploited for encoding information in chipless RFID. Next we will discuss two different methods of encoding information by using this behavior.

*Method I*
The first method is discussed in [11, 12, 16] where a discrete set of time lags produced at a given frequency encodes information bits in each group of C-sections in the DDS. Using more discrete levels of time delay for a particular frequency will increase the data capacity of an elemental encoding structure. Table 6.1 illustrates the coding principle used in [12] where the chipless tag consists of two groups of C-sections; each group encodes a single bit. The group delay for a particular frequency is changed by slightly increasing the length of a C-section. A larger group delay denotes a data bit "1," whereas a smaller group delay denotes a data bit "0." Furthermore the LSB is denoted using the lowest frequency and MSB is denoted using the highest frequency.

**Table 6.1**   Encoding Two Bits of Information Using Frequency-Group Delay Combinations

| C-Section 1 | C-Section 2 | Group Delay at $f_{L1}$ | Group Delay at $f_{L2}$ | Data |
|---|---|---|---|---|
| $L_1$ | $L_2$ | $T_1$ | $T_2$ | 00 |
| $L_1 + \Delta L_1$ | $L_2$ | $T_1 + \Delta T_1$ | $T_2$ | 01 |
| $L_1$ | $L_2 + \Delta L_2$ | $T_1$ | $T_2 + \Delta T_2$ | 10 |
| $L_1 + \Delta L_1$ | $L_2 + \Delta L_2$ | $T_1 + \Delta T_1$ | $T_2 + \Delta T_2$ | 11 |

Figure 6.19 illustrates the time of arrival of the output pulse for different modulation frequencies $f_m$ for the four different binary data combinations of the 2-bit chipless RFID tag. Here, the output is tested using two different interrogation input pulses in two steps where each input pulse is modulated with $f_{L1}$ and $f_{L2}$, respectively. The interrogation of the chipless RFID tag can also be done using one broadband pulse, which encompasses both frequencies of interest, $f_{L1}$ and $f_{L2}$. In

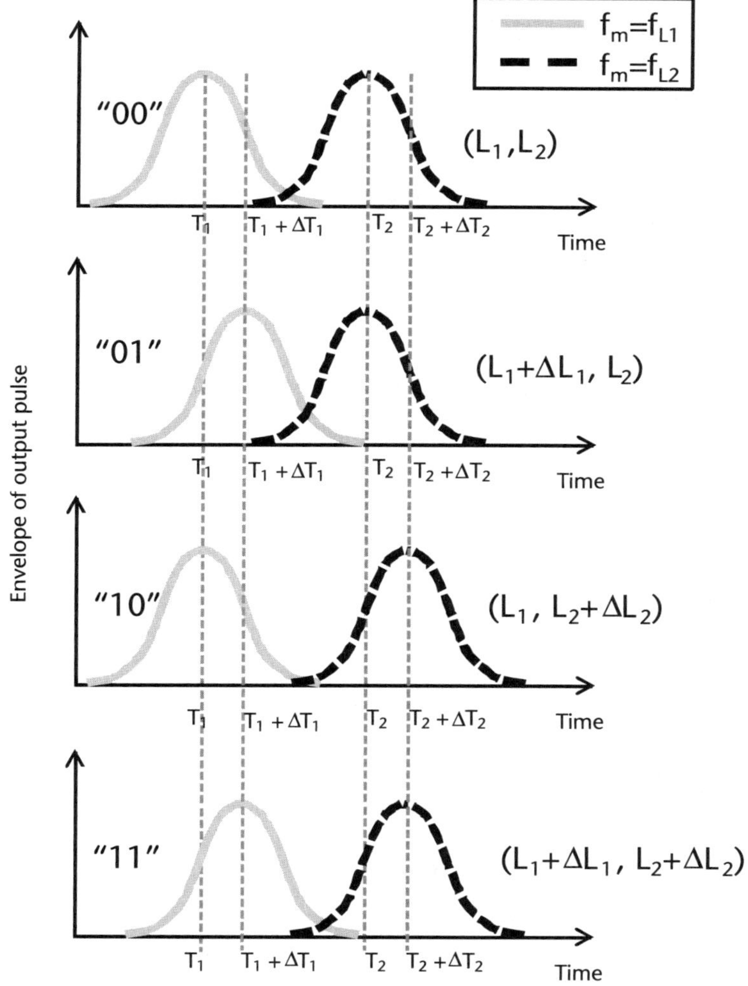


**Figure 6.19**   Four different combinations of time of arrival of the output pulse that encodes 2 bits of information.

such a case, further signal processing needs to be performed to isolate the time lag experienced by each frequency from the total output signal obtained.

*Method II*

The second method uses the DDS to pulse position modulate the interrogation signal in order to encode information [10, 15]. The interrogation signal is a train of pulses in which each pulse is modulated at a different frequency so that each pulse will experience a different delay when it passes through the DDS. Figure 6.20 illustrates the encoding principle of this method. The output pulse stream starts to appears after a propagation delay of Tp where each individual pulse in the stream is offset by slight amounts to the right or left of its intended location due to the frequency-dependent delays introduced by the DDS. The encoding principle is such that if a pulse is shifted to the right from its intended location, then it represents a bit "1"; if the shift is to the left, then it represents a bit "0."

Figure 6.21 depicts the complete chipless RFID tag consisting of the encoding elements (DDS) and tag antennas required for receiving and transmitting signals. The signal antenna design uses the same antenna for receiving and transmitting signals. In such a design the interrogation signal is first received by the antenna and then passes through the DDS structure where it gets transformed. Afterwards the signal reaches the end of the transmission line (which can be either open circuited or short circuited) and gets reflected where it travels back through the DDS for a second time and gets retransmitted by the same antenna to generate a backscattered signal. In the two-antenna design the signal captured by the receiving antenna only travels through the DDS once where it encounters the transmit antenna at the end of the transmission line. The two antennas are configured to be in orthogonal polarizations to reduce cross-talk between them. Either one of the methods discussed earlier for encoding information can be used for either of these tag designs. The actual signals received at the RFID reader through the wireless medium would be different than those illustrated in Figures 6.19 and 6.20 due to the presence of

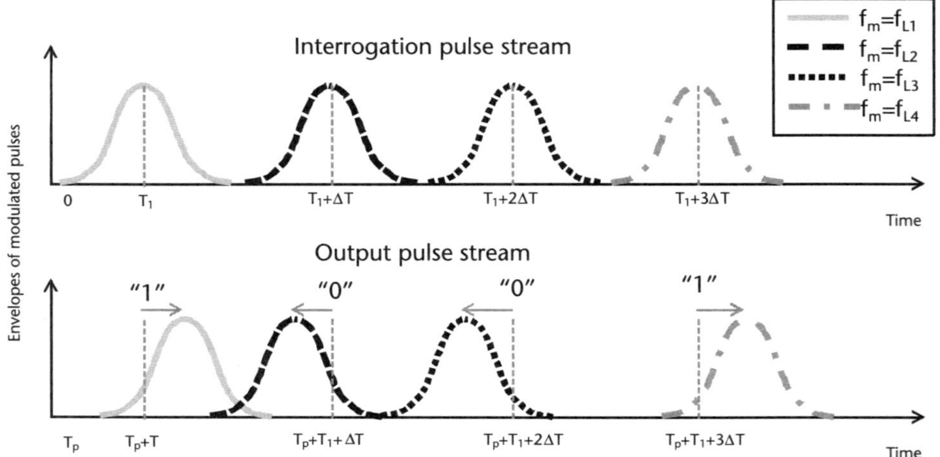


**Figure 6.20**  Pulse position modulation encoding technique for DDS based frequency-time chipless RFID [10, 15].

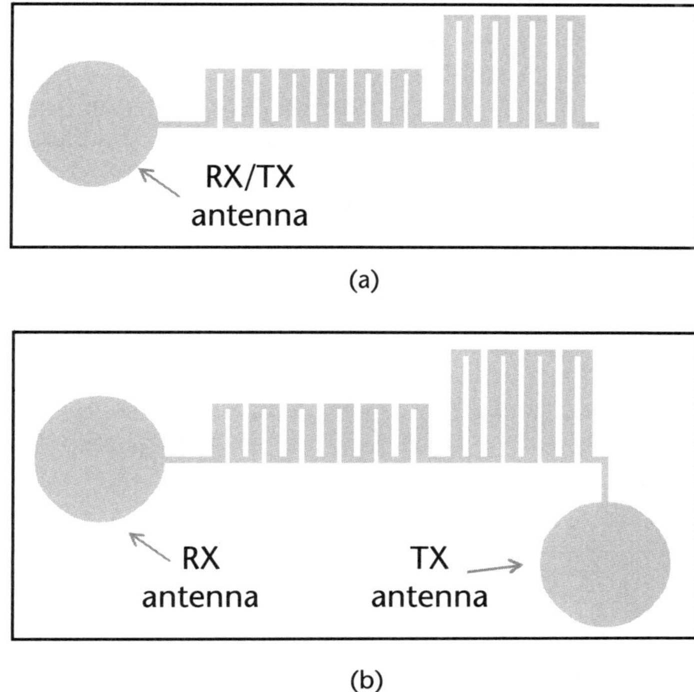



**Figure 6.21**   Complete chipless RFID used on C-section based DDS: (a) single antenna configuration and (b) two-antenna configuration.

multiple modes in the backscattered signal (antenna mode and structural mode [23]). However, the principle of operation would still be the same.

### 6.4.2   Non-Group-Delay Based Approaches for Hybrid Frequency-Time Chipless RFID

A non-group-delay based chipless RFID system is presented in [9] that relies on both time and frequency for its operation. The information is encoded in the frequency dimension as sharp changes in the amplitude using spiral resonators. In this design the dimension of time is not directly used in the process of encoding information; it is used as method of time division multiplexing for efficient spectrum utilization in encoding information. That is, information encoded in a given frequency spectrum is separated through the use of large time delays. Therefore, the total information that needs to be stored in the tag can be broken apart into smaller pieces and encoded in the same spectrum of frequency that is separated in time through delays as illustrated in Figure 6.22.

From the illustration shown in Figure 6.22, we can see that the frequency spectrum spanning $f_1$ to $f_3$ is used to encode 9 bits of data as three 3-bit groups separated in time. An example of a physical passive microwave structure capable of achieving this process is shown in Figure 6.23. This design was proposed in [9].

The chipless tag shown in Figure 6.23 consists of two transmitted antennas. The signal received by the tag's receiving antenna is divided into two parts by the power divider and sent through two paths. A meandering delay line is included

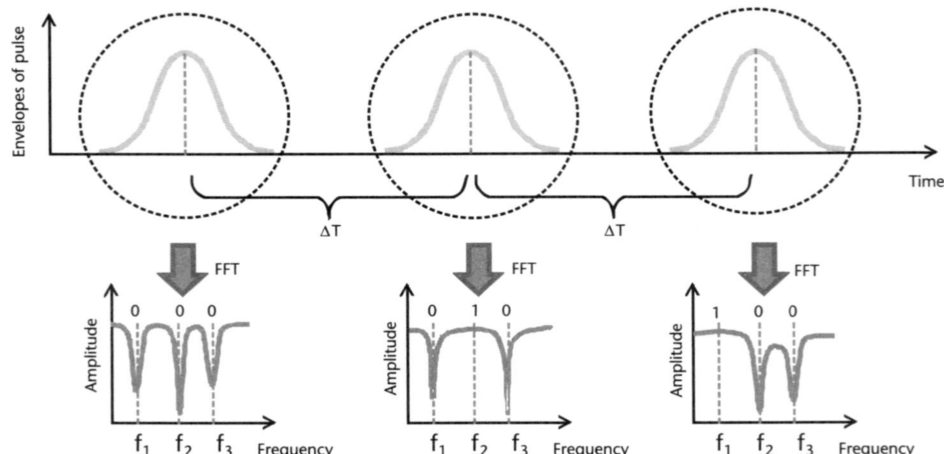



**Figure 6.22**  Time division multiplexed approach to encoding information in a given frequency spectrum.

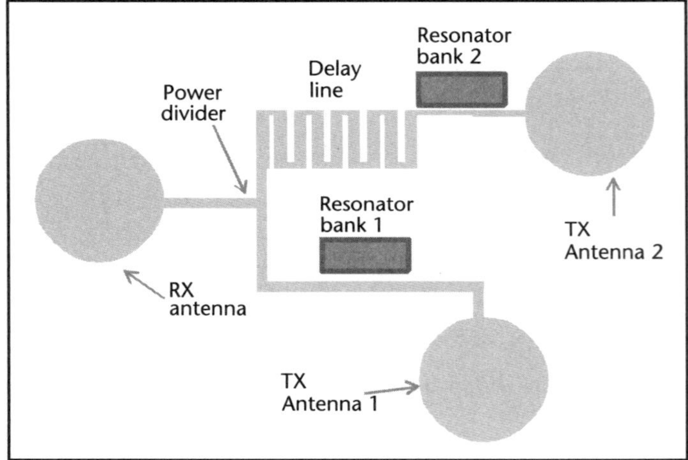



**Figure 6.23**  Frequency reuse based chipless RFID tag [9].

in one path in order to introduce an extra delay to the signal traveling through it. Each path is coupled to a bank of spiral resonators in order to encode information in the passing signal. The two signals are then retransmitted through the transmitting antennas. The two signals are delayed in time due to the extra delay introduced by the meandering transmission line.

### 6.4.3  Chipless RFID Reader for Hybrid Frequency-Time Based Chipless RFID

The design of the RFID reader required for reading hybrid frequncy-time chipless RFID tags can be either frequency domain based or time domain based. In either case the analysis needs to be performed in the time domain where windowed Fourier analysis of the time domain backscattered signal is required (particularly in the non-group-delay based approaches discussed in Section 6.4.2). Therefore, if a frequency domain based reader design is used, the time domain representation of the backscatter needs to be obtained through the inverse Fourier transform.

The reader shown in Figure 6.24 is based on a time domain architecture where an ultra-short-duration pulse is used for interrogating the tag. This reader design is aimed at reading the group delay based chipless RFID tags discussed in Section 6.4.1. The raw received backscatter is first sampled and digitized using a fast analog-to-digital converter, shown in inset (A). Next the strong coupled signal between the two antennas and the unwanted structural mode needs to be removed through windowing [24]. The resulting desired signal is shown in inset (B). This signal is then analyzed using discrete wavelet transform (DWT) methods in order to localize the position of specific frequency content along the time axis. An illustration of the resulting signal through an analysis using DWT for two frequency components in the backscatter is shown in inset (C). The time delay between these frequency components reveals the encoded information in the chipless RFID tag. The Fourier transform and its inverse (FFT and IFFT) are also two common DSP tools used in these types of readers.

## 6.5  Conclusion

Frequency spectrum and space occupied by a tag are two main resources in any chipless RFID tag design. Because of the limited availability of license-free bandwidth and the high cost of licensing frequency bands, spectrally efficient data encoding for chipless RFID is vital.

The typical chipless RFID application demands a tag to be comparable in size to existing chipped RFID tags or barcodes. Furthermore, the size of the tag dictates

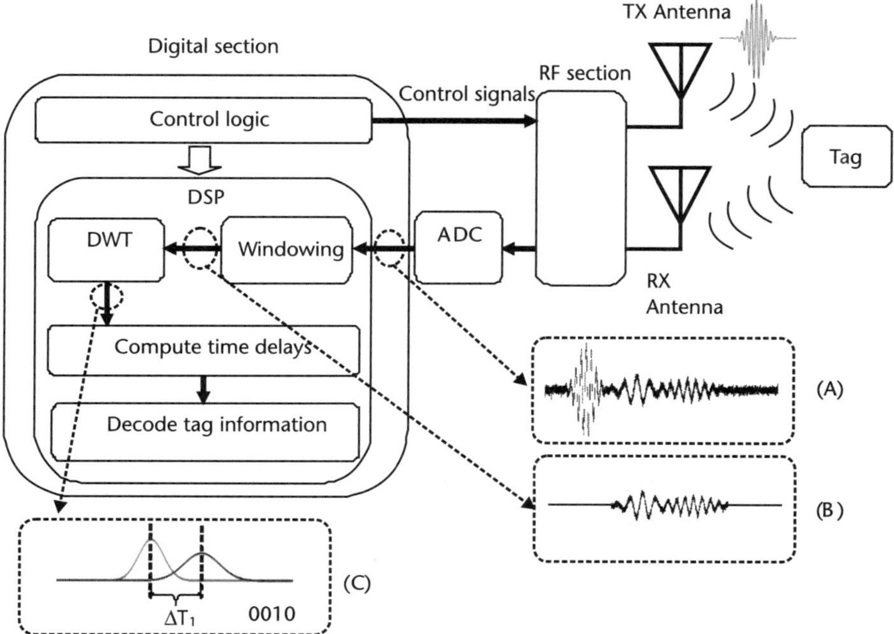


**Figure 6.24**  Time domain reader for hybrid frequency-time chipless RFID.

the cost associated with manufacturing it using conductive ink and substrate materials. Therefore, spatially efficient data encoding is also of great importance in chipless RFID.

Hybrid chipless RFID systems make use of more than one dimension for encoding information, enabling spectrally and spatially efficient data encoding in tags. The elemental encoding structures used in these systems are more efficient than elemental encoding structures used in chipless RFID systems that only use a single dimension, in terms of space and spectrum usage per encoded information bit. This is because they can simultaneously manipulate multiple dimensions to store information.

A comprehensive review of contemporary research on hybrid chipless RFID systems is presented where the working principles of different types of hybrid chipless RFID systems are discussed. The operation of encoding elements for the chipless RFID tags of hybrid systems using the dimensions of frequency-phase, frequency-polarization, and frequency-time are analyzed in detail. Through the use of novel planar microwave components (C-sections, SRRs, DDSs) and through the examination of their characteristic behaviors with different dimensions of interest (frequency, polarization, time, etc.) it is possible to innovatively utilize them for designing a hybrid chipless RFID tag design.

## Questions

1. What is the benefit of having multiple dimensions for encoding information in a chipless RFID system?
2. Briefly explain the working principle of the encoding element of a frequency-phased based system.
3. Which parameters determine the resonant frequency and the phase deviation bandwidth?
4. How are the locations of a resonant peak and resonant dip determined?
5. Explain how the bits are determined in a hybrid frequency-phase chipless RFID based on C-sections.
6. Explain why the encoding process utilizes the dimensions of both frequency and polarization?
7. How is the group delay related to the physical dimensions $L_i$, $g_i$, $d_i$, and $w_i$ of the C-section in a hybrid chipless RFID?
8. Explain how the DDS constructed using C-sections can be used to introduce a frequency-dependent time lag for a signal traveling through it.
9. Briefly explain the method in which the DDS is used to pulse position modulate the interrogation signal in order to encode information.
10. Explain the basic principle of the method in which a discrete set of time lags produced at a given frequency encodes information bits in each group of C-sections in the DDS.
11. In a two-antenna configuration, what step should be taken to reduce the cross-talk between the receiving and transmitting antennas?
12. Briefly explain the basic principle of a non-group-delay approach to hybrid frequency-time chipless RFID systems.

13. Develop reader architectures for (a) hybrid frequency-phase, (b) frequency-polarization, (c) frequency-group delay, and (4) frequency–non-group-delay based chipless tags.

14. Explain the signal processing steps used for an RFID reader reading frequency–non-group-delay based chipless RFID tags.

## Reference


[1] S. Preradovic et al., "A novel chipless RFID system based on planar multiresonators for barcode replacement," presented at the IEEE International Conference on RFID, 2008, Las Vegas, NV, 2008.

[2] S. Preradovic and N. C. Karmakar, "Design of fully printable planar chipless RFID transponder with 35-bit data capacity," presented at the 2009 European Microwave Conference (EuMC 2009), Rome, 2009.

[3] S. Preradovic, S. Roy, and N. Karmakar, "Fully printable multi-bit chipless RFID transponder on flexible laminate," in *2009 Asia-Pacific Microwave Conference (APMC 2009)*, pp. 2371–2374.

[4] Z. Linlin et al., "Design and implementation of a fully reconfigurable chipless RFID tag using Inkjet printing technology," in *IEEE International Symposium on Circuits and Systems (ISCAS 2008)*, pp. 1524–1527.

[5] Z. Lu et al., "An innovative fully printable RFID technology based on high speed time-domain reflections," in *Conference on High Density Microsystem Design and Packaging and Component Failure Analysis (HDP '06)*, pp. 166–170.

[6] B. Shao et al., "An ultra-low-cost RFID tag with 1.67 Gbps data rate by ink-jet printing on paper substrate," presented at the IEEE Asian Solid-State Circuits Conference (A-SSCC), Beijing, 2010.

[7] M. Schuler et al., "Phase modulation scheme for chipless RFID- and wireless sensor tags," in *2009 Asia-Pacific Microwave Conference (APMC 2009)*, pp. 229–232.

[8] I. Balbin and N. C. Karmakar, "Multi-antenna backscattered chipless RFID design," in *Handbook of Smart Antennas for RFID Systems*, N. C. Karmakar (Ed.), Hoboken, NJ: John Wiley & Sons, 2010, pp. 413–443.

[9] M. S. Bhuiyan, R. Azim, and N. Karmakar, "A novel frequency reused based ID generation circuit for chipless RFID applications," in *2011 Asia-Pacific Microwave Conference (APMC 2011)*, pp. 1470–1473.

[10] S. Gupta, B. Nikfal, and C. Caloz, "Chipless RFID system based on group delay engineered dispersive delay structures," *Antennas and Wireless Propagation Letters*, vol. 10, pp. 1366–1368, 2011.

[11] R. Nair, E. Perret, and S. Tedjini, "Novel encoding in chipless RFID using group delay characteristics," in *Microwave & Optoelectronics Conference (IMOC 2011)*, pp. 896–900.

[12] R. Nair, E. Perret, and S. Tedjini, "Temporal multi-frequency encoding technique for chipless RFID applications," in *2012 IEEE International Microwave Symposium Digest*, pp. 1–3.

[13] A. Vena, E. Perret, and S. Tedjini, "A compact chipless RFID tag using polarization diversity for encoding and sensing," in *2012 IEEE International Conference on RFID*, pp. 191–197.

[14] A. Vena, E. Perret, and S. Tedjini, "Chipless RFID tag using hybrid coding technique," *IEEE Trans. on Microwave Theory and Techniques*, vol. 59, pp. 3356–3364, 2011.

[15] S. Gupta, B. Nikfal, and C. Caloz, "RFID system based on pulse-position modulation using group delay engineered microwave C-sections," in *2010 Asia-Pacific Microwave Conference (APMC 2010)*, pp. 203–206.

[16]    R. Nair, E. Perret, and S. Tedjini, "Chipless RFID based on group delay encoding," in *2011 IEEE International Conference on RFID—Technologies and Applications (RFID-TA)*, pp. 214–218.

[17]    M. A. Islam and N. Karmakar, "Design of a 16-bit ultra-low cost fully printable slot-loaded dual-polarized chipless RFID tag," in *2011 Asia-Pacific Microwave Conference (APMC 2011)*, pp. 1482–1485.

[18]    M. A. Islam and N. C. Karmakar, "A novel compact printable dual-polarized chipless RFID system," *IEEE Trans. on Microwave Theory and Techniques*, vol. 60, pp. 2142–2151, 2012.

[19]    S. Gao, A. Sambell, and S. S. Zhong, "Polarization-agile antennas," *Antennas and Propagation Magazine*, vol. 48, pp. 28–37, 2006.

[20]    M. G. Pettus, "RFID system utilizing parametric reflective technology," U.S. Patent 7460014 B2, 2008.

[21]    T. K. Ishii, *Handbook of Microwave Technology*, Boston: Elsevier Science, 1995.

[22]    S. Gupta et al., "Group-delay engineered noncommensurate transmission line all-pass network for analog signal processing," *IEEE Trans. on Microwave Theory and Techniques*, vol. 58, pp. 2392–2407, 2010.

[23]    P. Kalansuriya and N. C. Karmakar, "Time domain analysis of a backscattering frequency signature based chipless RFID tag," in *2011 Asia-Pacific Microwave Conference (APMC 2011)*, {AU: page numbers?}

[24]    P. Kalansuriya, N. C. Karmakar, and E. Viterbo, "On the detection of frequency-spectra-based chipless RFID using UWB impulsed interrogation," *IEEE Trans. on Microwave Theory and Techniques*, vol. 60, pp. 4187–4197, 2012.

## CHAPTER 7

# Antennas for Chipless RFID Readers

## 7.1  Introduction

The preceding chapters have presented frequency, time, and hybrid domain chipless RFID tag reader systems. In all cases these systems encode and decode chipless RFID tags in UWB frequency bands. Therefore, the antennas for both tags and readers must be designed to cover the entire frequency band of the system. This chapter presents such antenna developments.

Various types of reader antennas for chipless RFID systems are presented in the chapter. Antennas are classified in terms of form factors, beamforming types, and operating frequency bands. Both planar and wire antennas are used in the UWB technology based reader systems. They are disk-loaded monopoles, leaf dipoles, log-periodic dipole antennas (LPDAs), log-periodic dipole array antennas (LPDAAs), and horn antennas. Theories of antenna elements and array analysis help in the development of these antennas. The designs and results of these antennas are presented for unlicensed instrumentation, scientific and medical (ISM) microwave and millimeter-wave frequencies bands. Antennas are designed targeting reading the chipless RFID designed in UWB microwave frequencies of 3 to 10.6 GHz and millimeter-wave frequencies of 22 to 26.5 GHz. All frequencies fall in unlicensed UWB bands. Antenna input return losses, radiation patterns, and gain are presented to augment the required specifications of readers. Finally, the antennas developed are discussed in terms of their applications to low-cost UWB chipless RFID tagging systems.

Antennas are an integral part of a wireless system. They are the "eyes" of electronics that allow them to communicate with the outer world. An antenna is a conduit between free space and the guided structure. In that sense antennas are also called *spatial filters* or *transducers*. Antennas are fully passive circuits. However, they have gain due to their characteristics of forming radiation patterns in different shapes and directions. Antenna gain is directly added to the gain and loss factors of microwave active and passive devices, respectively, in the link budget calculations for any wireless system. Therefore, effective antenna design is the prime goal for an efficient system.

Antennas can be designed to be reconfigurable and active. The reconfigurability is realized in terms of frequency and beamforming. A part of the antenna can be tuned electronically to obtain frequency reconfiguration. A beamforming network in an antenna array can be added to obtain beam agility. Antennas can be made active as well by adding an amplifying device integrated into the antenna as is done in active filters. Antennas are also used to scavenge DC power by adding a Schottky diode and a lowpass filter as the rectifying and filtering circuits, respectively. This type of rectifying antenna is called a *rectenna* [1].

All of these advanced features of different antennas can be incorporated in RFID readers to establish efficient wireless links between the tag and the reader and to energize some parts of the reader electronics if required[1]. Readers can refer to a previous work on various smart antennas for RFID systems [2]. For the particular application with chipless RFID systems developed by the authors' research group—Monash Microwave Antennas, RFID and Sensors (MMARS) Laboratories at Monash University—the developed tag and reader antennas can be classified into the categories illustrated in Figure 7.1. As can be seen in the figure, the developed antennas can be broadly classified into their physical configurations—planar and nonplanar, beam agility, and operating frequency bands. In terms of form factors, antennas are developed mainly in microstrip antenna configurations. Some antennas are designed in modified wired and waveguide configurations such as UWB leaf dipoles, wideband ridge horns and similar structures. These antennas are 3D in shape.

The efficacy of an RFID reader is highly dependent on the performance of the reader antenna. An efficient antenna controls the link margin and the interference for a wireless system. This is very vital for an RFID system in which the transmitted

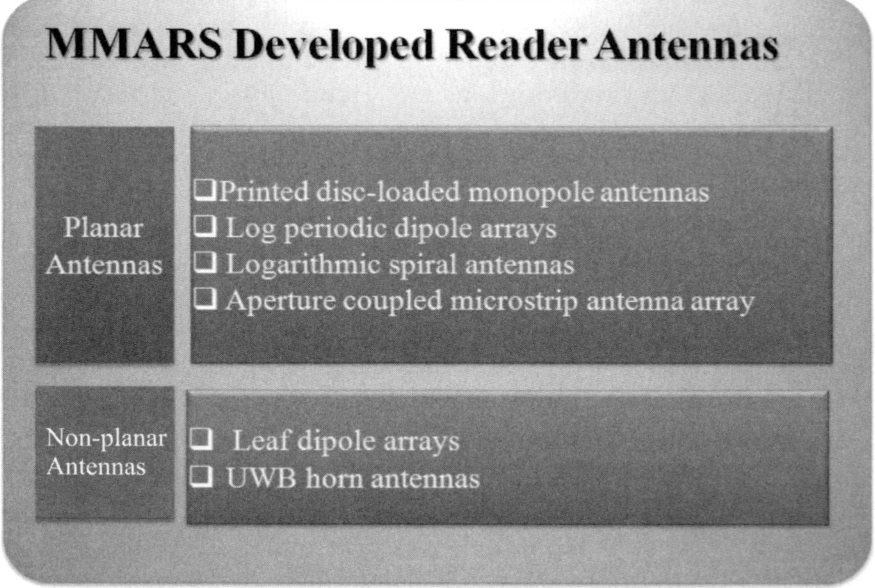


**Figure 7.1** Monash Microwave Antennas, RFID and Sensors (MMARS) Laboratories categories for antennas developed for tags and readers.

---

1.   Some low-powered ICs are possible to be energized with rectennas.

signal from a chipless RFID tag, which is a minute radar target, is returned from the tag with a fourth-power reduction in magnitude of the reading distance ($R^{-4}$), where $R$ is the distance between the tag and the reader. Therefore, a slight improvement in antenna gain and directivity plays a significant role in improving reading range, accuracy, and localization of the tag. The RFID tag and reader antennas can be broadly classified according to three important parameters: (1) form factors, (2) beamforming, and (3) frequency bands, as shown in Figure 7.2. In terms of the form factors, the antennas are divided into planar and nonplanar configurations. Planar antennas are printed circuit board (PCB) based antennas such as microstrip patch antennas. Nonplanar antennas are dipoles and their arrays and horn antennas, which are made of wires, coaxial cables, metals, and waveguides, respectively. In terms of their beamforming capabilities, antennas are classified into fixed-beam and agile-beam antennas. A fixed-beam antenna can be a single-element antenna or it can have an array of elements connected to a fixed feed network. Agile-beam antennas are phased arrays, switched beam arrays, and adaptive array antennas. A detailed account of such antennas can be found in [2].

The fixed-beam antenna for the RFID reader is commonplace, and a plethora of antenna vendors are selling RFID antennas to cater to the RFID market demand. The fixed-beam antenna has a unique radiation pattern. Panel antennas, which are microstrip patch antenna arrays, are used as the reader antennas for the most part. Most UHF and microwave band RFID readers are equipped with omnidirectional or wide-beamwidth antennas in order to include as much area as possible in their interrogation zones. Several fixed-beam antennas are also used, and can be commonly found in Alien Technology [3] and Omron [4] readers. The fixed-beam antennas are easy to install and do not need any switching electronics and associated

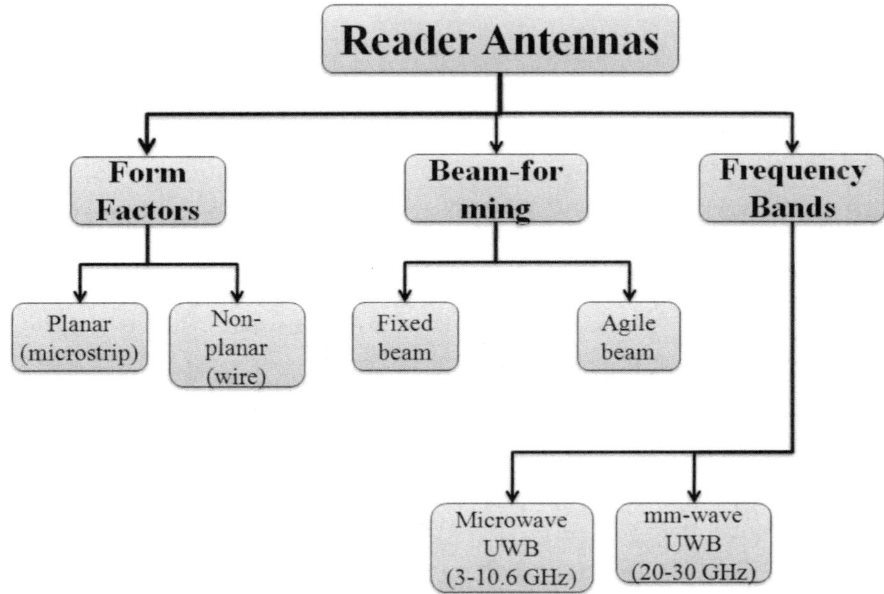


**Figure 7.2**   Broad classifications of RFID tag and reader antennas.

logic control to steer their beams. However, they pick up multipath signals and interference and are incapable of localizing the tag's position when receiving transponders' backscattered signals. This situation may lead to reading errors during interrogations. The smart antenna for an RFID reader can mitigate these problems by offering a directional high-gain beam toward the desired tags and nulls toward the interferers. Multipath interference makes the reading unreadable even if it is within the reading range of the reader.

A smart antenna can solve the interference problem by electronically controlling the main beam emitted from the reader's antenna toward the desired tags and nulls toward the interfering signals. The smart antenna technology incorporated in RFID readers reduces reflections from surroundings and thus minimizes degradation of system performance due to multipath and other undesired effects. A $2 \times 3$-element phased array antenna for a UHF universal RFID system was introduced in [2]. However, considering the trade-off between cost and performance, fixed-beam antennas are preferred over agile-beam smart antennas. This chapter covers various UWB fixed-beam antennas for chipless RFID readers.

The chapter is organized as follows (also see Figure 7.3). Section 7.2 introduces fundamental antenna parameters such as radiation patterns, gain and directivity, bandwidth, polarization and axial ratio, input impedance, and array factor. Section 7.3 reviews the various broadband antennas. Section 7.4 presents design specifications for chipless RFID reader antennas and practical design procedures. Section 7.5 presents the design and results of various reader antenna types, followed by applications and discussion of results in Section 7.6 and a conclusion in Section 7.7.

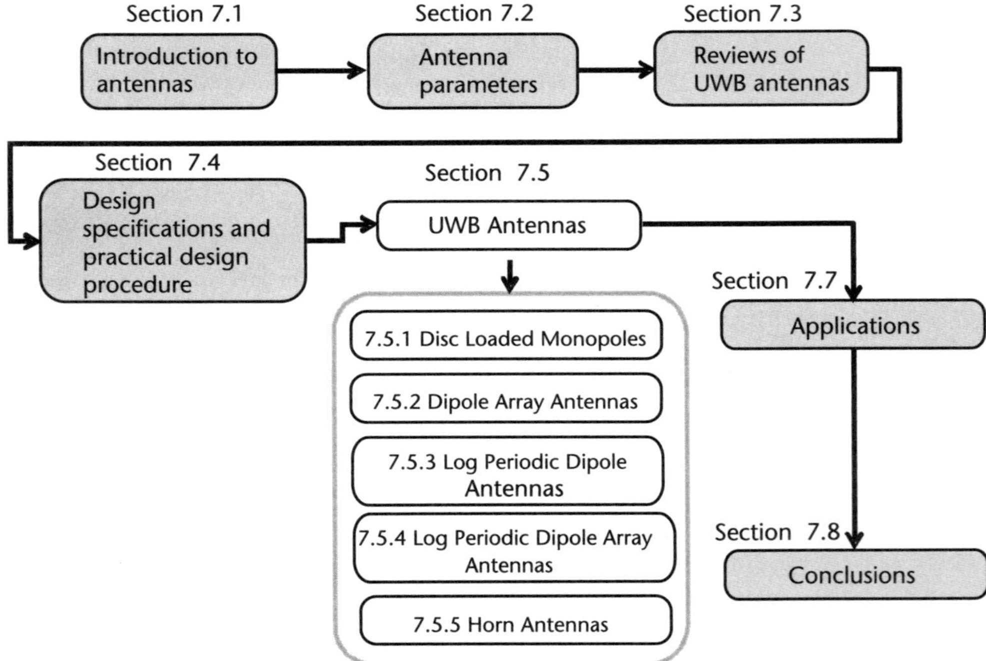


**Figure 7.3**   Organization of Chapter 7.

## 7.2   Antenna Parameters

The characteristics of an antenna are defined by a set of parameters. The important basic parameters of an antenna are: (1) radiation patterns (beamwidth, front-to-back ratio, sidelobe levels), (2) directivity and gain, (3) bandwidth, (4) polarization and axial ratio, (5) input impedance return loss and voltage standing wave ratio (VSWR) and (6) the array factor of an antenna array. The basic antenna parameters are defined below.

*Radiation Patterns*

The radiation pattern of an antenna is defined as the antenna power and field patterns emanating out of the antenna in its radiating modes. An antenna pattern represents the 3D influential zone of radiation when the antenna is active in both the transmit (Tx) and receive (Rx) modes. To enhance this understanding, antennas are also defined in terms of 3D solid angle (cone angle instead of 2D angular concept). From the solid angle one can easily predict the antenna directivity and its patterns.

Figure 7.4 shows a typical radiation pattern for an antenna. As can be seen in the figure, the antenna's mainlobe has a finite half power (3-dB) beamwidth that defines the directional property of the antenna. Antennas also have sidelobes, which are undesirable in many applications especially for our chipless RFID readers. Sidelobes are inherent properties of antennas and pick up unwanted signals from the surrounding interferers. Therefore lower sidelobes are better to avoid such unwanted signals being picked up by the antenna[2]. Antennas also have back lobes. For wideband design of slot-type antennas and broadband aperture coupled patch antennas, the dimensions of the slots become comparable to half the wavelength of the design frequency. The slots become resonant with the patch antenna to obtain a large bandwidth from these antennas. These slots radiate equally in both the front and back directions; hence back lobes become comparable to the front lobes. The ratio between the mainlobe and back lobe power levels is defined as the *front-to-back-ratio* (F\B). Antennas also produce nulls in particular intervals in

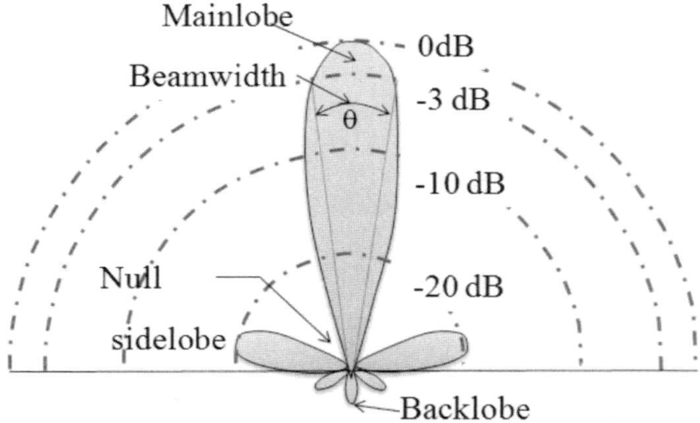


**Figure 7.4**  Antenna radiation pattern.

---

2.   Very small terminals (VSATs) use very low sidelobe antennas to avoid receiving unwanted signals from surrounding satellites.

angular positions. If adaptively controlled, such nulls can be used in advantageous situations to avoid interference at known positions of interferers. Thus, adaptive antennas improve the signal-to-noise-plus-interference ratio (SNIR) of a system.

### Directivity and Gain

Directivity represents the directional properties of an antenna and depends primarily on its radiation characteristics. Directivity is defined as the focusing capability of an antenna compared to an isotropic radiator. An isotropic radiator radiates power uniformly in all directions. However, antennas have losses. The losses make antennas less efficient. Some antenna types are quite highly efficient. As an example, microstrip patch antennas etched on a low-loss laminate enjoy efficiencies as high as more than 90%. The antenna gain is the product of the directivity and the efficiency. For an antenna design, the gain is more important to observe than the operating frequency band.

### Bandwidth

The bandwidth of an antenna is defined as the frequency band of operation over which the antenna operates satisfactorily. Some antennas have a large radiation pattern and gain bandwidth compared to the input impedance bandwidth. The gain bandwidth is defined as the frequency band over which the gain does not drop a certain level compared to its peak value at the center frequency. Usually a 1-dB gain drop limits the gain bandwidth. Because most low-profile and planar antennas enjoy a much larger gain and radiation pattern bandwidth than the input impedance bandwidth, a designer first attempts to achieve a satisfactory input impedance bandwidth. The input impedance bandwidth is defined in terms of a 10-dB return loss bandwidth or a VSWR of 2 over the operational frequency band. Therefore, for a UWB antenna, a 10-dB return loss and/or 2 VSWR are ensured over the frequency band of 3 to 10.6 GHz. Other bandwidth definitions are also available but less common. They include the axial ratio bandwidth (appropriate for the circularly polarized antennas), co- and cross-polar level bandwidth, 3-dB beamwidth bandwidth, and F/B ratio bandwidth.

### Polarization and Axial Ratio

Polarization of an antenna is defined as the orientation of the antenna's radiated electrical field vector with time. Simple wire antennas such as dipoles and waveguide apertures have very definitive polarization. The polarizations of these antennas depend on their physical orientations. For example, if a dipole antenna is positioned vertically with respect to the ground, the antenna is vertically polarized. The same rules are applied to the waveguide. If the feed pin of the waveguide is vertically centered along the wider dimension of the rectangular aperture, the waveguide antenna is vertically polarized. Insight regarding the electrical field distribution of an antenna aperture explains its polarization. When both Tx and Rx antennas are in the same polarization, they receive maximum power, and their wireless communication is at its most effective. If the antennas are not positioned at the same orientation, the antenna loses power in proportion to the square of the cosine of the angle formed between the antennas. This loss term due to the mismatch of polarization between two antennas is defined as the polarization loss factor.

All antennas are inherently elliptically polarized, meaning the electric field vector in one orientation is larger than that for another orientation. The *axial ratio* is the ratio between these two fields. As defined earlier, linear polarization and circular polarization are subsets of elliptical polarization. In the linear case the field is always in one orientation, as seen, for example, in a vertically polarized antenna, where the vertical electric field is most pronounced compared to the field in another orientation. For circularly polarized antennas, the field vector is continuously rotated either in a clockwise or counterclockwise direction. In the ideal case both fields have the same amplitude and 90° phase offset for a circularly polarized antenna. Therefore, the axial ratio is 1 (0 dB). However, in reality a 0-dB axial ratio is very hard to maintain over the bandwidth of operation. Therefore, the axial ratio is defined as the 3-dB axial ratio bandwidth over the operating frequency band. This means the power of one oriented field (say, $E_x$) is double that for the field oriented in the other direction ($E_y$).

*Antenna Input Impedance*
When a practical antenna is designed, a designer always looks at the system perspective to determine how the antenna will be used in the particular system. The designer always has the goal of making an efficient radiator for both transmission and reception. In that respect, impedance matching of an antenna is the prime goal of antenna design. As stated earlier, when a designer designs an antenna, the first parameter he/she looks at is the antenna input impedance bandwidth. In this regard an antenna is always designed with a matching section as an integral part of the antenna. The matching section can be as simple as the feed line with a different impedance from the antenna's input impedance or a delicate matching network. Details of antenna matching can be found in any antenna textbook. Since antennas are reciprocal devices, they behave similarly in both the Tx and Rx modes. Therefore, the same matching section can be used for both transmission and reception. A simple equivalent circuit model of an antenna in a transceiver system is shown in Figure 7.5. As can be seen the antenna has an inherent input impedance $z_{Ant}$ (also called antenna driving point impedance). The antenna needs to be matched with a transceiver of inherent impedance $Z_o$. A matching section is added between the two to match the antenna with the transceiver. Once perfect matching of the antenna

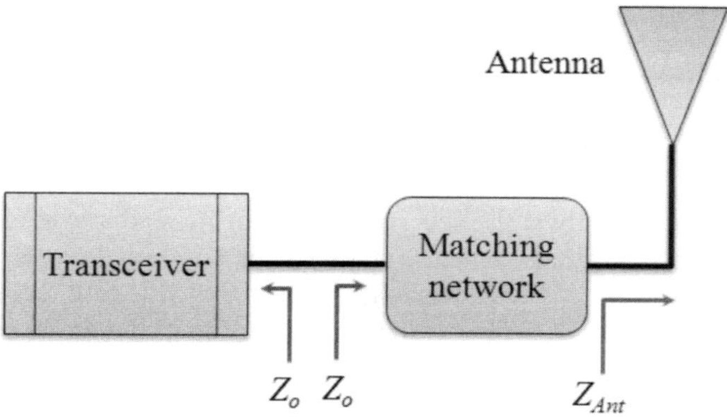


**Figure 7.5**   Antenna in transceiver system.

over the frequency band of interest is obtained, the designer looks at other parameters such as radiation patterns, gain, directivity, and polarization.

*Antenna Array Factor*
In many applications, high gain and directivity antennas are needed. In those cases an antenna element is severely limited by its low gain and broad beamwidth. The antennas can be arranged in linear, rectangular, circular, and spherical grids to make arrays. For the chipless RFID reader, a $2 \times 2$-element broadband dipole array was designed to enhance the gain and radiation performance of the antenna [5, 6].

For an $M \times N$ number of elements planar array, the normalized array factor ($AF$) (to its maximum amplitude) is represented [7] by:

$$AF(\theta, \varnothing) = \left\{ \frac{1}{M} \frac{\sin\left(\frac{M}{2}\psi_x\right)}{\sin\left(\frac{\psi_x}{2}\right)} \right\} \left\{ \frac{1}{N} \frac{\sin\left(\frac{M}{2}\psi_y\right)}{\sin\left(\frac{\psi_y}{2}\right)} \right\} \tag{7.1}$$

where $\psi_x = kd_x \sin\theta \cos\varnothing + \beta_x$

$$\psi_y = kd_y \sin\theta \cos\varnothing + \beta_y \tag{7.2}$$

where

$d_x$ = interelement spacing along the $x$ axis
$d_y$ = interelement spacing along the $y$ axis
$k$ = free space wave number ($2\pi/\lambda$)
$\beta_x$ = progressive phase difference between antenna elements along the $x$ axis
$\beta_y$ = progressive phase difference between antenna elements along the y axis.

Generally, to avoid grating lobes, element spacing is kept less than the wavelength of the operating frequency ($\lambda$) [7]. However, to optimize mutual coupling of radiating elements, element spacing is kept as 0.7 times the wavelength ($0.7\lambda$) [8, 9].

The total radiation pattern of the array can be expressed as:

$$E_{total} = E \text{ single element} \times AF \tag{7.3}$$

Figure 7.6 shows the representative normalized single-element pattern, the array factor of a $2 \times 2$-element array and the combined pattern of the $2 \times 2$-element array antenna. A 6-GHz wire dipole antenna of length 2.5 cm yields the element pattern (only a 0–90°-elevation angle is shown in the figure). The array factor (AF) is calculated based on a $2 \times 2$ rectangular grid array of point sources with interelement distances dx and dy equal to 3.5 cm. The combined array antenna pattern

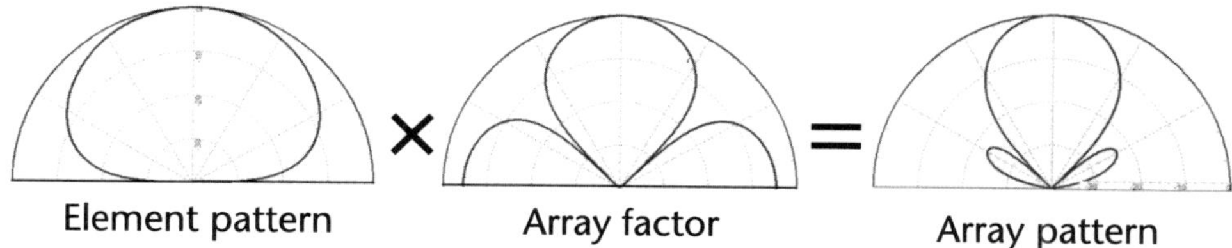



**Figure 7.6**   Normalized single-element pattern, array factor, and combined pattern of 2 × 2-element 6-GHz wire dipole array antenna.

is a narrow beamwidth high-gain directional beam with two sidelobes of 13.9 dB below the peak of the main beam.

## 7.3   Review of UWB Antennas

Since the inception of UWB communication technology in February 2002, UWB communications devices have gone through tremendous development. UWB antennas have occupied significant space in the development. Various UWB antennas are under development. The UWB antennas developed so far can be broadly classified in terms of the gain performance: low gain antennas and high gain antennas. Disk-loaded monopole and printed dipole antennas are low-gain antennas. The dipole arrays, log-periodic dipole antennas and arrays, and tapered ridged horn antennas are relatively high-gain antennas. Figure 7.7 shows the classifications. The antenna reviews are discussed next.

A historical background and recent development can be found in [10–12]. There are many kinds of UWB antennas, including printed disk-loaded monopole [13–15], bowtie [16], TEM horn [17], spiral [18], Vivaldi, and biconical [19, 20]. Biconical UWB antennas are the earliest type used in wireless communication systems. There are many variations of biconical antennas, such as finite biconical

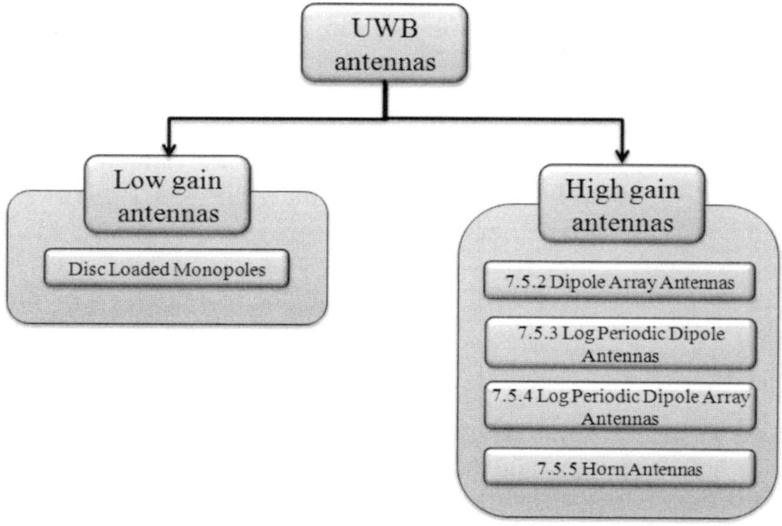



**Figure 7.7**   Classification of UWB antennas.

antennas and discone antennas [21]. Although they exhibit a broad impedance bandwidth [22, 23], due to their metallic structure and bulky size, they are mainly used in laboratories.

Theoretically frequency-independent antennas that provide a constant performance over the operating frequency band can be applied to broadband antenna design. Log-periodic antennas such as log-periodic slot antennas, log-periodic dipole arrays, and log spiral antennas are some of the popular frequency-independent antenna types used in practice [24]. However, when such frequency-independent antennas are used for communicating UWB pulses, they show unstable phase centers dependent on frequency. This distorts the radiated pulse waveform [25].

Elliptical-shaped monopole [26, 27] and dipole planar antennas [28, 29] are the most common models for UWB communication applications in terms of suitable gain, low cost, and low profile, as well as high stability in input impedance [23]. A comprehensive study of planar elliptical dipole antennas working from 0.5 to 6 GHz has been carried out in [27–30]. Triangular and rectangular slots have been used to improve the bandwidth performance of semicircular bowtie antennas [31, 32]. A planar elliptical dipole has been designed for UWB wireless communication applications working in the frequency range of 0.75 to 6 GHz [33].

The elliptical-shaped dipoles reported in [9, 26, 27] show broadband behavior and have a low profile. Several techniques are used to improve the operating bandwidth of the antennas. In [31, 34] exact elliptical shapes are used with the optimized ratio of the radii of the major axis and the minor axis. Nazli et al. [26] use elliptical-shaped slots inside the elliptical-shaped dipole arms to achieve improved bandwidth. Cerny and Mazanek [35] report that elliptical-shaped dipoles exhibit a better impedance match with higher impedance feeds (greater than $50\Omega$). Using tapered feed lines and a shorting bridge across dipole arms improves the bandwidth [36]. However, most of these antennas have omnidirectional radiation patterns and the gains of these antennas are relatively low at approximately 0 to 3 dBi.

When these antennas are attached to walls or metals, their performance can be degraded owing to their omnidirectionality. If directional antennas are utilized, the degradation of performance due to omnidirectionality can be avoided [37]. Further, less power transmission and higher speed can be achieved by the communication system.

By using antenna arrays, a higher gain and narrower beamwidth can be achieved [17, 18]. Several designs have comparatively high gain with a directional radiation pattern and show broadband characteristics. One antenna array uses leaf-shaped dipole elements and has an operating bandwidth of 4.1 to 10 GHz and fairly high gain over the frequency band [38]. However, its physical size is not suitable for a chipless RFID reader. A modified elliptical-shaped dipole array reported on in [39] uses a balun transformer between the feed network and each dipole element. Thus it has a complex structure, although it has a gain of 9 dBi over the frequency range of 3.1 to 8.5 GHz.

The printed antenna arrays discussed in [39, 40] use a broadband feed network with an integrated balun transformer to feed each dipole element. They have achieved more than 200% fractional operating frequency bandwidth. Although they have 32 elements, the size of the antenna array is fairly small compared to those discussed in [35]. However the minimum operating frequency is 6 GHz. Due to the fact that broadband dipoles are better matched with higher impedance [35], the designs reported in [41–43] use integrated baluns [44] to feed dipole elements. All of the dipole array antennas discussed above use a metallic reflector to achieve unidirectional radiation characteristics, thus showing high gains.

From the above review we can see that there is a broad range of selections for UWB antennas for chipless RFID reader systems. In the following section the design specifications and practical design procedure for UWB chipless RFID reader antennas are presented.

## 7.4   Practical Design Procedure for UWB Antennas

The antenna parameters discussed earlier are used in this section to form the design specifications and practical design procedure for an antenna for a specific system, in this case a chipless RFID system. A few salient features of a chipless RFID tag and\ or a reader antenna are as follows [45]:

- The antenna should exhibit a bandwidth large enough to accommodate as many frequency signatures as are needed to encode large numbers of data bits.

- The radiation pattern of the antenna should be as uniform as possible over the entire frequency band of operation.

- The antenna radiation pattern should be omnidirectional to cover the reading range.

- The antenna should offer polarization purity for optimum wireless communication an sufficient isolation between the Tx and Rx links.

- The antenna should exhibit a compact, low-profile, planar layout for ease of fabrication.

Table 7.1 lists the design specifications for UWB antennas used with a chipless tag and its reader. These design specifications provide guidelines for design engineers as they plan for an appropriate antenna design. First, based on the deployment environment, the antenna topology is selected.

Figure 7.6 shows a practical design guideline for an antenna. Once the antenna application area has been determined and the antenna type selected, the bills of materials are formulated. For planar patch antennas, the microwave laminates and input connectors are the two prime components to be selected first. For nonplanar antennas, wire gauges, metals sheets, and connector assemblies are important. In both cases the design guidelines (i.e., the dimensions and antenna configurations) are maintained with high precision. High-end software tools with a full-wave electromagnetic (EM) solver such as CST Microwave Studio™, Agilent Momentum™, Zealand IE3D™, and Ansoft HFSS™ are some of the most popular EM solvers in the market. They come with nice graphical user interfaces (GUIs) for precision 2.5D to 3D design suites for planar and nonplanar antennas. At MMARS we have been using CST Microwave Studio™ for all of our passive design requirements.

Once the antenna configuration and bills of materials have been chosen, the next phase is to do the appropriate design based on the theoretical framework and associated design parameters such as length, width, and height of various sections

**Table 7.1**    UWB Chipless RFID Antenna Design Specifications

| Antenna Parameters | Specifications | Remarks |
|---|---|---|
| UWB frequency bands | 3–10.6 GHz (microwave) <br> 22–26.5 GHz (millimeter wave) | No license fee is required as long as regulatory requirements are met. |
| UWB transmitted power | Approximately 10–15 dBm | Power can vary with antenna gain and beamwidth. |
| Antenna form factor | Planar\interleaved\nonplanar | Wire and horn antennas are nonplanar antennas. |
| Tag antenna gain | Approximately 0 dB | |
| Tag antenna beamwidth | Omnidirectional (for short range) <br> Directional (for long range) | The beamwidth limits the reading zone and distance. |
| Reader antenna gain | >8 dBi preferred for short range <br> >22 dBi for long range | Gain varies with transmission power and reading distance. |
| Reader antenna beamwidth | Approximately 40° | A narrower, steerable beam is preferred for long range and direction finding and anticollision. |
| Antenna polarization | Linear or dual polarization | Depends on tag design requirement. |
| Reading range | 1–5 cm (short distance) | For authentication and security applications |
| | 5–50 cm (intermediate range) | For retail applications |
| | >2m (long range) | For inventory control, vehicular applications, etc. |

of an antenna. Various layers of the antennas are designed with high precision, and appropriate connectivity between layers is examined before the simulation is performed. The input ports (sources) of the antennas are connected and appropriate excitation methods suitable for the simulation types are added. The simulation goals based on the design specifications are determined in terms of input impedance (10-dB return loss and/or 2 VSWR), bandwidth, radiation patterns, beamwidth, peak gain, gain bandwidth, and isolation between ports for dual polarized antennas before the simulation is started.

Once the design of the antenna and the simulation setup are ready, the design is simulated with the software tool and the design outcomes are extracted in terms of S-parameters versus frequency, radiation patterns, gain versus frequency, and surface current plots on the antenna structures. The S-parameters versus frequency plots give the antenna input impedance match, which is vital for examining antenna performance. The input return loss, which is defined as the negative log of the input reflection coefficient ($S_{11}$), is observed in the first instance. If the input return loss is more than 10 dB over the frequency band of interest, the design passes the first requirement specification.

Then the other parameters such as radiation patterns in both 2D and 3D, co-polar and cross-polar components, gain versus frequency, axial ratio versus frequency are observed in the design frequency band. If these parameters are not satisfactory, the surface current plot of the antenna can be analyzed to determine why the antenna is not behaving well. Rectification and modifications are made to the design to mend the problem and then further simulations are performed. The complexity of the simulation increases with the complexity of the antenna geometry and high-end requirement specifications. A systematic approach and appropriate planning help to ease the simulation cycles of the antenna design. As shown in Figure 7.8, surface current plots help to identify problems with the design. Therefore,

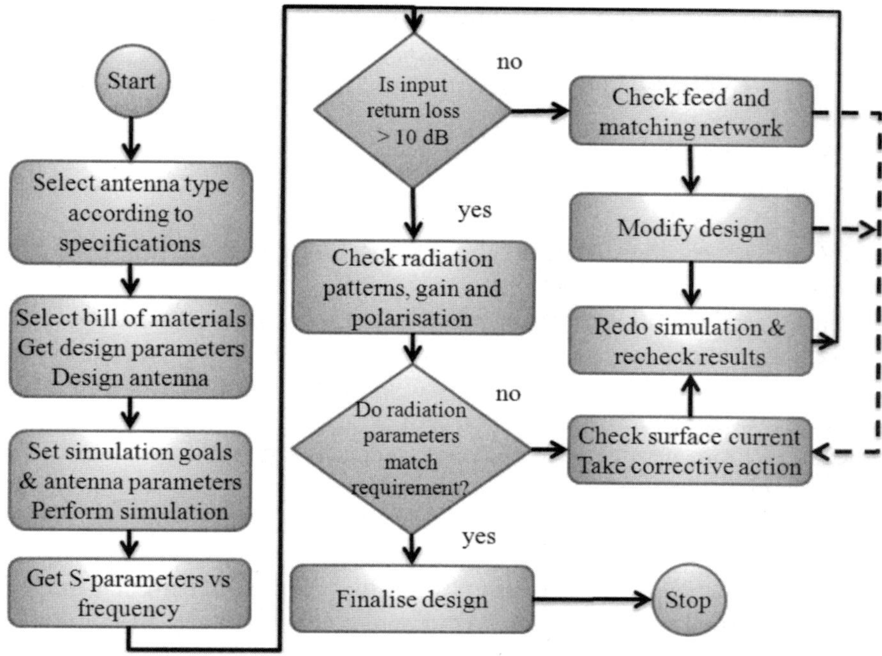



**Figure 7.8**   Design procedure for chipless RFID UWB reader antennas.

it is a good practice to check the surface current plots at times when the desired outcome in both input impedance and radiation characteristics is not obtained.

## 7.5   UWB Antenna Development

In this section various UWB antennas developed for chipless RFID readers are presented. Sequentially they are: (1) disk-loaded monopoles, (2) UWB dipole arrays, (3) log-periodic dipole antennas (LPDAs), (4) log-periodic dipole array antennas (LPDAAs), and (5) broadband horn antennas. All antennas provide sufficiently large bandwidth to sustain multibit tag reading in various UWB frequency bands.

### 7.5.1   UWB Disk-Loaded Monopole Antennas

The most popular UWB antenna is the disk-loaded monopole antenna. It has a simple layout and exhibits extremely large bandwidth and a figure-eight radiation pattern [13–15]. A ground plane in the vicinity of the disk-loaded monopole emulates dipole-type behavior. UWB dipole antennas have been reported on by researchers that have radiation properties and bandwidth similar to those found using UWB monopoles. Both types of antennas are fully printable and initially appear suitable for deployment as chipless RFID tag antennas. According to the definition of the Federal Communications Commission (FCC), antennas having a bandwidth greater than 500 MHz or a fractional bandwidth greater than 0.2 are considered UWB antennas [11, 12]. In this regard a disk-loaded monopole outperforms the

FCC regulation and offers a bandwidth over the entire UWB microwave frequency band from 3 to 10.6 GHz.

The chipless RFID tags [45] use circular disk-loaded monopole antennas as the Rx and Tx antennas. These monopole antennas have low gain (near 0 dB) and omnidirectional radiation patterns. Because the tag antenna gain is low, the retransmitted signal with spectral signatures is also weak. If directional antennas are utilized as reader antennas, the sensitivity of the reader system can be enhanced, and better reading distance can be achieved by transmitting the same amount of power. Therefore, the development of UWB antennas with directional radiation characteristics and high gain is highly desirable. Antennas with omnidirectional radiation patterns can also be used as reader antennas, but are suitable only for applications that require a very short reading distance (less than 10 mm), for example, secured document authentication. For conveyor belt applications, a clearance of at least 100 mm between the conveyor belt and antenna is required because a larger reading distance of more than 100 mm is needed.

As stated earlier, UWB disk-loaded monopole antennas are the most popular antenna due to their simple configuration and uniform performance over the entire UWB frequency band. Because of their low gain, UWB circular and elliptical disk, and deformed versions, loaded monopole antennas meet all favorable criteria and are used as short-range reader antennas. The only disadvantage is that the antenna has a figure-of-eight radiation pattern and losses half the transmitted power in an unwanted direction.

Figure 7.9 shows the layouts for three types of disk-loaded monopole antennas. As can be seen in the figure, the antenna consists of three important components, which play a significant role in antenna performance. They are a circular disk with radius $R$, a feed line of length $L_{feed}$ and width $W_{feed}$ connecting the disk and the source input port, and a ground plane at the bottom side of the substrate with length $L_{Gnd}$ and $W_{Gnd}$. The gap between the disk edge and the ground plane edge influences the input impedance matching of the antenna drastically. Therefore, the gap length is a significant design parameter. Both the ground plane and the antenna are supported with a substrate of height $h$, dielectric constant $\varepsilon_r$, and loss

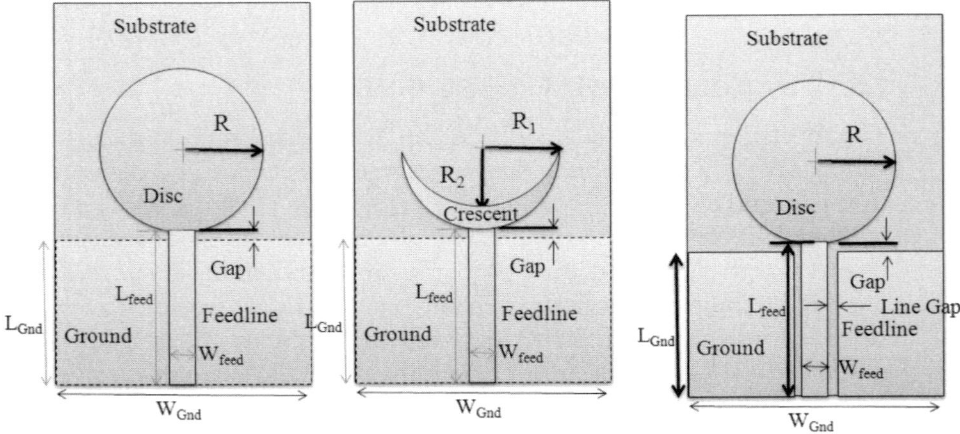


**Figure 7.9** Configuration of a disk-loaded monopole antenna: (a) full disk, (b) crescent, and (c) CPW feed full disk.

tangent tan $\delta$. The antenna radius $R$ is determined based on its fundamental mode frequency. A crescent antenna can be designed to reduce the size of the antenna. For compatibility with MMIC designs and to achieve low-cost simple fabrication, a CPW line feed antenna can be designed. This monoplane design is preferred for ease of integration with RF\microwave electronic circuits.

The UWB disk-loaded monopole antenna operates over its fundamental and higher order modes as shown in Figure 7.10. As can be seen, the resonances of different modes are closely distributed over the UWB frequency band [13–15, 45]. The frequency responses of various modes are overlapping each other and the combined result is a UWB VSWR response at the input port of the antenna. Although the fundamental mode frequency of the disk-loaded monopole determines the radius $R$ of the disk and provides an undistorted figure-of-eight omnidirectional radiation pattern, the higher order modes distort the E-field patterns in higher frequencies than that of the fundamental frequency. However, the overall performance is broadbeam omnidirectional radiation patterns with a moderate gain of 1 to 2 dBi.

Figure 7.11 shows the input impedance return loss versus frequency plot [45]. As can be seen, the input return loss is more than 10 dB over the frequency band. Figures 7.12(a) and (b) shows the radiation patterns. Both plots show omnidirectional properties. Figure 7.13 shows the gain versus frequency plots of the antenna.

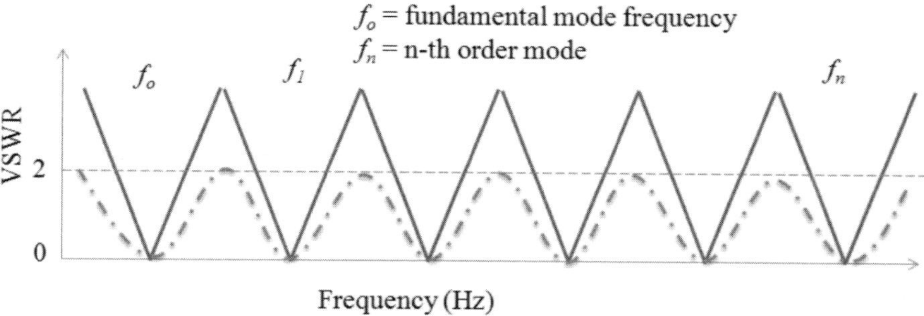


**Figure 7.10**   Frequency response of UWB disk-loaded monopole antenna.

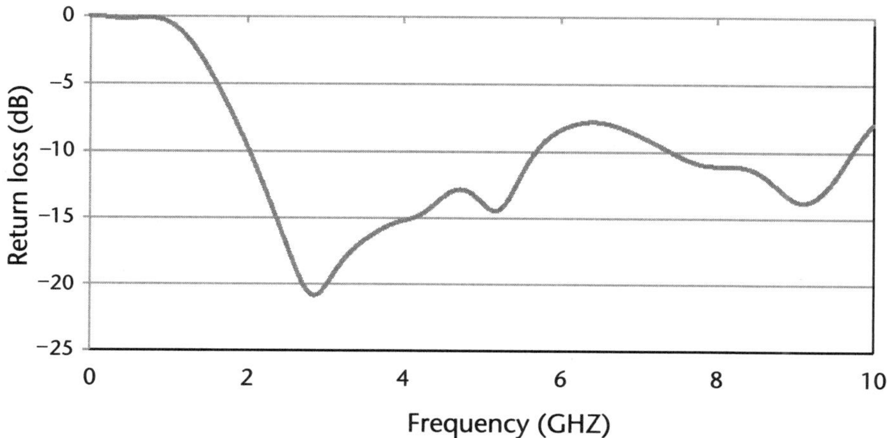


**Figure 7.11**   Input return loss versus frequency response of UWB disk-loaded monopole antenna. (From S. Preradovic, "Chipless RFID system for barcode replacement," PhD Dissertation, Monash University, December 2009.)

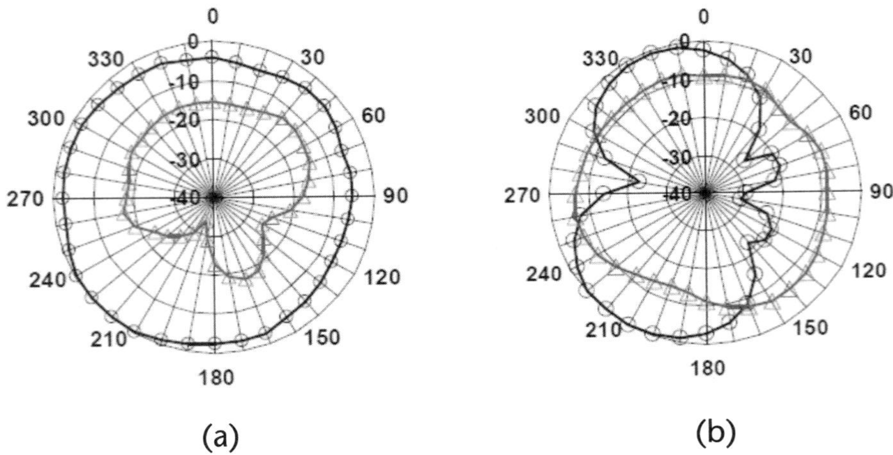

(a)                                              (b)

**Figure 7.12** Three-gigahertz radiation patterns of dipole antennas: (a) E-plane and (b) H-plane (circle: co-polar: triangle: cross-polar). (From S. Preradovic, "Chipless RFID system for barcode replacement," PhD Dissertation, Monash University, December 2009.)

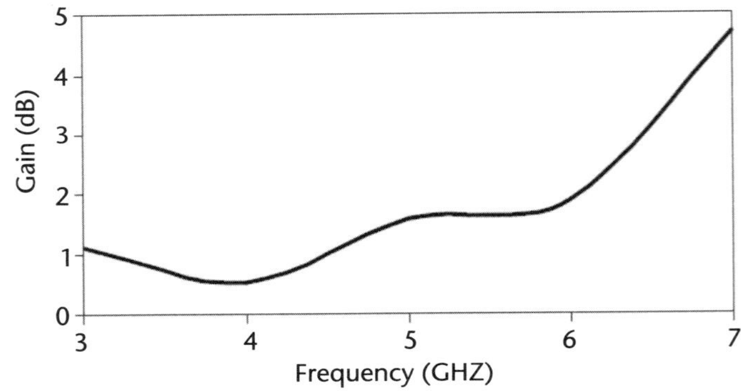

**Figure 7.13** Gain versus frequency plot of disk-loaded monopole antenna. (From S. Preradovic, "Chipless RFID system for barcode replacement," PhD Dissertation, Monash University, December 2009.)

*Crescent-Loaded Monopole Antenna*

There is always pressure to design compact antennas that meet stringent requirements, including low cost. In addition, for a UWB impulse radio (UWB-IR) based reader, a low structural mode RCS is desirable. A crescent-loaded monopole antenna is designed to address these issues. Figure 7.14 shows such an antenna. As can be seen, the results and performance parameters of the antenna are very similar to those of a disk-loaded monopole antenna. Some gain patterns need to be sacrificed in H-plane radiation patterns and cross-polar level due to the asymmetry of the configuration of the antenna.

*CPW Feed Disk-Loaded Monopole on Polymer*

A CPW feed disk-loaded monopole antenna has been designed and fabricated as shown in Figure 7.15(a). Specially made plastic jigs were developed at Monash

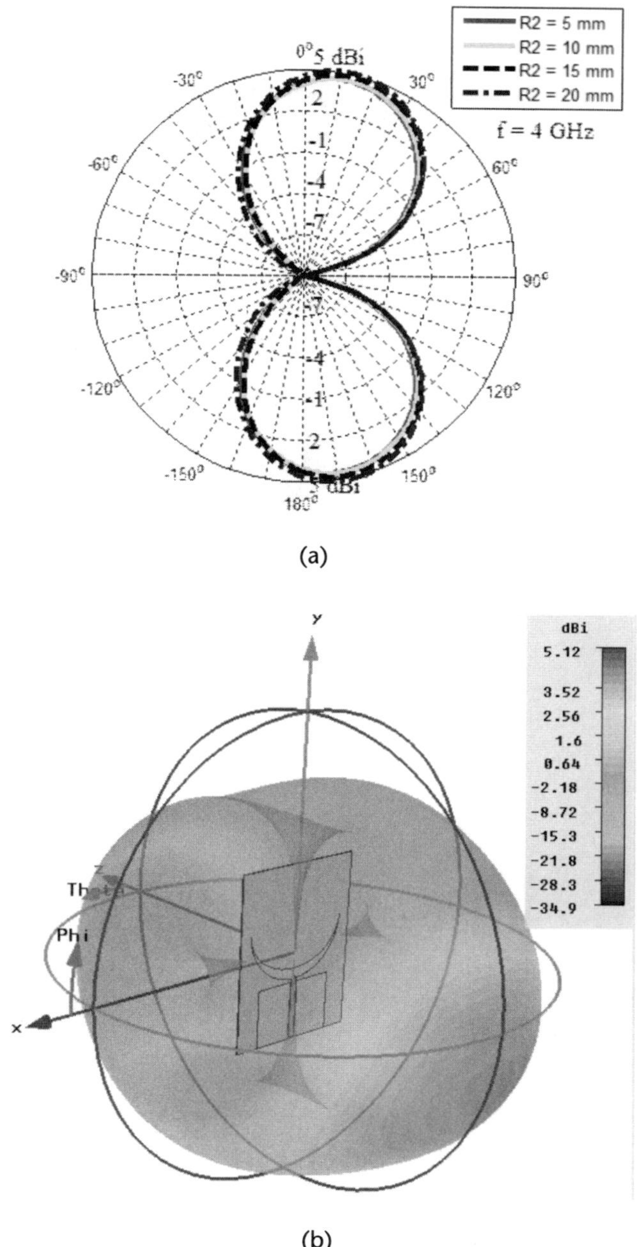



(a)

(b)

**Figure 7.14**  (a) Crescent-loaded monopole radiation patterns and (a) 2D patterns with R2 as parameter and 3D radiation patterns with R2 = 20 cm, (b) f = 3 GHz, and (c) f = 4 GHz.

University's mechanical workshop to hold the flexible polymer antennas to measure their return loss, gain, and radiation patterns. All of the jigs were made of Perspex (nonmetallic, clear plastic-like substance) to avoid unwanted reflection. Prior to application of the conductive epoxy on the CPW monopole, the applicant should make sure that the layer to which the epoxy is being applied is the conductive layer. This can be ensured by checking the electrical conductivity using a simple multimeter. After the procedure, conductive epoxy is applied in small amounts to both

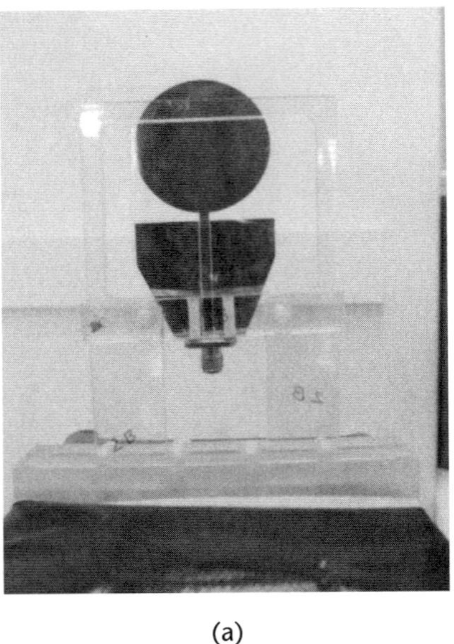

(a)

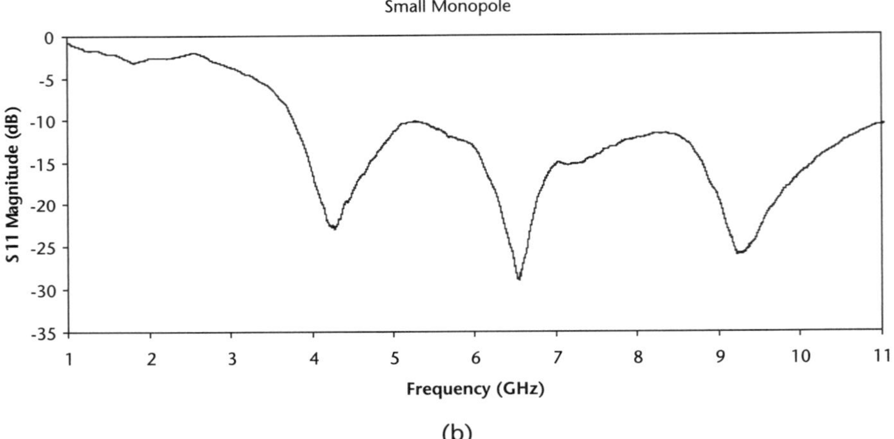



(b)

**Figure 7.15** (a) A CPW monopole on polymer substrate mounted on a Perspex jig and (b) return loss versus frequency plot.

the ground planes to ensure smooth and continuous connectivity with the ground of the SMA connector. Figure 7.15(a) shows a photograph of the polymer antenna and Figure 7.15(b) shows the input impedance return loss versus frequency of the antenna. As can be seen in Figure 7.15(b), the antenna meets the 10-dB return loss bandwidth requirement from 3.5 to 11 GHz. The antenna exhibits a maximum gain of 1.5 dBi with a minimum gain of −2 dB due to excess loss of the polymer substrate, epoxy glue, and Perspex fixture. The radiation patterns are very similar to those for the microstrip line feed disk-loaded monopole antenna.

### 7.5.2   UWB Dipole Array Antennas

The monopole antennas discussed above are very low gain antennas. They are mainly used for tag design and short-range (<5 cm) reader design. More details about the design principles, parametric studies, and operations of these antennas can be found in [45].

In this section, the detailed design for two linearly polarized, high-gain, low-profile, UWB *dipole reflector antenna arrays* is presented. During the process of antenna design, two antenna prototypes were developed: (1) the elliptical-shaped dipole reflector antenna array, operating in the 4.0- to 6.0-GHz bandwidth and (2) the elliptical-shaped printed dipole reflector antenna array, operating in the 3.8- to 10-GHz bandwidth. Each prototype has a different operating bandwidth and is suitable for reading chipless RFID tags operating in different frequency bandwidths.

The quality factor ($Q$) of a resonator is defined as the ratio of inductive reactance to the resistance ($j\omega L\backslash R$). Resonators with a high quality factor yield narrowband operations, and resonators with a low quality factor yield wideband operations. Balanis [7] provided excellent accounts of resonant dipole antennas. The thin wire half-wave dipole antenna is a high-Q resonator hence yields narrow bandwidth. However, the Q decreases with the wire radius of the dipole. The bandwidth versus dipole length to diameter ($l/d$) ratio of the antenna has been studied [7]. As an example, a dipole antenna with a ratio of $l/d = 5000$ has an acceptable bandwidth of about 3%, which is a small fraction of the center frequency. An antenna of the same length but with a ratio of $l/d = 260$ has a bandwidth of about 30%. The phenomenon with the Babinet and Rumsey's principles [7] helps develop the concept of designing frequency-independent antennas.

Figure 7.16 shows a generic block diagram for a dipole array antenna in an RFID reader system. The RF source (reader electronics) is connected to the antenna with a SMA connector and a feed network is used to feed the four elements of the array. The design and results of the antenna element, feed network, and the 2 × 2-element dipole array antenna are explained in the following subsections. CST

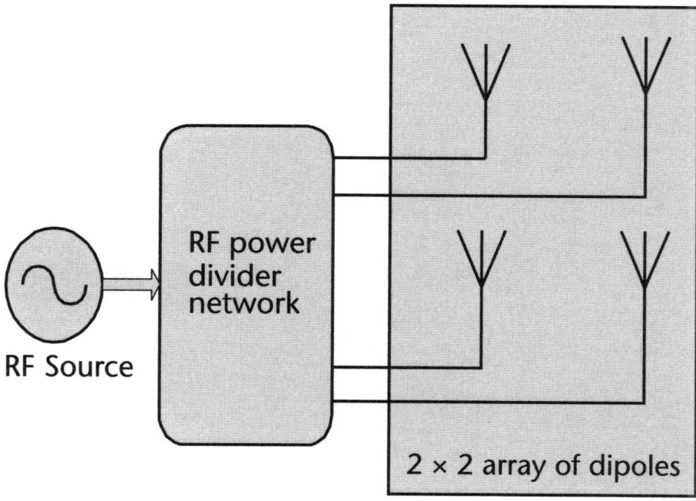


**Figure 7.16**   Block diagram of dipole reflector antenna array.

Microwave Studio 2010 (CST MWS) software was used for the simulation of feed networks and antennas.

### 7.5.2.1  Design of Elliptical Leaf Dipole

The design of the elliptical leaf dipole requires the optimum dimensions of the leaf to be known in order to achieve the best radiation characteristics and antenna performance. The survey of dipole antennas has shown that modified ellipse shapes perform better than exact elliptical-shaped dipole elements. A single-dipole reflector antenna was designed and developed to study the effects of different dimensions. To find the optimum values for the dimensions of the dipole elements, a single-dipole reflector was designed first and parametric study of the dipole elements was carried out.

Figure 7.17(a) shows the CST Microwave Studio simulation model used for the parametric study to find the optimum values for the dimensions of the dipole leaves. Models of leaves were created using 1-mm-thick brass plates. The gaps between the two leaves were kept at 2 mm and they were fed with a 75Ω discrete port. A metal plate was used as the reflector with $h$ = 12 mm (assuming a center frequency of 6 GHz since the required bandwidth is 4.0 to 8.0 GHz) behind the modeled dipole elements, as shown in the figure. Figure 7.16(b) shows the important design parameters: length ($L$) and radius ($R$) of the dipole leaves, with $h$ being the reflector distance from the dipole leaves. Figure 7.16(c) shows the prototype leaf dipole antenna backed by a reflector.

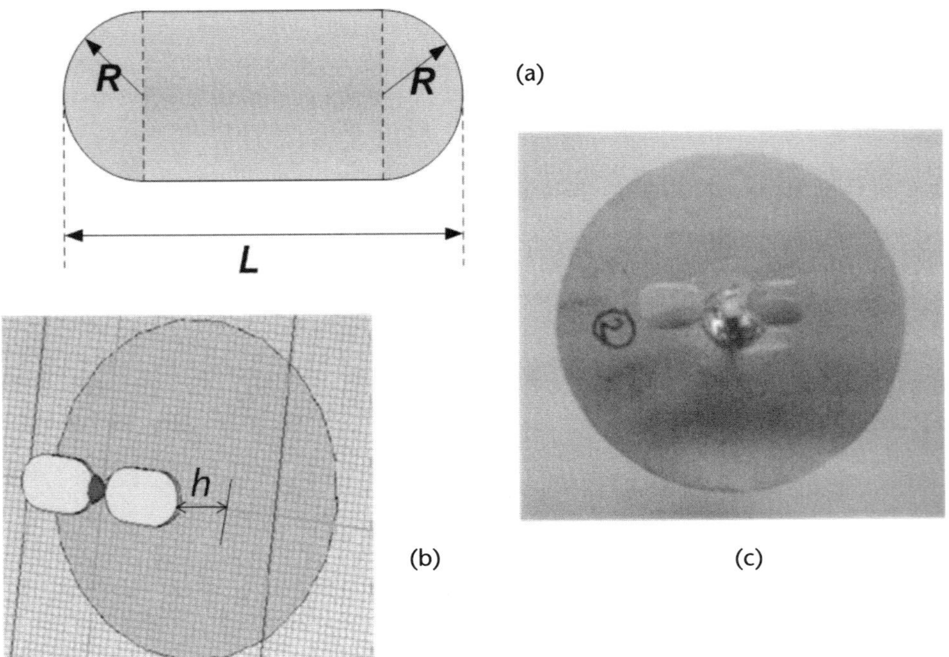


**Figure 7.17**  (a) Important parameters of the leaf-shaped dipole arm, (b) CST MWS simulation model, and (c) dipole reflector prototype.

A comprehensive parametric study of the dipole antenna transforms the antenna to its best performance. The following inferences have been extracted from the study:

1. Resonant frequency decreases linearly with $L$. Input impedance matching also gets better at lower frequency with $L$.
2. Radius $R$, the width of the dipole leaf, has no influence on the resonant frequency. However, $R$ has a significant influence on the input impedance matching of the antenna. The 10-dB return loss bandwidth increases almost linearly and then is saturated at $R = 3$ mm ($0.06\lambda$) at a 6-GHz center frequency.
3. Reflector distance $h$, the antenna height, significantly influences the input impedance matching[3]. At $h = 0.4\ \lambda$, the best matching is achieved and it plateaus. Otherwise the variation is linear on both sides of $h = 0.4\ \lambda$.

Table 7.2 shows the optimum design parameters for the antenna after the comprehensive parametric study of the antenna. To validate the optimized results, a physical dipole reflector antenna was built with the optimized dipole arms and reflector distance as shown in Figure 7.17(c). The simulation and measured input impedance return loss versus frequency results are shown in Figure 7.18. Measured results are in good agreement with simulation results as shown in the figure.

### 7.5.2.2   Broadband Feed Networks

After achieving satisfactory performance for the antenna element, the broadband feed network is designed. For a $2 \times 2$-element array antenna, a four-way corporate feed network with equal power distributions for individual elements was designed. For the applications that require miniature size and large operating bandwidth, T-junction power dividers with quarter-wavelength tapered line transformers are the most appropriate. A compact size and a wide operating bandwidth for the feed network were achieved with T-junctions having tapered transmission lines. The feed network is a 1-to-4 equal broadband power splitter. The input impedance of the feed network is $50\Omega$ and output ports are transformed to $75\Omega$, as shown in Figure 7.19. The feed network was designed on Taconic TLX-8 substrate ($\varepsilon r = 2.50$; thickness t = 0.5 mm, and loss tangent $\tan\delta = 0.0019$). The 10-dB input impedance

**Table 7.2**   Optimized Design Parameters of the Dipole Reflector Antenna Used for Validating Optimization

| Parameter | Required/Optimized Value |
| --- | --- |
| Designed frequency bandwidth | 4.0–6.0 GHz |
| Optimized reflector distance | 20 mm |
| Semirigid cable characteristic impedance | $50\Omega$ |
| Input port characteristic impedance | $50\Omega$ |
| Optimized value for $L$ | 11.2 mm |
| Optimized value for $R$ | 3.6 mm |

---

3.   Reflector distance also influences the mode of excitation and, hence, the beamwidth and radiation patterns.

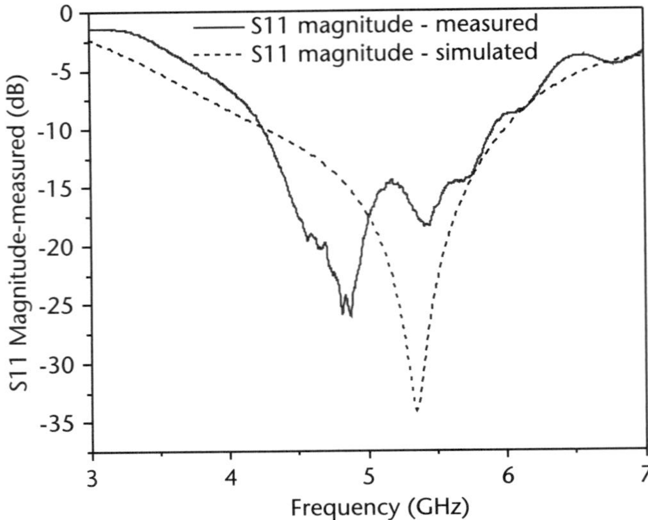

Figure 7.18   Simulated and measured input impedance return loss versus frequency of dipole reflector antenna.

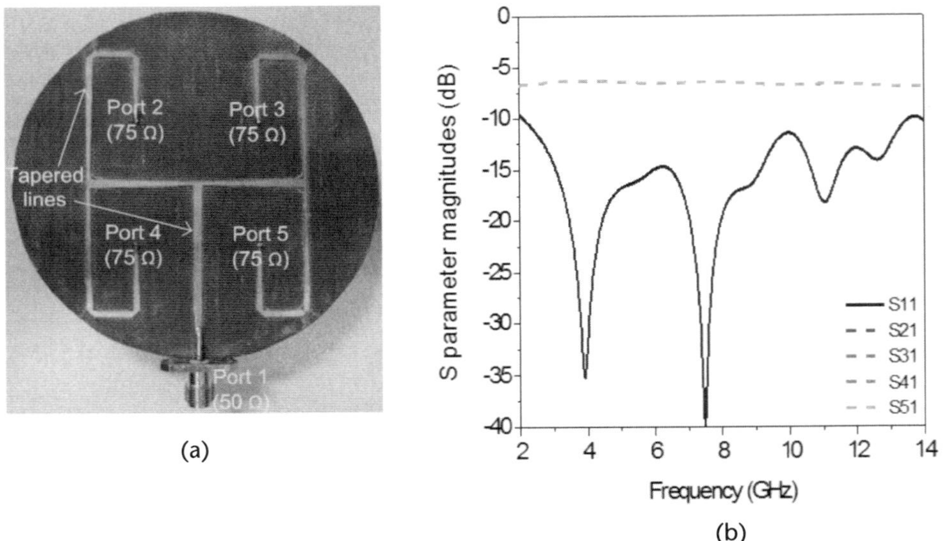



(a)

(b)

Figure 7.19   (a) Feed network and (b) S-parameters versus frequency plots for four-way feed network on TLX-8.

return loss bandwidth of 2 to 14 GHz was achieved in both simulations and measurements. Figure 7.19(a) shows the feed network and Figure 7.19(b) shows the S-parameters versus frequency plots of the feed design. As can be seen from the figure, the feed network yields very uniform insertion loss (S21 and S31 in decibels) between the input and output ports and more than 10-dB input impedance return loss over the frequency band of 2 to 14 GHz. The comprehensive design of the feed network for the antenna is discussed in Chapter 8.

## 7.5.2.3  Elliptical Leaf Dipole Array

After achieving a satisfactory design and results for the single element and the $2 \times 2$-way corporate feed network, the two were combined to achieve the $2 \times 2$-element array. Array analysis was performed to determine the optimum interelement distance and optimum mutual coupling between the elements. Once satisfactory results from the parametric study and array analysis were achieved, the complete array antenna was designed in CST Microwave Studio™, fabricated, and finally tested to determine its best performance.

The parametric study revealed the following facts about the mutual coupling and grating lobe behaviors of the array antenna:

1.  The 10-dB input impedance return loss bandwidth (BW) decreases linearly from BW = 6 GHz at interelement distances $d_x = d_y = 0.52\,\lambda$ to BW = 1 GHz at interelement distances $d_x = d_y = 0.68\lambda$ and then saturates.
2.  Input impedance matching also worsens from (44-dB return loss) to 12 dB in that interval. However, the input impedance return loss of the array antenna never goes below 10 dB.
3.  At around the height of the antenna $h = 0.2\lambda$, BW = 6 GHz and the best matching of a 22-dB input impedance return loss is obtained. At $h = 0.36\,\lambda$ the return loss reduces to 6 dB and therefore no further study was made.

Based on the parametric studies, the optimum design parameters are listed in Table 7.3. When the dipoles were fed with $50\Omega$ ports, the antenna continued to show 2-GHz, $-10$-dB S11 magnitudes. However, the center frequency of the operating bandwidth changed and new values for the interelement distance and reflector distance should be used for the new array. Two dipole reflector arrays were designed with 2- and 5.5-GHz operating bandwidths. Optimized values for each design parameter for the two versions of the antennas are shown in Table 7.3. Figure 7.20 shows the simulated input impedance return loss (S11 in decibels) and realized gain for each design.

**Table 7.3**  Optimized Design Parameters for Two Versions of the Dipole Reflector Antennas

| Parameter | Required/Optimized Value | Required/Optimized Value |
|---|---|---|
| Designed frequency bandwidth | 4.0–6.0 GHz | 4.0–10 GHz |
| Optimized interelement distance | 36 mm | 26 mm |
| Optimized reflector distance | 19 mm | 12 mm |
| Optimized value for $L$ | 11.2 mm | 11.2 mm |
| Optimized value for $R$ | 3.6 mm | 3.6 mm |
| Semirigid cable characteristic impedance used to feed each element | $50\Omega$ | $75\Omega$ |
| Input port characteristic impedance of the antenna array | $50\Omega$ | $50\Omega$ |

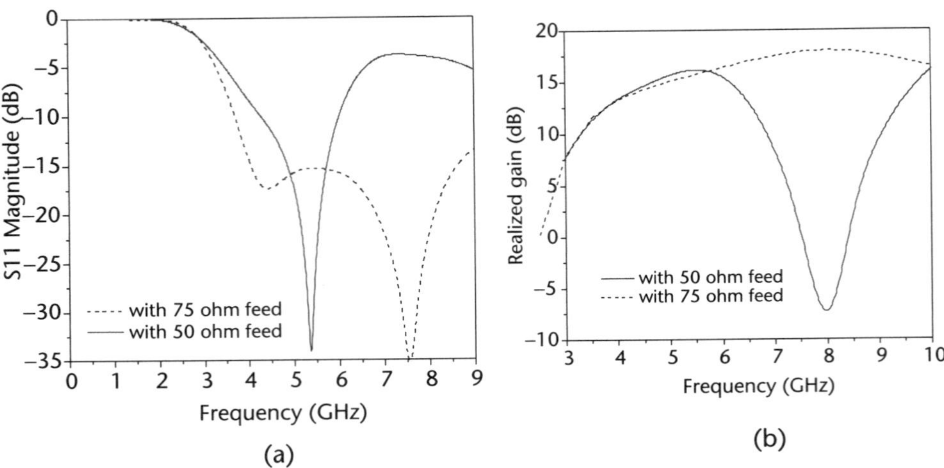

**Figure 7.20** Simulated (a) input impedance return loss and (b) realized gain versus frequency of dipole reflector arrays with 50Ω and 75Ω feeds.

### 7.5.2.4 Mechanical Assembly of Antennas

The feed network design is presented in Chapter 8. Here, the summary of design specifications for the feed network is shown in Table 7.4. In this section the construction of the antenna is presented.

The mechanical construction of a 50Ω fed elliptical dipole array was carried out as shown in Figure 7.21(a). A prototype antenna is shown in Figure 7.21(b). As shown in Figure 7.21(a), the ground plane of the feed network acts as the reflector. Due to the complexity of the mechanical construction of the leaf dipoles (they are soldered separately with the feeder coaxial cables), the leaf dipoles were etched on a thin laminate. Each element is fed with 75Ω semirigid coaxial cable. The feeder cables act like supporting posts for the feed network PCB and the dipole array PCB. Both PCBs were etched on 0.17-mm, thin Taconic TLX-8 ($\varepsilon_r = 2.4$, tan $\delta = 0.0019$) substrate. The complete antenna and the plan view of the assembly are shown in Figures 7.21(c) and (d), respectively.

**Table 7.4** Summary of Design Specifications for the Feed Networks

| Specifications | Elliptical-shaped Leaf Dipole Reflector Antenna with 50Ω Feed | Elliptical-shaped Leaf Dipole Reflector Antenna with 75Ω Feed |
|---|---|---|
| Input impedance of the feed network | 50Ω | 50Ω |
| Output impedance for each dipole element | 50Ω | 75Ω |
| Required minimum operating bandwidth | 4.0–6.0 GHz (2 GHz) | 4.0–8.0 GHz (4 GHz) |
| Substrate used | Taconic TLX-0 | Taconic TLX-8 |
| Operating bandwidth achieved in simulations | 5.5 GHz | 8 GHz |

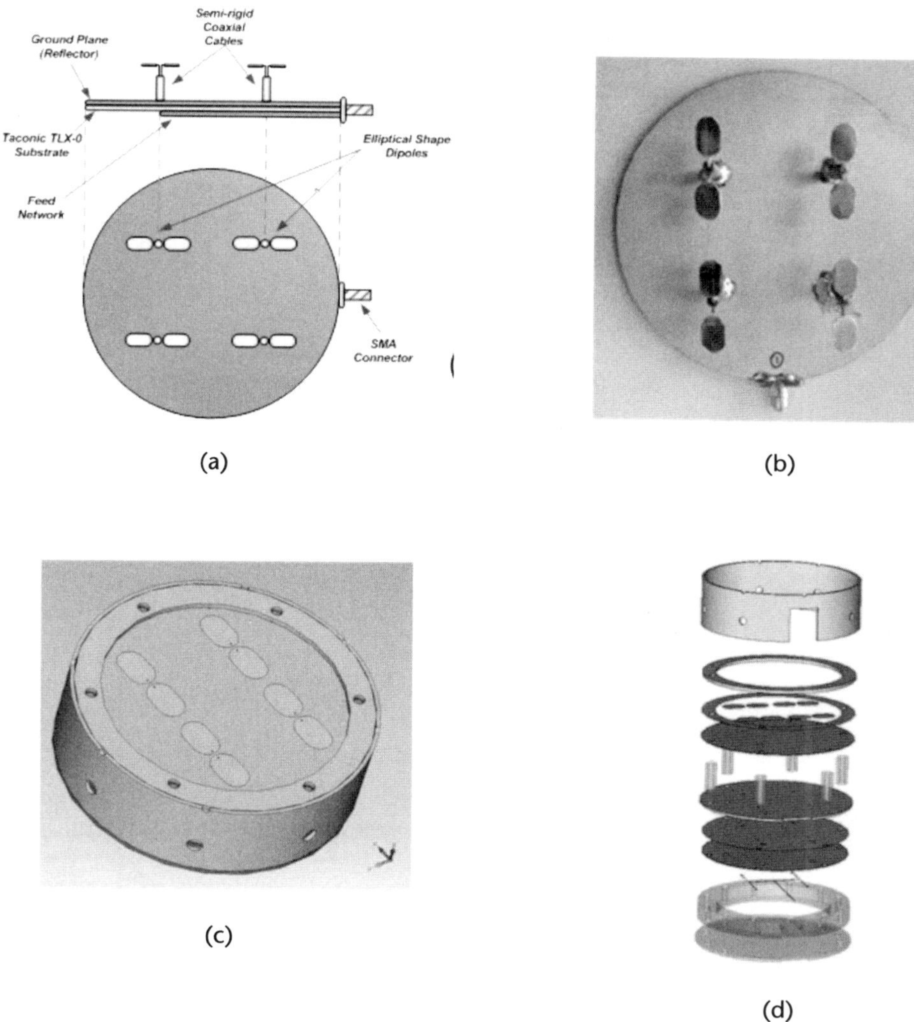


**Figure 7.21**   Mechanical assembly of antennas: (a, b) antenna assembly with 50Ω feed and (c, d) antenna assembly with 75Ω coaxial cable feed.

### 7.5.2.5   Results

In this section, we present the CST MWS simulated and measured results for the antenna designs.

*Results for an Elliptical-Shaped Dipole Reflector Antenna Array with 50Ω Feed*
The measured E- and H-plane radiation patterns of the antenna shown in Figures 7.22(a) and (b) for 4.0, 5.0, and 6.0 GHz. The main beams show sum patterns and have a 25° to 35o half-power beamwidth over the frequency range. Sidelobe levels are below 15 dB from the peak value of the main beam for the frequency range of 4.0 to 6.0 GHz. Figure 7.23 shows the simulated and measured return loss and gain versus frequency of the antenna. Both measured results have good agreement with the simulated results. The measured gain is about 2 dB less than the simulated value

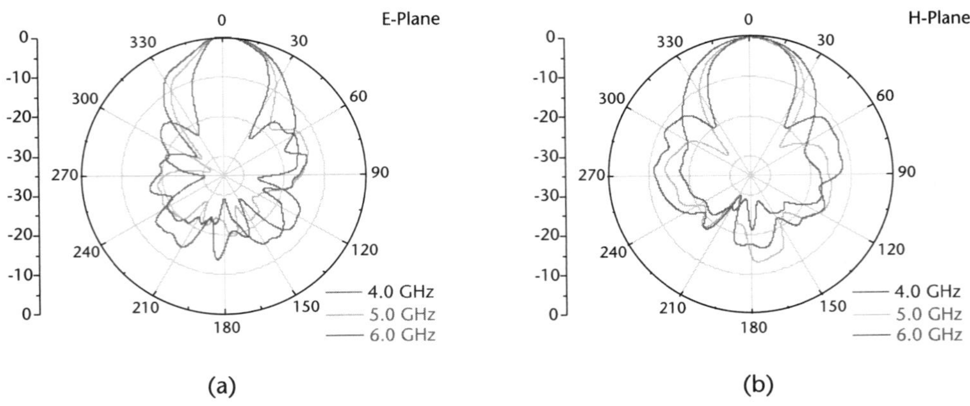



**Figure 7.22**  Measured radiation patterns of elliptical-shaped dipole antenna array.

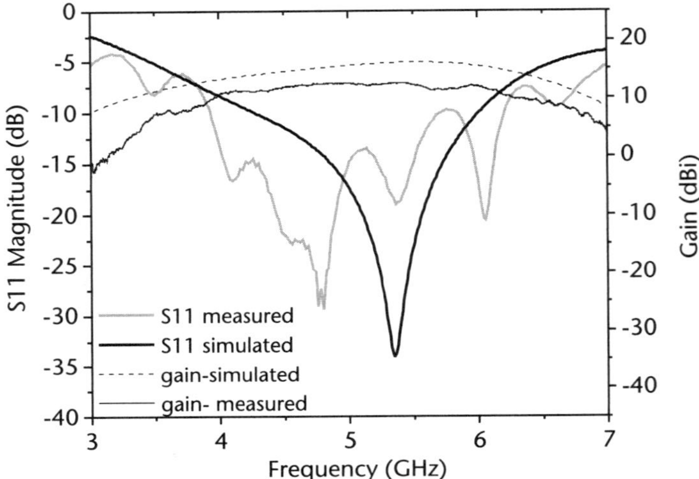



**Figure 7.23**  Measured and simulated return loss and gain versus frequency plots for elliptical-shaped dipole reflector array with 50Ω feed.

for the whole frequency range. This loss is probably introduced by the feed network and coaxial cable assemblies used to feed each element.

*Results for an Elliptical-Shaped Leaf Dipole Reflector Antenna Array with 75Ω Feed*
The E- and H-plane radiation patterns of the antenna are shown for frequencies 4.0, 6.0, 8.0, and 10 GHz in Figure 7.24. The sidelobe level becomes – 5 dB at 10 GHz; and for frequencies below 10 GHz, sidelobe levels are below – 10 dB in both the E- and H-planes. The 10-dB return loss bandwidth is 3.8 to 12.5 GHz.

Figure 7.25 shows the simulated and measured input impedance return loss and gain versus frequency plots of the antenna. As shown in the figure, the measured 10-dB input impedance return loss bandwidth is 4.3 to 9.5 GHz, and over a 9-dBi gain bandwidth is 3.9 to 9.3 GHz. The measured and simulated realized gain show good agreement up to 8 GHz. However, a large discrepancy beyond 9 GHz

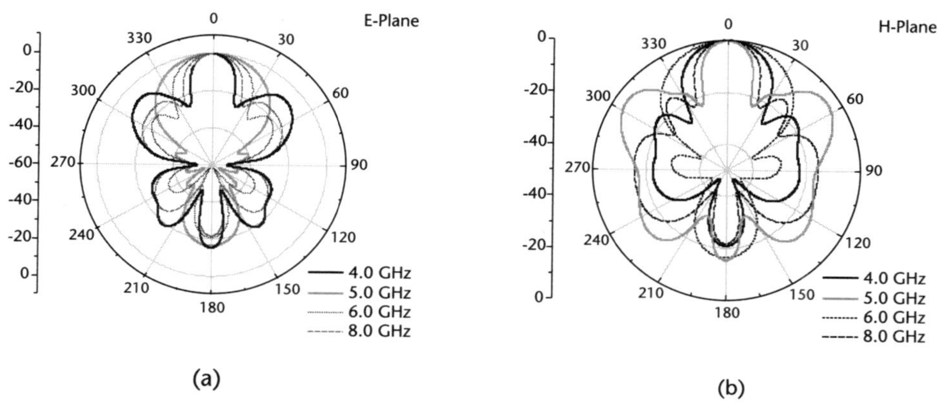



**Figure 7.24**   Simulated results of elliptical-shaped dipole reflector antenna with 75Ω feed.

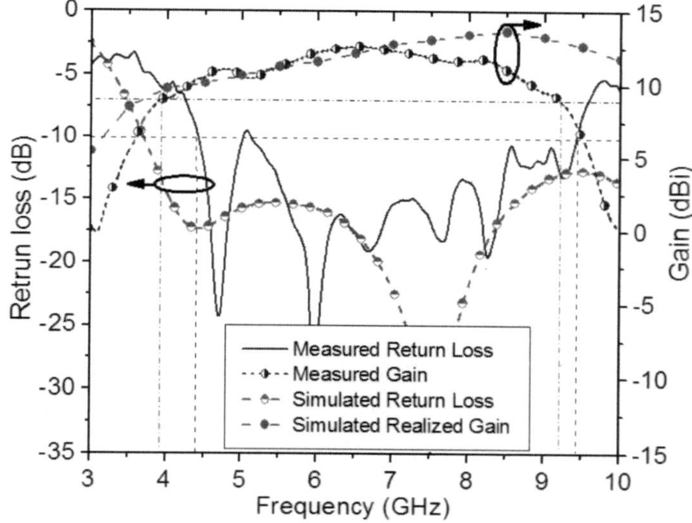



**Figure 7.25**   Simulated and measured return loss and gain versus frequency plot of elliptical-shaped dipole reflector antenna  with 75Ω feed.

between the measured and simulated results is observed. Using a balanced and unbalanced (balun) feed, this discrepancy in gain performance is minimized [7, 44].

*Surface Current Distribution*

For a wire dipole antenna, the bandwidth of the antenna is limited by its fundamental mode and higher order mode currents. Therefore, different patterns are observed when the antenna operates above its fundamental mode. For the broadband dipole antenna, this is not the case. The operating principle can be explained with the help of Rumsey's principle [7]. The elliptical dipole antenna geometry is a function of angle and helps overcome the frequency dependency. However, the ground plane reflector under the broadband elliptical dipole antenna makes the antenna directional. Boresight radiation patterns are obtained due to the reflection of the radiated figure- eight field pattern of the leaf dipole from the ground reflector. As stated before, the distance of the ground reflector from the antenna element has significant influence on the input impedance match and radiation patterns. To understand the broadband characteristics of the antenna, a comprehensive study on the surface current distribution of the antenna elements and the ground plane was made. The broadband radiation mechanism is interpreted with the aid of the surface current distributions in the leaf dipoles and the ground plane at difference frequencies.

Figure 7.26 shows the CST Microwave Studio simulated surface current distributions on dipole elements and the reflector plate. With simulated return loss and radiation patterns, satisfactory operation of the antenna was observed over 3.8 to 10 GHz. The sum patterns of the main beam and broadband matching over a bandwidth of more than 6 GHz can be explained with the aid of surface current distribution on dipoles and the reflector plate as follows:

1. The surface current distributions on the elliptical-shaped dipole elements confirm their broadband behavior. At lower frequencies (4 to 6 GHz), a high current density in a large portion of the dipole elements can be observed. The direction of current on the ground plane (reflector) at 4 GHz is opposite that of the currents in dipole elements. However, since the current density of radiating elements is high compared to that for the ground plane, this opposing current does not distort the radiation pattern.

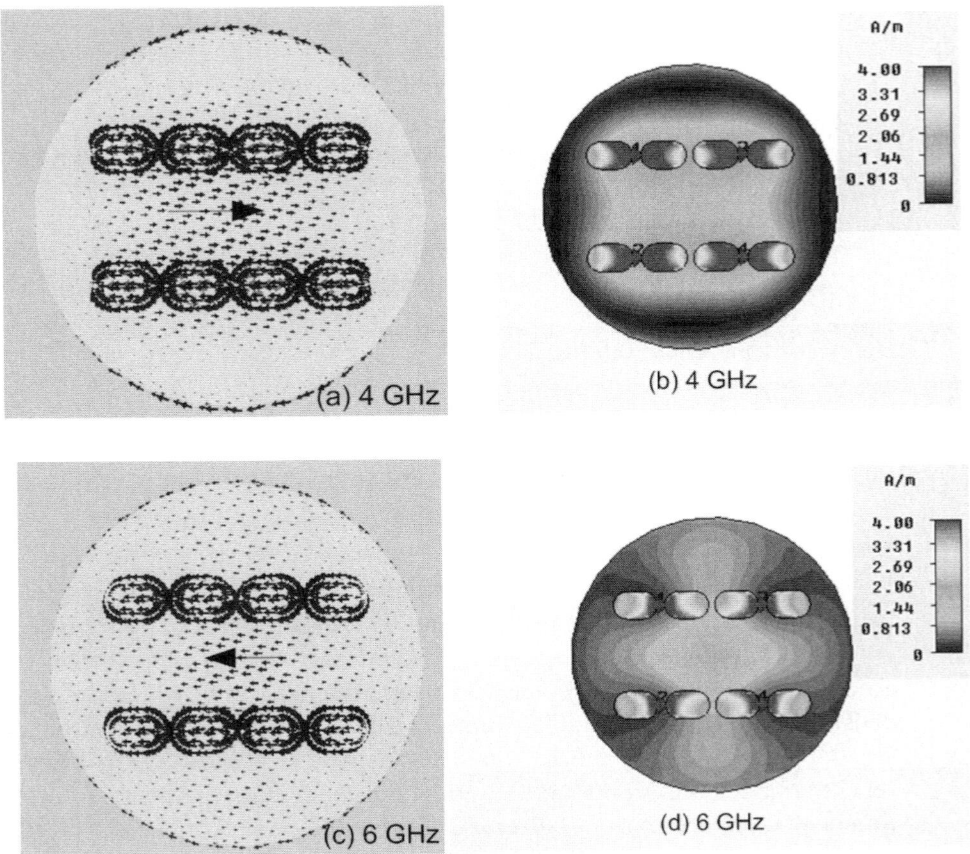


**Figure 7.26**    Surface current distributions of antenna.

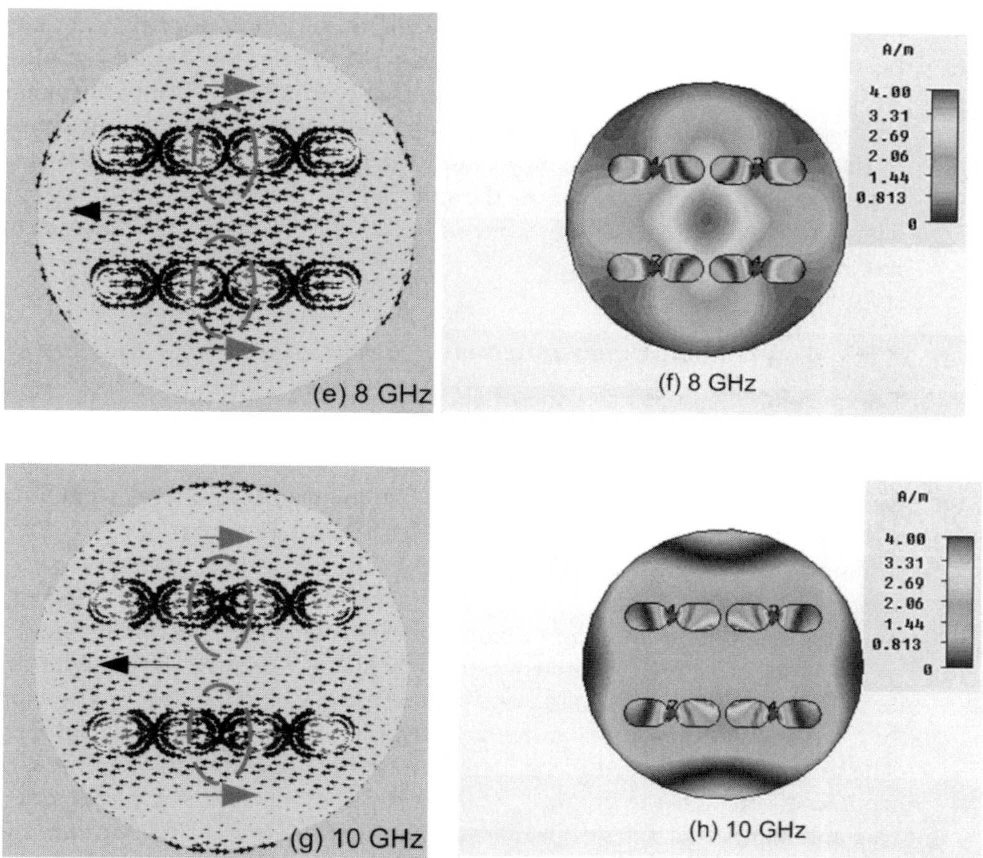



**Figure 7.26**   (continued)

2.   At higher frequencies (8 to 10 GHz), the current distribution is confined to the outer edges of the dipole elements, as shown in Figures 7.26(e) and (g), and the current is opposite in the areas marked with ovals. At this stage, dipole elements are not radiating as a conventional $\lambda/2$ dipole. The surface current distribution on the ground plane is in the direction marked with a black arrow. It is confined in the area surrounded by dipole elements and contributes to the radiation of dipole elements. The current distributions of the grounded reflector plate show that it is also acting like a radiating element of the antenna. Therefore, a sum pattern main beam and excellent impedance match with moderately high gain over a large frequency range were achieved with this compact profile antenna.

### Discussion of UWB Dipole Array

Two new antennas that have two different operating bandwidths, 4.0 to 6.0 GHz and 3.8 to 10 GHz, have been designed. Two antennas were fabricated and their input impedance return losses and gain were measured. With the utility of surface current distribution on the dipole elements and reflector plate, the operation of the antennas over a wide bandwidth has been explained. Each antenna shows a gain greater than 10 dBi over the operating bandwidth and no discontinuity in the main beam over the below 10-dB return loss bandwidth (S11 in decibels). The half-power beamwidth for each antenna is in the range of 15 o to 35o over the operating bandwidth. All antennas are linearly polarized and suitable for use with the multiresonator based chipless RFID tags mentioned in Section 7.6. Using antennas working at the 4.0- to 6.0-GHz frequency range, the frequency signature of the chipless tag was measured wirelessly to show the suitability of the antennas as reader antennas. These antennas were used with one of the chipless RFID reader designs described in Chapter 10.

### 7.5.3 Log-Periodic Dipole Antennas

Log-periodic antennas are one of the most popular UWB antennas. They have a low-profile configuration with moderately high gain and uniform radiation characteristics over the frequency band of interest. Since MMARS' first retransmission based chipless tag [45] needed precise alignment with the reader antenna, a very compact low-profile antenna was investigated and a compact circular log-periodic antenna was the selected design candidate. The advantages of the log period antennas are: (1) very compact in size, (2) low profile, (3) light weight, and (4) moderately high gain with uniform radiation patterns over the UWB frequency band.

A circular LPDA antenna was designed that operates over a wide frequency band (4 to 6 GHz). The log periodic dipole array (LPDA) antenna is supposed to be a frequency-independent antenna [7] operating over a wide range of frequencies. The antenna is supposed to be a comparatively high gain antenna, linearly polarized, and fitted to the chipless RFID reader. The basic principle of operation of this antenna is that a few closely resonant dipoles are placed in an array in such a way that their resonances overlap; hence, the antenna resonates over a much wider frequency band. A microstrip line feed circular planar log period antenna is shown in Figure 7.27.

The main design parameters of a log-periodic antenna are its flared angle $\alpha$, the stem angle $\beta$, and the radii of the complementary teeth $r_n$, $R_n$, and $R_{n+1}$. Using uniform periodic teeth, the geometrical aspect ratio of the antenna as shown in Figure 7.27 is [7]:

$$\tau = \frac{R_n}{R_{n+1}} = \frac{f_1}{f_2}; f_2 > f_1 \tag{7.4}$$

The width of the antenna slot is defined by the second geometric ratio:

$$\zeta = \frac{r_n}{R_{n+1}} \tag{7.5}$$

Based on this design equation, an LPDA that covers a bandwidth of 3 to 10.6 GHz can be achieved. Figure 7.28 shows the antenna. As can be seen, the antenna is fed at one end of the log-periodic dipole arm, and a tapered matching section is extended to the end of the antenna to facilitate the assembly of the coaxial feed line.

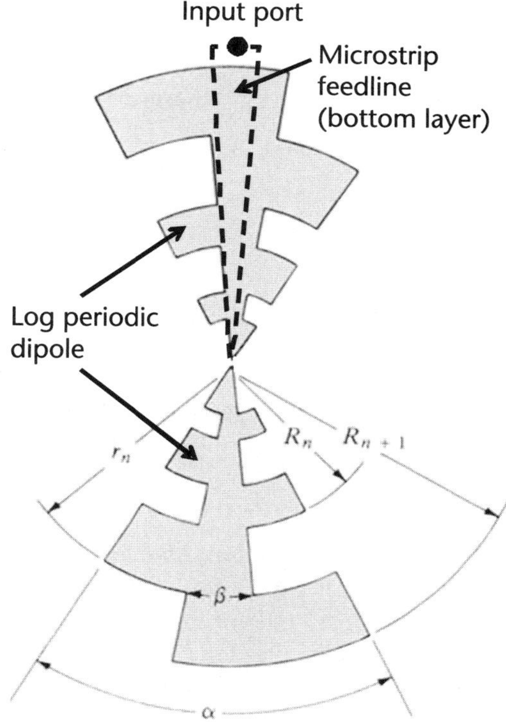



**Figure 7.27**   Configuration of a microstrip feed line circular log-periodic antenna.

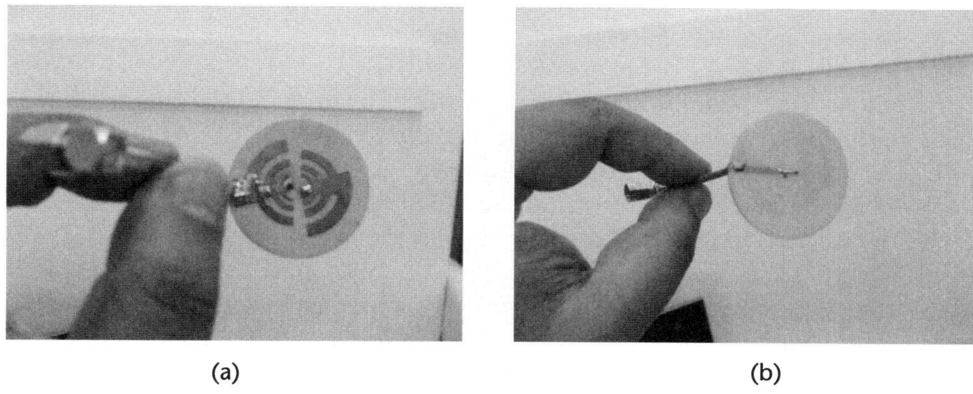

(a)                                                                              (b)

**Figure 7.28**   A circular LPDA: (a) front side and (b) bottom view showing the tapered matching section and coaxial feed line (as shown in Figure 7.27).

Figure 7.29 shows the simulated input impedance return loss versus frequency of the antenna. The antenna shows a 10-dB return loss over a frequency range from 4 to 7 GHz. The antenna was designed on 1.5-mm-thick FR4 substrate with a dielectric constant of 4.2. Figure 7.30(a) shows the CST-generated coordinate system and Figures 7.30 (b), (c), and (d) show 3D gain patterns from 4 to 6 GHz in 1-GHz steps. As can be seen, the antenna exhibits omnidirectional radiation pat-

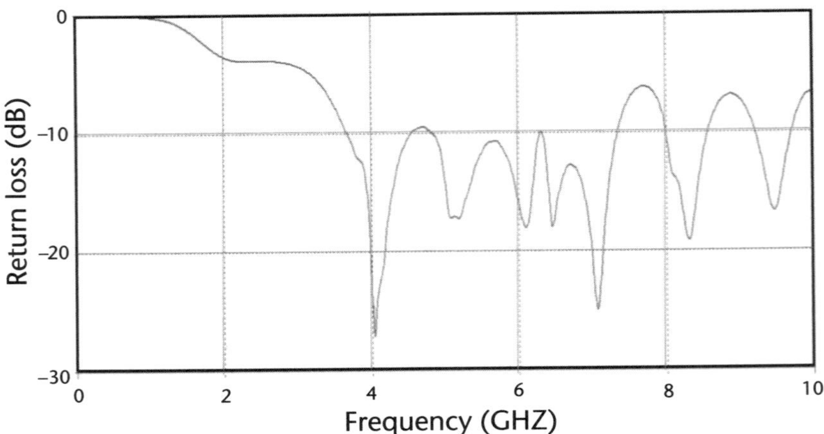

**Figure 7.29**  CST Simulated return loss (S11 parameter) for the antennas shown in Figure 7.28.

terns (figure-eight) in both the front and back sides with nulls in the elevation zero angle (more pronounced in the 5-GHz gain pattern).

Because the antenna loses half of its power in an unwanted direction, similar to the dipole reflect array, the circular LPDA antenna is placed in a conical cavity to suppress the back lobes and direct the beam in the boresight direction of the antenna. The conical cavity backed reflector provides different reflective distances for different wavelengths in a smooth transition and thereby a directional and high-gain beam is obtained in the front side of the antenna.

Figure 7.31 shows the two types of cavity backed circular LPDAs. This complementary design was done to accommodate a right-angle SMA connector while creating minimum obstruction to the radiative parts of the antenna and also to prevent the existence of any unknown radiative part (e.g., the cable junction below the antenna), especially inside a reflective cup. Figures 7.31(a) and (b) show the back and front views of the fabricated antenna PCB of the circular LPDA, respectively. As can be seen in Figure 7.31(a) the tapered matching section is connected to half of the dipole. The center pin of the SAM connector is soldered at the end of the tapered matching section from the other side of the PCB. As shown in Figure 7.31(b), the thick complementary tapered line at the center of the other half of the dipole (top view) acts as the ground plane of the tapered matching section on the bottom side of the antenna. The stem of the SMA connector, which is the ground of the connector, is soldered with the tapered line. The aluminum conical back reflector housing is shown in Figure 7.31(c) and the complete assembled antenna is shown in Figure 7.31(d). The whole structure was designed and simulated with a discrete input port in CST, as shown in Figure 7.31(e). This completes the fabrication and assembly of a type 1 antenna with conical back reflector housing.

The type 2 antenna was a compact circular LPDA with tight element spacing for much closer frequency resonances. The antenna is backed by a flared cylindrical metal cup reflector. Figure 7.31(f) shows the compact assembled antenna.

Figure 7.32(a) shows the CST simulated return loss and measured S-parameters (return loss and insertion loss) of the cavity (cup) backed circular LPDA. In both simulation and measurement, the circular LPDA shows good return loss figures over a wide frequency band similar to the case for the antenna with the cavity backed

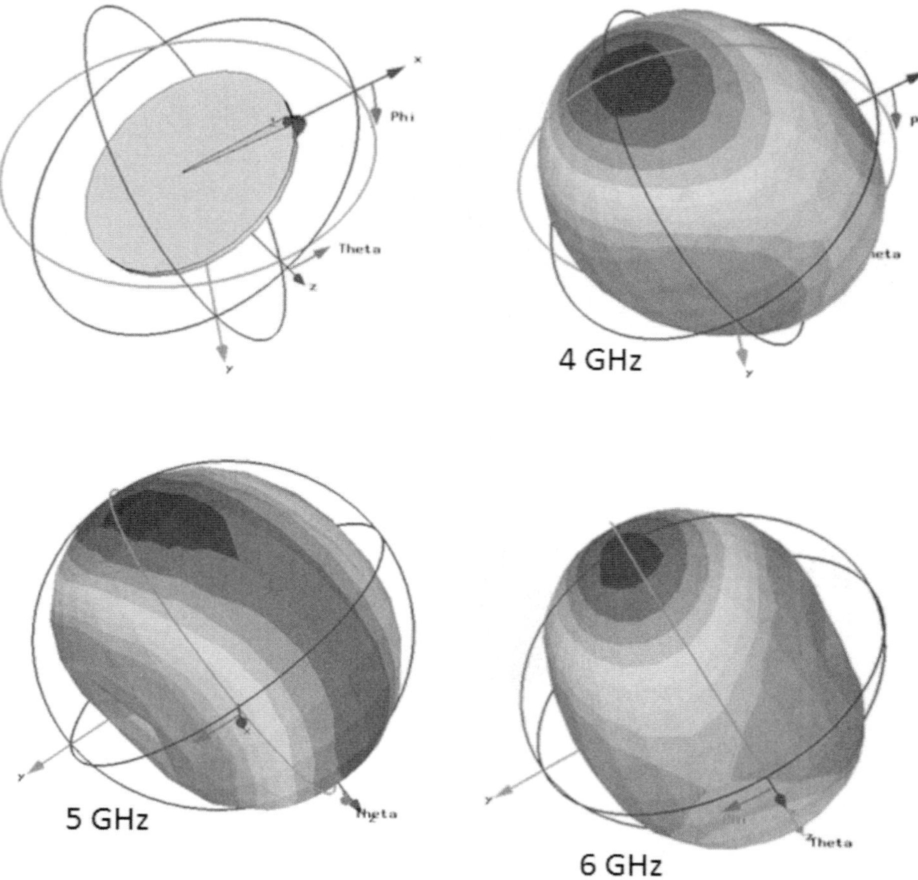



**Figure 7.30**   CST simulation results of circular log-periodic antenna on FR4 substrate: (a) CST-generated bottom layout of the circular LPDA showing the coordinate system. CST simulated 3D gain pattern with maximum gain of (b) 3.4 dBi at 4 GHz, (c) 3.3 dBi at 5 GHz, and (d) 3.9 dBi at 6 GHz.

reflector in Figure 7.29. Figure 7.32(b) shows the measured S-parameters—both return loss $S_{11}$ and $S_{22}$ of a pair of tuned cavity backed circular LPDAs and insertion loss—forward transmission coefficient $S_{21}$ and reverse transmission coefficient $S_{12}$. The tuning was done for the first time and showed a rise in the return loss centered on 5 GHz. The two antennas were placed face to face while attached to the VNA and their insertion loss was also measured. The rise in the return loss was reflected by a dip in the insertion loss. Further tuning may solve the slight mismatch and consequential transmission issues.

Figure 7.33 shows the simulated and measured radiation patterns for the cavity backed circular LPDA antenna. The patterns yield directional symmetry in both the E- and H-planes with an average 3-dB beamwidth of approximately 60°. Figure 7.34 shows the gain versus frequency plots for the LPDA with the cavity connected with the chassis ground. In both cases the antenna exhibits an average 7-dBi gain over the frequency band. The antenna was used to encode a 5-bit retransmission based chipless RFID tag. The encoded data are presented in Section 7.6.

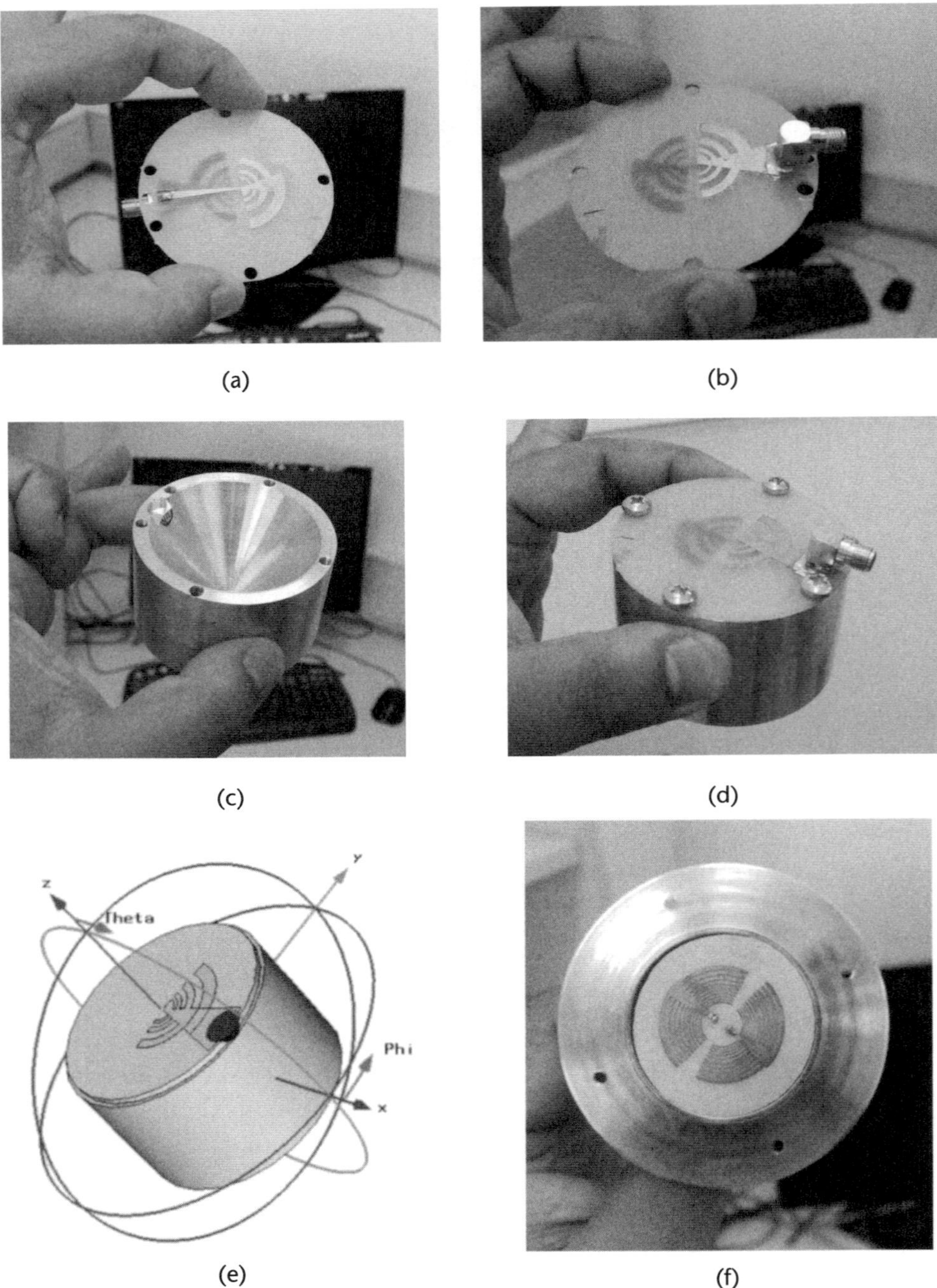

(a)                                        (b)

(c)                                        (d)

(e)                                        (f)

**Figure 7.31**   Two types of cavity backed circular LPDAs. Type 1: Complementary LPDA design with right-angle SAM feed: (a) bottom view, (b) top view, (c) conical metallic back reflector, (d) assembled circular LPDA fitted to conical metal cup reflector, and (e) CST-generated 3D layout of circular LPDA fitted to conical metal cup reflector. Type 2: (f) Assembled circular LPDA fitted to a flared cylindrical metal cup reflector with tight element spacing for much closer frequency resonances.

## 7.5.4   Log-Periodic Dipole Array Antennas

In the preceding section two types of cavity backed circular LPDAs were presented. The construction of the antenna is complex and nonplanar due to the aluminium

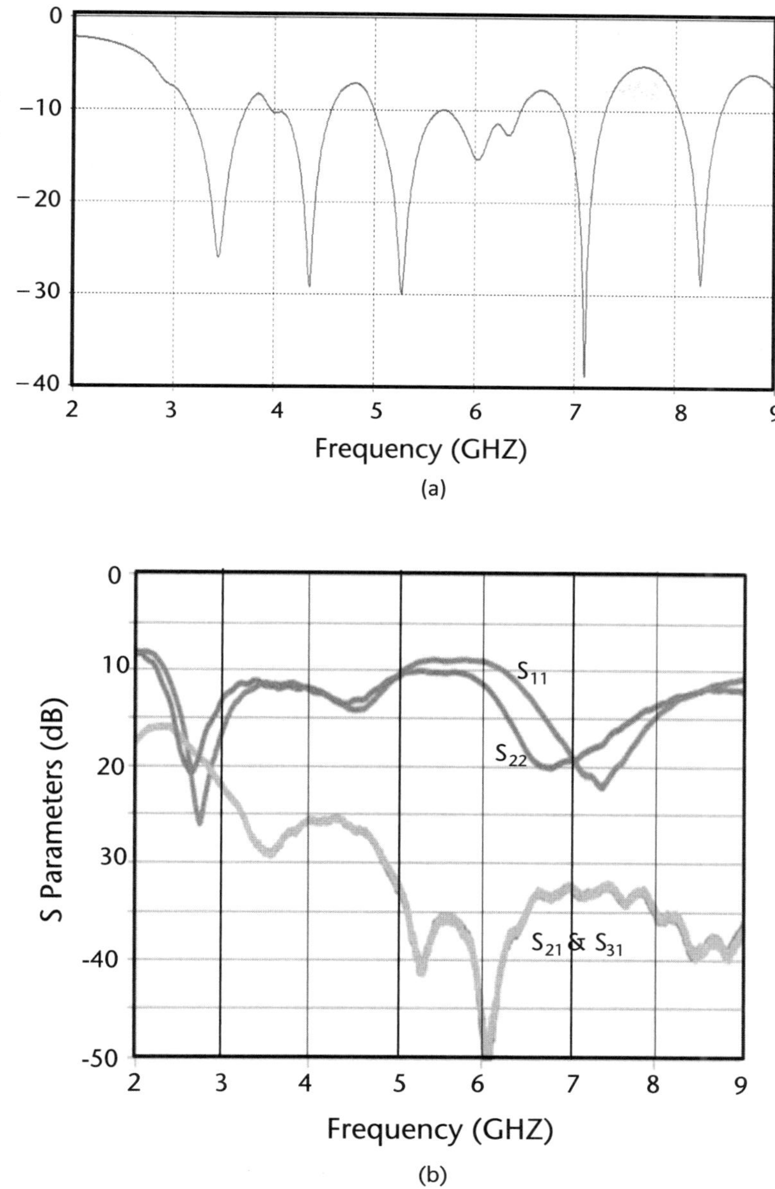



**Figure 7.32**   (a) Simulated return loss and (b) measured S-parameter versus frequency plots of the circular LPDA with cavity (cup) backing.

cavity housings. In chipless RFID reader systems, lightweight, low-profile, planar antennas are preferred. In this section a couple of linear log-periodic dipole array antennas (LPDAAs) for both microwave and millimeter-wave UWB bands are presented. LPDAAs are very common linearly polarized broadband antennas with relatively high gain and favorable radiation characteristics. Since the antenna radiates at its end fire direction, it has found many applications in broadcasting and receiving operations.

Another advantage of LPDAAs is that both input impedance match and radiation performances (radiation patterns, gain and front-to-back ratio) are essentially

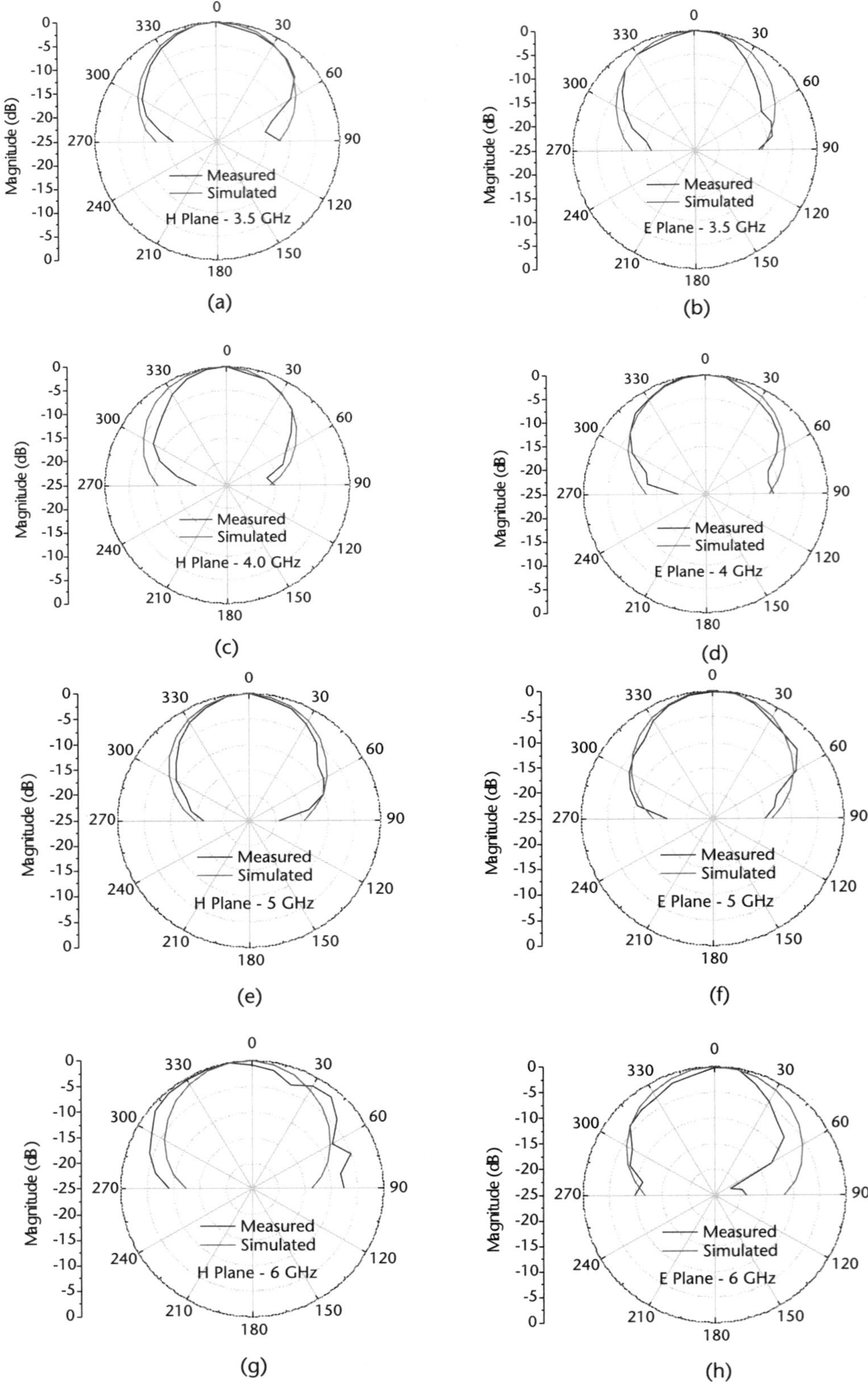


**Figure 7.33**　E-plane and H-plane radiation patterns for a cavity backed circular LPDA.

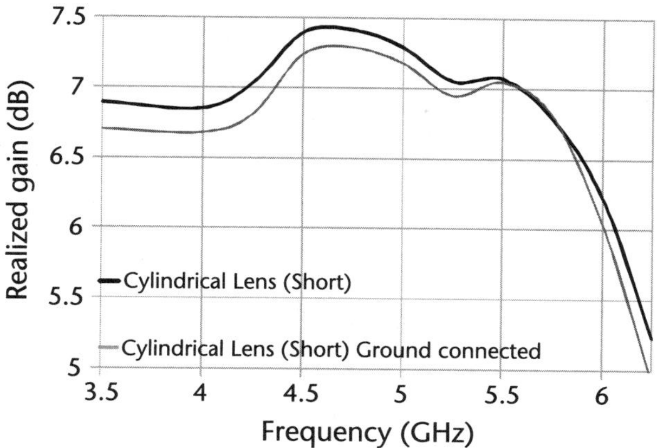



**Figure 7.34**  Gain versus frequency plot of cavity backed circular LPDA.

frequency independent over the frequency band of operation. The basic configuration of a log-periodic antenna is similar to any fishbone antenna such as a Yagi Uda antenna. The antenna elements gradually increase with a flared angle and are fed by a two-wire transmission line, as shown in Figure 7.35. An LPDAA design requires three design parameters [7]: geometric ratio $\tau$, angle factor $\alpha$, and spacing factor $\sigma$, as shown in Figure 7.35. By defining the desired directivity, the geometric ratio $\tau$ and spacing factor $\sigma$ may be easily found from [7].

The design equations of an LPDAA are:

$$\tau = \frac{L_{n+1}}{L_n} = \frac{R_{n+1}}{R_n}\frac{f_{n+1}}{f_n}; \text{ for } \quad f_{n+1} > f_n \tag{7.6}$$

$$\sigma = \frac{R_{n+1} - R_n}{2L_{n+1}} \tag{7.7}$$

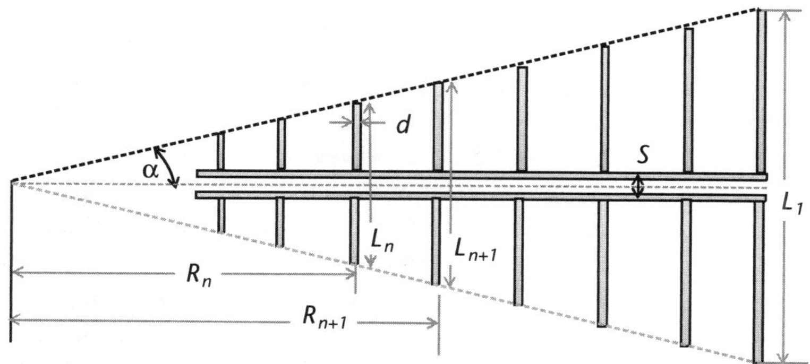



**Figure 7.35**  Configuration of log-periodic dipole array antenna.

$$B_s = \frac{f_n}{f_1}\left[1.1 + 7.7\left(1 - \tau\right)^2 \frac{4\sigma}{1 - \tau}\right] \tag{7.8}$$

$$\alpha = \tan^{-1}\left[\frac{1 - \tau}{4\sigma}\right] \tag{7.9}$$

$$N = 1 + \frac{\log B_s}{\log\left(1 / \tau\right)} \tag{7.10}$$

$$L_{total} = \frac{\lambda_{max}}{4}\left(1 - \frac{1}{B_s}\right)\frac{1}{\tan\alpha} \tag{7.11}$$

$$Z_a = 120\left[\ln\left(\frac{L_n}{d_n}\right) - 2.25\right] \tag{7.12}$$

$$S = d\cosh\left(\frac{Z_o}{120}\right) \tag{7.13}$$

where $f_n$ and $f_{n+1}$ are the resonant frequencies of the nth and $(n + 1)$th dipole elements, respectively. The spacing factor $\sigma$ defines the separation between the dipole elements in order to achieve the desired bandwidth Bs and directivity Do. The terms $R_n$ and $R_{n+1}$ represent the separation between the $n$th and $(n + 1)$th dipole elements, respectively, and $L_n$ and $L_{n+1}$ are the lengths of the two dipole elements of the $n$th and $(n + 1)$th orders as shown in Figure 7.35. The above design equations and design procedure give the dimensions of an LPDAA. Figure 7.36 shows the practical design procedure for an LPDAA.

Transferring the design into a microstrip PCB introduces a new chapter of practical realization of the antenna. The practical design procedure (see Figure 7.37) is performed in a CAD tool such as CST Microwave Studio™, Agilent Momentum™, or LineCalc™ and similar EM CAD tools. Design optimization in CAD tool needs both fundamental knowledge of antenna theory and specific skills sets for those specific CAD tools. In this regard, it is not expected that the optimum design and performance benchmark will be attained in one go. Design optimization needs a set of goals to be achieved and that should be incorporated in the simulation engine. An iterative process is the common mode of CAD design culture. If the initial iteration does not meet the goal, the tuning tools of some advanced level software package such as CST Microwave Studio can be used or manual trial-and-error

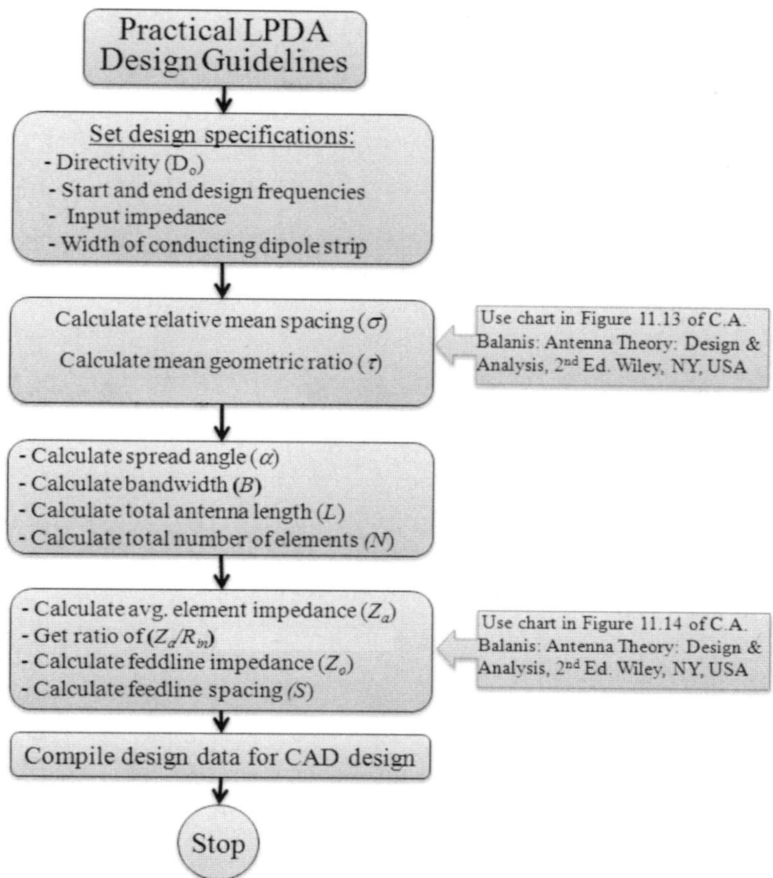



**Figure 7.36**   Design procedure for an LPDAA.

observations and tuning of design parameters based on the skill set and knowledge can be applied. Surface current plots on the designed layout can also be a useful tool to verify the root cause of the problem (see, e.g., Figure 7.26). The design can be modified based on the perception of surface current plots, and the design can be optimized to achieve the required goal. Once the design is optimized, the layout can be transferred to Gerber or dfx format for fabrication and production.

Based on the preceding design equations and design procedures, a comprehensive parametric study was done on the width of the dipoles $d$, the gap between feed lines $S$, and the geometric ratio $\tau$, leading to the following conclusions:

1. Reducing the geometric ratio $\tau$ results in a better return loss performance. However, it also reduces the antenna gain and the bandwidth.
2. Increasing the gap between feed lines $S$ results in shifting the center frequency to a lower value (shifts to the left), which gives a poor performance in the desired frequency band of operation.
3. Increasing the width of the dipoles results in an increase in the bandwidth. Based on the parametric study performed on various design parameters, a geometric ratio $\tau$ of 0.8 to 0.95 yields the most desirable operating parameters and the best compromise between them.

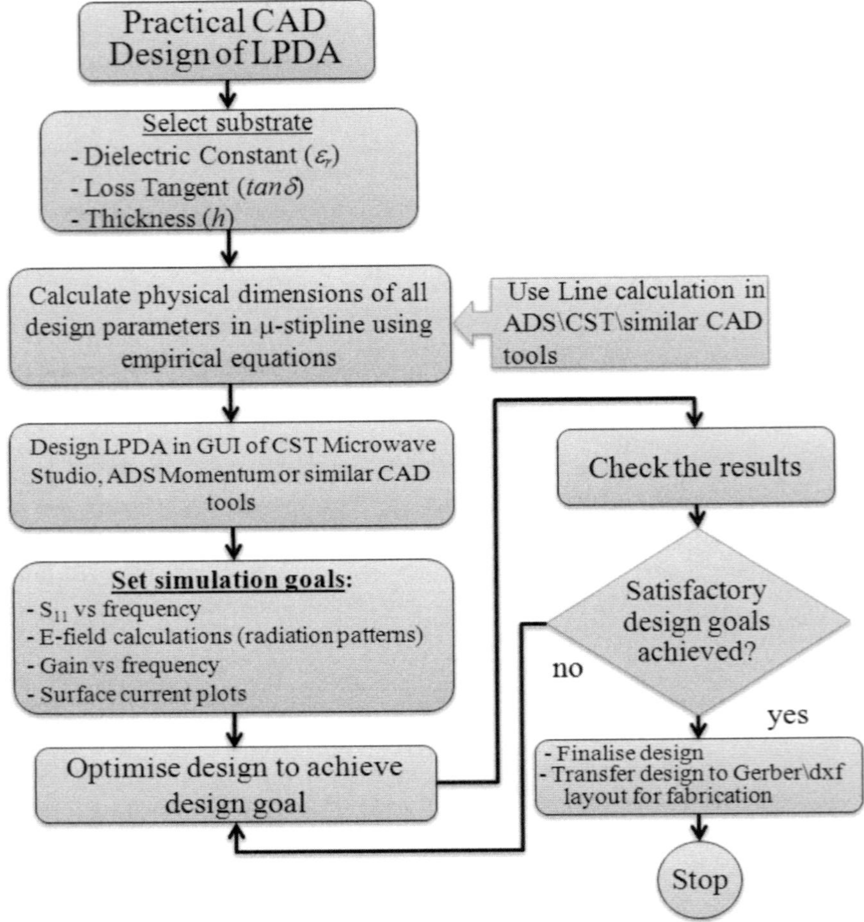


**Figure 7.37**    CAD design procedure for printed LPDAA.

4.  Increasing the width of a dipole $d$ will only have a positive effect on performance. The gap between feed lines $S$ is dictated by the fabrication capabilities of the milling machine, which is 0.15 or 0.1 mm in our available resource (LKPF Milling Machine).

### 7.5.4.1   Design of Microwave LPDAAs

Based on the design guideline given earlier, two LPDAAs were designed for RFID chipless tag readers in the microwave frequency band (2 to 3 GHz) and the millimeter-wave frequency band (20 to 30 GHz). Figure 7.38 showsa photograph of an LPDAA designed for the microwave frequency band. The design parameters for the microwave LPDAA are as follows:

- Geometric factor $\tau = 0.93$
- Starting dipole length L1 = 2.99 cm
- Width of dipole elements w1 = 2 mm

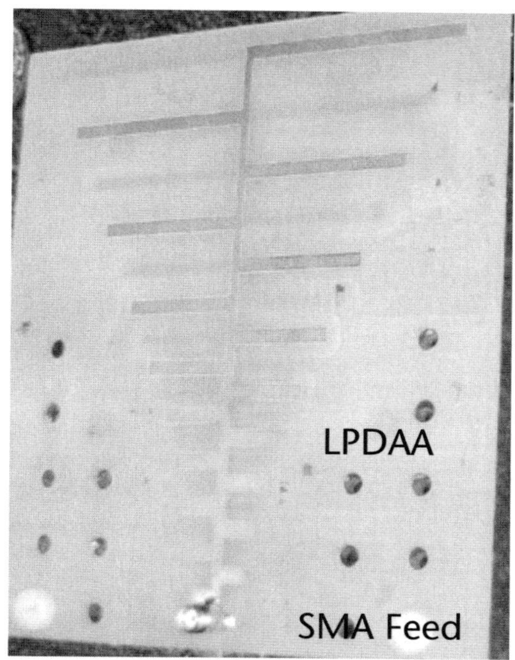


**Figure 7.38**   Photograph of an LPDAA designed at the microwave frequency band form 2- to 3-GHz.

- Gap between the two arms of the LPDA w2 = 0.4 mm
- Number of elements = 9
- Substrate FR4, $\varepsilon_r = 4.3$, tan $\delta = 0.025$.

*Results for Microwave LPDAA*

Measurements were performed for the input impedance return loss versus frequency plot, gain versus frequency plot, and measured co-polar radiation patterns of the microwave LPDAA operating in the 2- to 3-GHz frequency band. The antenna has a 10-dB input impedance return loss from 2 to 3 GHz and beyond. The average gain of the antennas is 6 dBi and the 3-dB beamwidth is approximately 60° and 40° at 2.2 and 2.9 GHz, respectively. Significant sidelobes and back lobes are pronounced due to the truncated ground plane and edge diffraction of the antenna. More details for similar microwave LPDAAs can be found in [45].

### 7.5.4.2   Design of Millimeter-Wave LPDAA

Here the design and results for a millimeter-wave LPDAA are presented. Figure 7.39 shows a CST-generated layout for a millimeter-wave LPDAA. The antenna was designed on microstrip laminate Taconic TLX0, which has a dielectric constant of 2.45 and thickness of 0.5 mm. The dimensions for the 2- to 3-GHz operational LPDAA [45] were scaled down by a factor of 10 to obtain the initial structure of the 20- to 30-GHz LPDA. The dimensions of the antenna are 1 cm × 2.5 cm × 1.6 cm. The connector is a 2.4-mm air core flange coaxial connector.

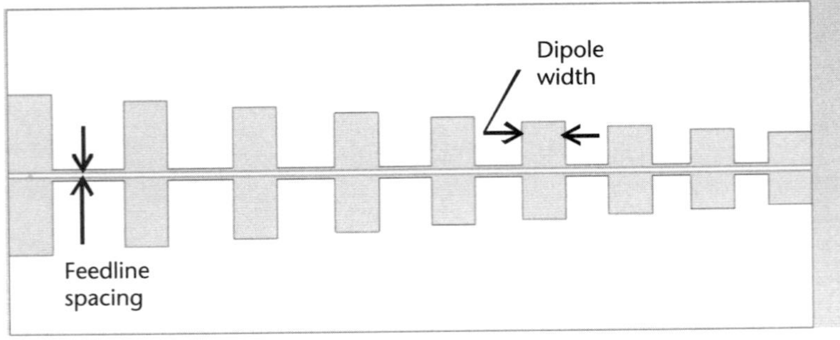



(a)

(b)

**Figure 7.39**   Millimeter-wave LPDAA on Taconic TLX0, which has a dielectric constant of 2.45 and thickness of 0.5 mm: (a) CST-generated layout and (b) prototype.

The prototype LPDAA has the following design parameters:

- Geometric ratio $\tau = 0.9$
- Width of dipoles d = 1.1 mm
- Gap between feed lines S = 0.15 mm
- Number of elements N = 9.

*Results of Millimeter-Wave LPDAA*

The antenna was measured with an Agilent E8361A PNA with full two-port error correction calibration. This instrument can measure microwave and millimeter-wave signals from 10 MHz up to 67 GHz with high precision. Figure 7.40 shows the input impedance returns versus frequency plot of the antenna. As can be seen in the figure, the prototype antenna shown in Figure 7.39 yields a 10-dB return loss bandwidth throughout the design operating range of 20 to 30 GHz.

Figure 7.41 shows the CST simulated gain patterns for 20, 25, and 30 GHz. The realized gain of the antenna gradually increases from 6 to 10 dBi from 20 to 30 GHz. Due to the unavailability of a rotary joint at that frequency range, it was not possible to measure the gain pattern in the anechoic chamber. However,

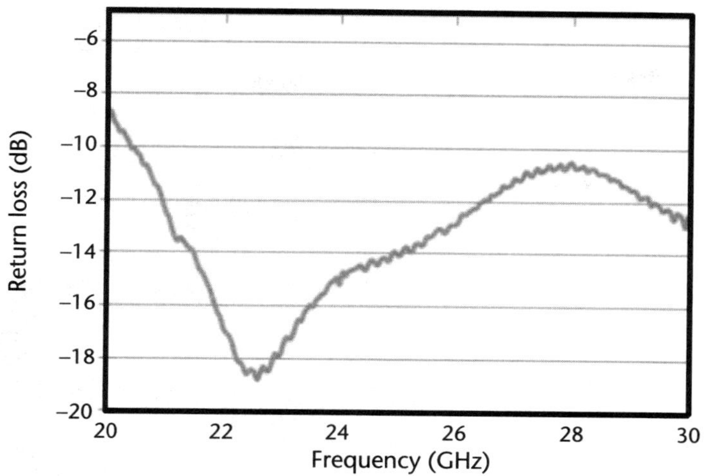

Figure 7.40   Measured input impedance return loss of millimeter-wave LPDAA on Taconic TLX0 substrate.

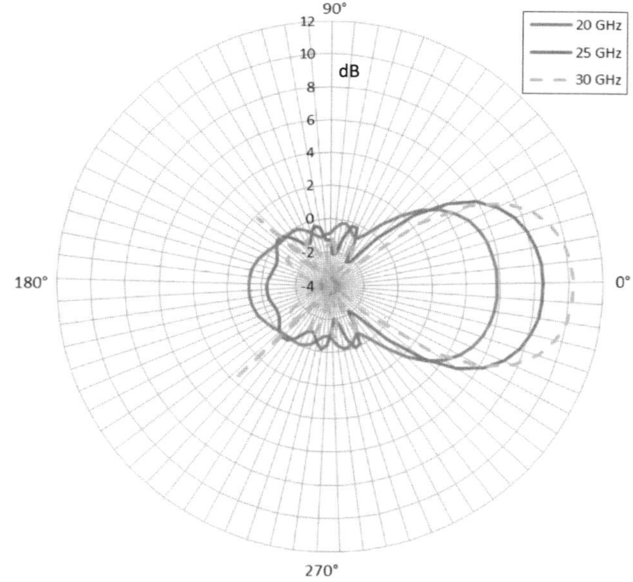


Figure 7.41   CST simulated radiation patterns of millimeter-wave LPDAA.

the forward transmission of the antenna was measured with a millimeter-wave 4 × 4-element microstrip feed aperture coupled patch antenna array in the Monash University anechoic chamber, as shown in Figure 7.42(a). The realized gain derived from the results of the forward transmission measurement is shown in Figure 7.42(b). Taking measurements in the reflection-free anechoic chamber ensures the true performance of the LPDAA. As can be seen in Figure 7.42(b), the gain gradually increases with frequency.

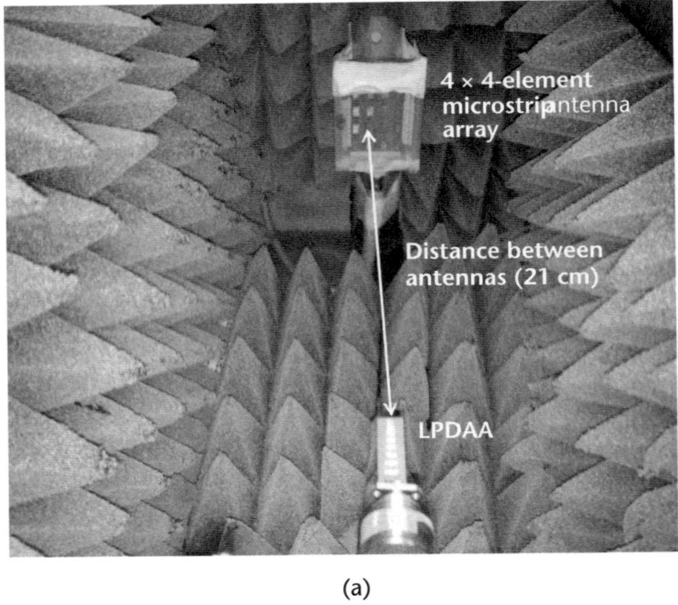


(a)

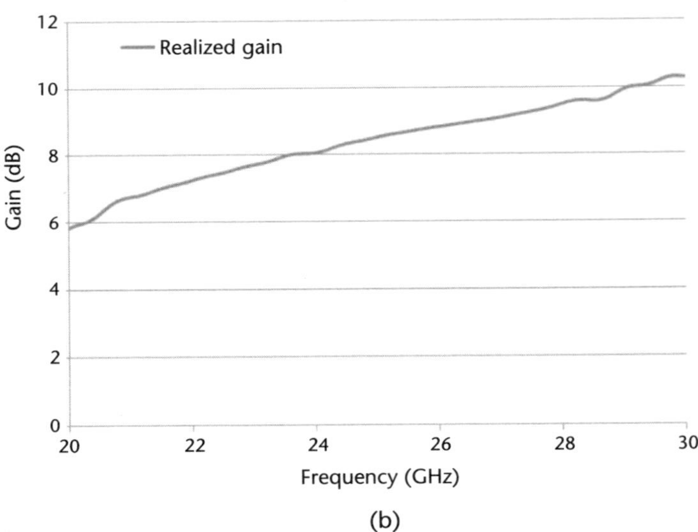


(b)

**Figure 7.42**    (a) Experimental setup for gain measurement and (b) forward transmission gain versus frequency of millimeter-wave LPDAA.

*Concluding Remarks for Millimeter-Wave LPDAA*

The millimeter-wave LPDAA was designed as a scaled-down version of the micro-wave frequency LPDAA (also reported in more details in [45]). The antenna provides consistent input impedance return losses, antenna gain patterns, and realized gains over the design frequency. A few important parametric studies over the design parameters help to convert the design to its optimum performance. The antenna exhibits an average 8-dBi gain over the frequency band from 20 to 30 GHz.

### 7.5.5   UWB Horn Antennas

An exponentially tapered ridged horn antenna is present in this section. The benefits of UWB horn antennas over other antennas are mainly their superior performance in terms of wider frequency bandwidth, high gain, and directivity. However, the manufacturing process of a horn antenna can be complicated due to the mechanical properties of the metal (K-factor and bending radius of the tapered ridges), which will affect the performance of the antenna due to the shift in its overall dimension. Furthermore, the high-precision manufacturing process is both costly and time consuming. The UWB horn antenna comprises mainly three parts: (1) a waveguide launcher, (2) a pair of tapered ridges, and (3) the horn itself. Figure 7.43 shows a cross section of the antenna, with these three parts showing. This section provides the design of an antenna that operates over the 3- to 12-GHz frequency band.

#### 7.5.5.1   Design of Exponentially Tapered Ridge

The design of the tapered ridge is the most vital part for achieving the desired UWB performance. Each part of the tapered ridge provides different impedances as it progresses, starting with 50Ω at the feeding point to 377Ω at the end of the antenna. Thus a wideband matching over the desired frequency band is achieved.

The variation of the impedance can be expressed using the following equations [7, 46, 47]:

$$Z(y) = Z_0 e^{ky} \; ; \text{ for } 0 \leq y \leq L \tag{7.14}$$

$$k = \frac{1}{L} \ln\left(\frac{Z_L}{Z_o}\right) \tag{7.15}$$

where $L$ is the maximum length of the ridged, $Z_L = 377\Omega$ and $Z_o = 50\Omega$.

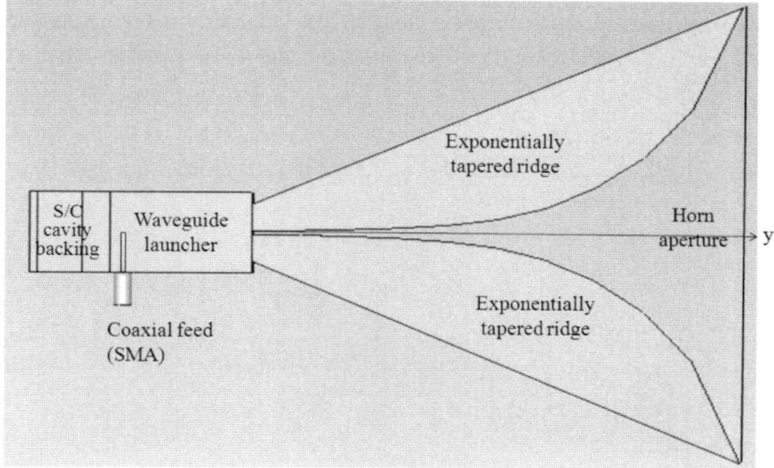


**Figure 7.43**   Wideband exponentially tapered horn antenna.

In the current design $L = 0.4\lambda$ is selected, where $\lambda$ is the smallest wavelength corresponding to the highest frequency of 12 GHz. Based on the design calculation, an exponentially tapered section is designed. Figure 7.44 shows the geometrical layout of the tapered section with dimensions. The total length of the section is 70 mm. Each taper has a uniform width of 7 mm. The corresponding lengths of the sections in millimeters are tabulated in Figure 7.44. The exponential contour of the taper section is also shown in the figure.

### 7.5.5.2  Design of Waveguide Launcher and Cavity Back

The design of the waveguide launcher is the means for excitation of the horn antenna. The wideband matching of the launcher ensures efficient transfer of energy from the microwave source to the antenna. The short circuited cavity backing is very crucial for providing the perfect match with the horn antenna [46–49]. The overall dimensions of the waveguide launcher with its feed assembly (coaxial connector) and the short circuited cavity backing yield the perfect match for the antenna. A common practice to achieve the lowest possible input return loss from the transition from the feeder to waveguide is to apply a cavity backing, as shown earlier in Figure 7.43. Therefore, it is vital that the overall length of the short circuited cavity backing be optimized. The ridge thickness will influence initial impedance, $Z_0$. Therefore, it is important to match its impedance with the feeder rod impedance (the inner conductor of the coaxial cable), which is 50$\Omega$, to ensure the highest possible input impedance return loss, which results in perfect input impedance matching of the antenna.

To simplify the fabrication, the waveguide and the cavity backing are retained in the traditional rectangular shape. The overall dimensions of the waveguide are $a = 40$ mm, $b = 10$ mm, thickness of the ridge $w = 3.5$ mm, and total length of the waveguide launcher $l = 20$ mm. Figure 7.45(a) shows the cross section of the waveguide launcher, and Figure 7.45(b) shows the isometric 3D view with the length of the waveguide launcher that ends at the beginning of the tapered ridges.

Based on simulation, the optimum distance for the short circuited cavity backing to waveguide, $L_{cb}$, is 10.5 mm, which is approximately $\lambda/4$ of the center frequency of 7.5 GHz spanning the 3- to 12-GHz frequency band. This distance of short circuited end provides a perfect match for the reflected signal for the entire band and reduces reflection loss during the transition from feeder rod to waveguide. Figure 7.46 shows the isometric view of the launcher with cavity backing.

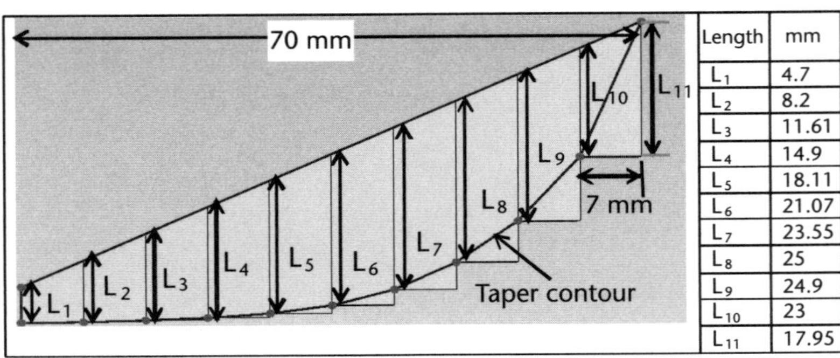


| Length | mm |
| --- | --- |
| $L_1$ | 4.7 |
| $L_2$ | 8.2 |
| $L_3$ | 11.61 |
| $L_4$ | 14.9 |
| $L_5$ | 18.11 |
| $L_6$ | 21.07 |
| $L_7$ | 23.55 |
| $L_8$ | 25 |
| $L_9$ | 24.9 |
| $L_{10}$ | 23 |
| $L_{11}$ | 17.95 |

**Figure 7.44**  Geometrical configuration of exponentially tapered ridge with uniform width of 7 mm.

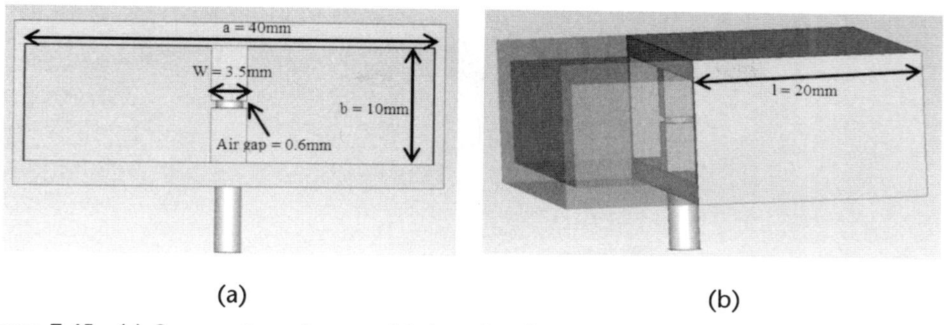

(a)                                                        (b)

**Figure 7.45**   (a) Cross section of waveguide launcher (front view) and (b) 3D isometric view showing length of waveguide launcher and beginning of tapered ridges.

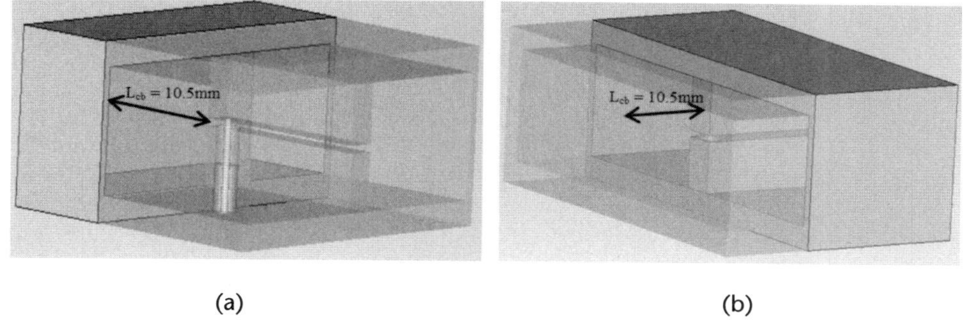

(a)                                                        (b)

**Figure 7.46**   Isometric rear view showing distances of (a) feed pin to waveguide end and (b) feed pin to short circuited end.

The antenna was manufactured in the Monash University ECSE workshop. The tapered ridges were welded with the horn. The cavity backing is an integral part of the horn antenna[4]. A SMA connector is added at the end of the horn. Figure 7.47 shows the manufactured horn antenna.

### 7.5.5.3   Results

The complete antenna with the launcher and the horn was designed in CST Microwave Studio. Figure 7.48 shows the CST simulated input impedance return loss versus frequency plot of the antenna. As can be seen in the figure, the horn antenna has a 10-dB return loss bandwidth of 3 to 12 GHz with a slightly lower return loss at a couple of frequency points. Improvement of the input impedance bandwidth can be achieved by tuning the launcher of the horn antenna. Usually, tuning screws in the launcher waveguide are used to tune the input impedance of the horn antenna.

To examine the operational bandwidth, the radiation characteristics of the horn antenna were thoroughly investigated in a CST simulation. Figure 7.49 shows the E- and H-plane radiation patterns from 3 to 12 GHz with a 3-GHz interval. As can be seen in the figure, the antenna exhibits directional radiation patterns except at 9 GHz with pronounced high sidelobes. Due to these sidelobes, in both the E- and

---

4.   Such complex fabrication is possible with a 3D printer.

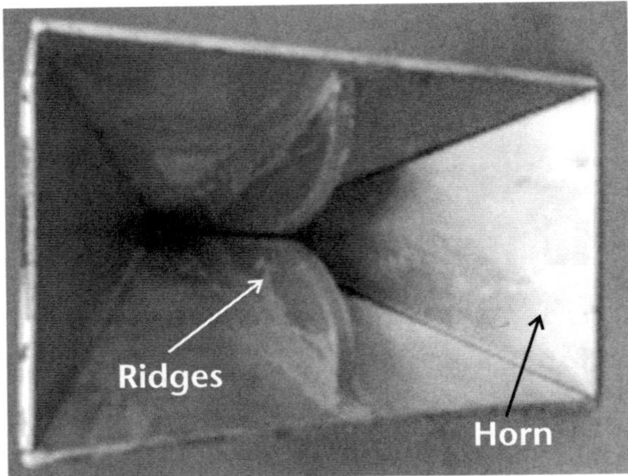



**Figure 7.47**   Manufactured version of exponentially tapered ridge UWB horn antenna.

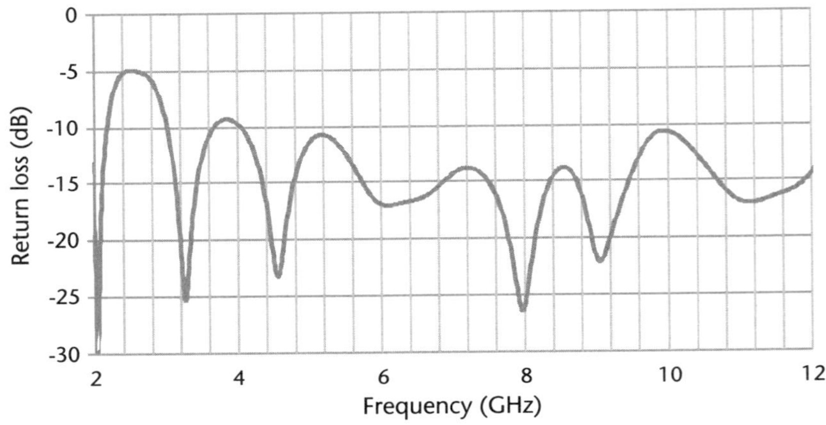

**Figure 7.48**   Input impedance return loss of exponentially tapered ridge horn antenna.

H-planes the antenna suffers a drop in gain. This reduced gain and sidelobes can be improved with the modification of side walls (ribbed side walls) on the horn antenna. Figure 7.50 shows CST-generated 3D radiation patterns for those frequencies.

Figure 7.51 shows the realized gain versus frequency plot. The realized (with 50Ω input impedance match) gain of the horn antenna varies from 5.5 to 10.5 dBi with a dip to 7 dBi at 9 GHz due to the spread of the mainlobe in the E-plane and the sidelobes in the H-plane.

A few commercial tapered ridged horn antennas with excellent antenna performance have been acquired by the authors' research group that work from 1 to 18, 18 to 40, and 50.0 to 75.0 GHz. The 1- to 18-GHz antenna suffers from similar sidelobe and gain issues at the higher end of the frequency band, whereas the other two types of antennas provide excellent directional radiation patterns and gain. More details on the antenna performance can be found at the manufacturer's website ( www.ainfoinc.com).

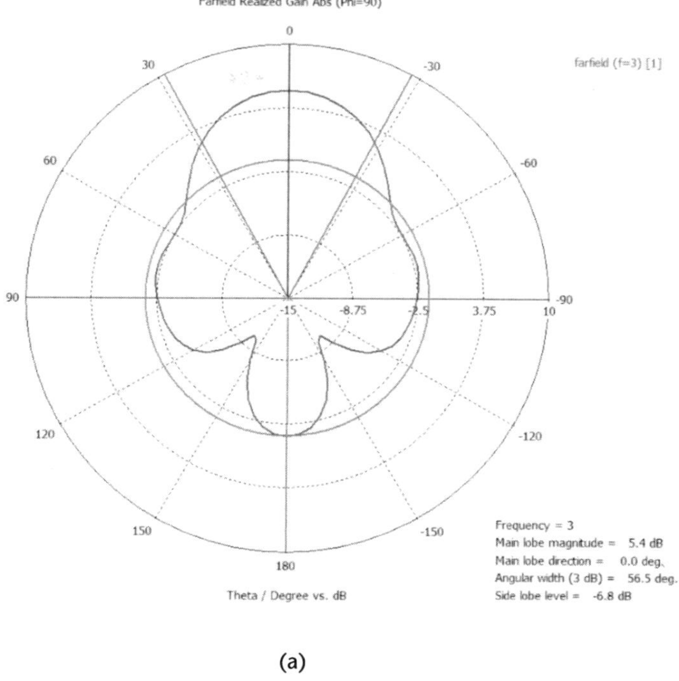


(a)

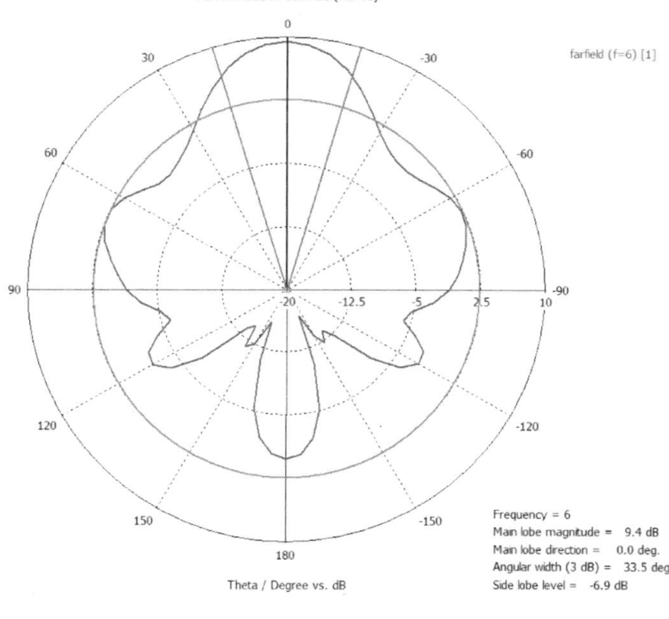


(b)

**Figure 7.49**   2D gain patterns of exponentially tapered ridged horn antenna: (a–) E-plane and (e–h) H-plane. (Frequencies are indicated in the plots.)

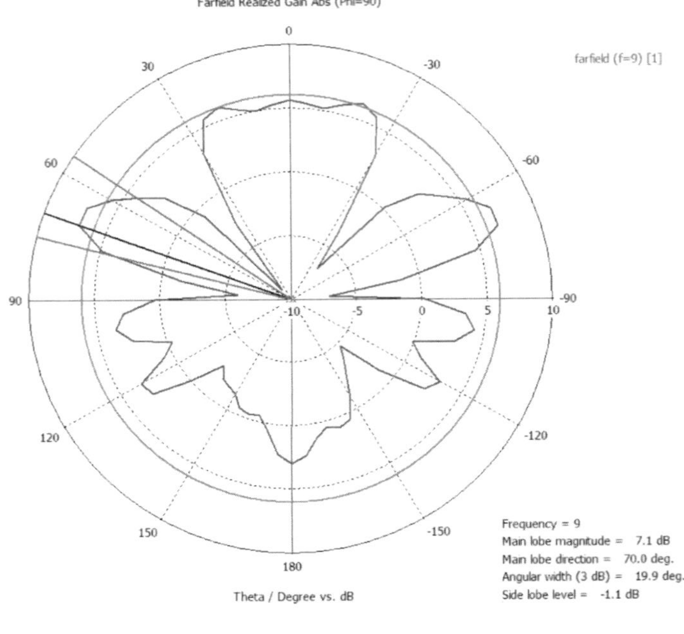



(c)

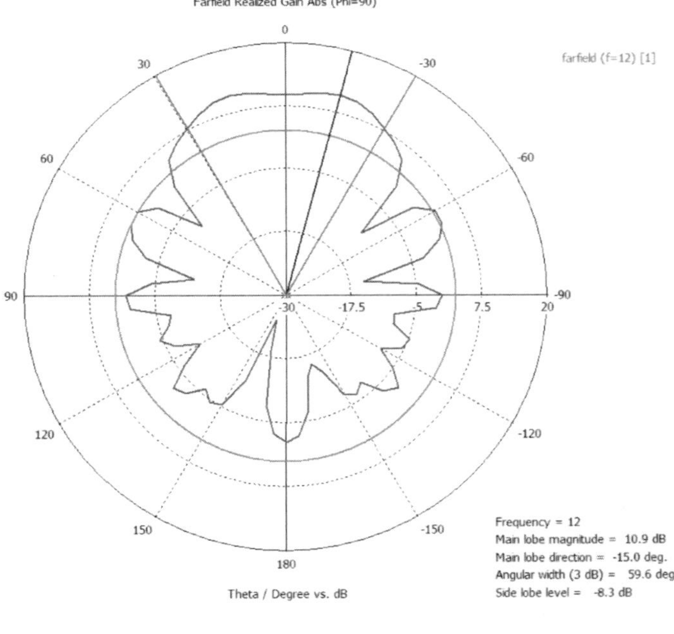



(d)

**Figure 7.49**   (continued)

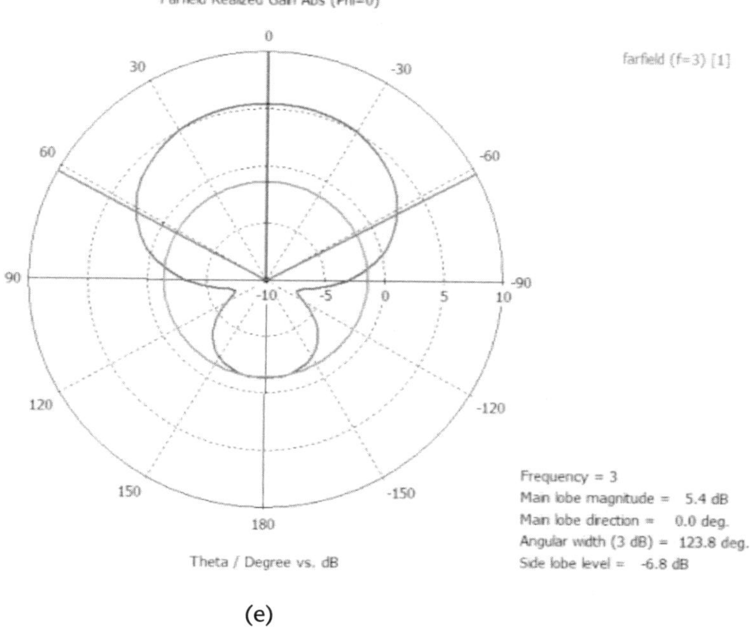


(e)

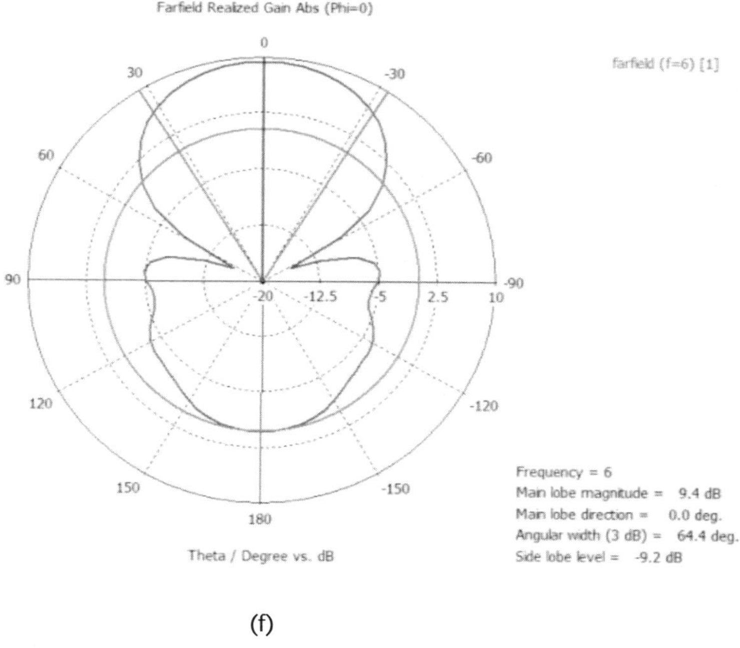


(f)

**Figure 7.49**   (continued)

### Conclusions for Horn Antennas

Though horn antennas are bulky, metallic, nonplanar antennas, due to their excellent radiation performance they are used in the laboratory setting to read chipless

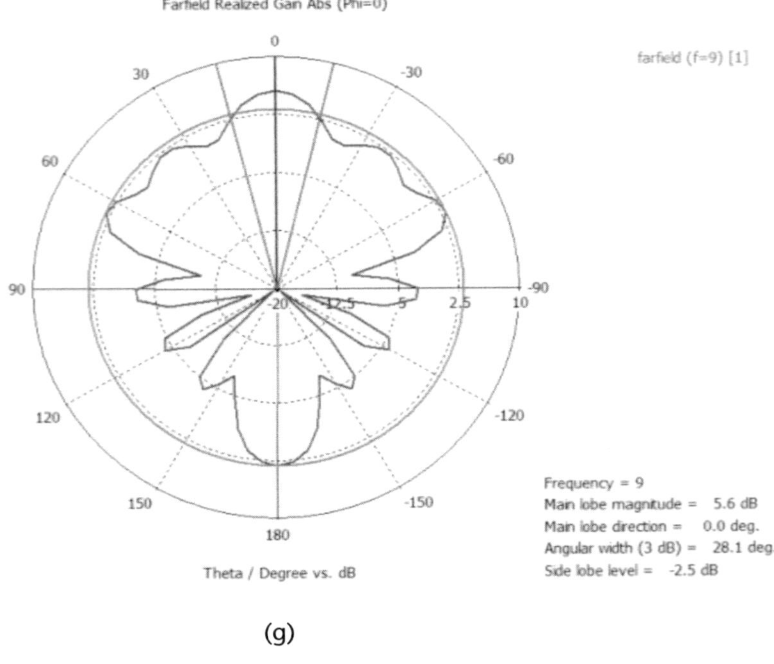



(g)

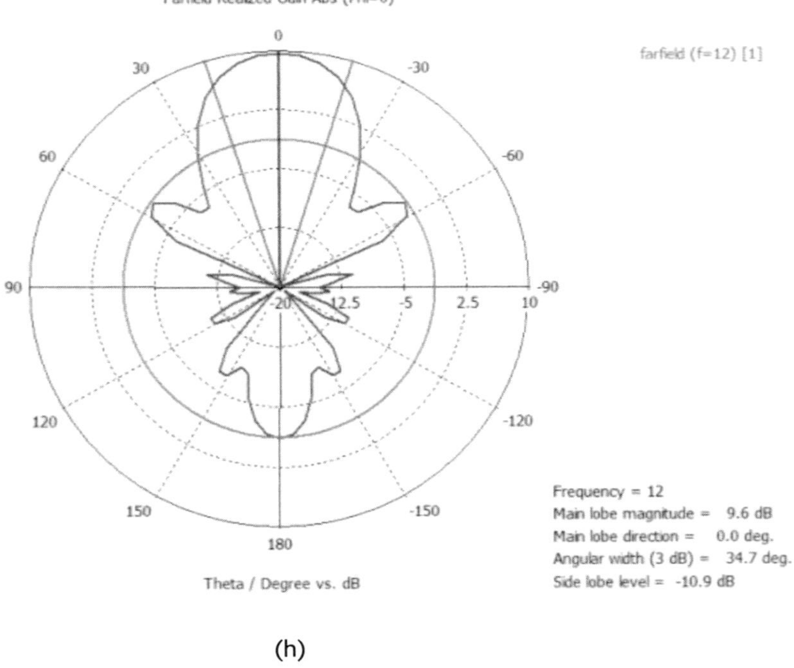



(h)

**Figure 7.49**  (continued)

RFID tags. The antennas enjoy UWB input impedance bandwidth, moderately high gain directional radiation patterns, extremely favorable isolation between the Tx and Rx chains, and a uniform gain profile over the frequency band of interest. These advantageous features of horn antennas result in error-free and robust readings of

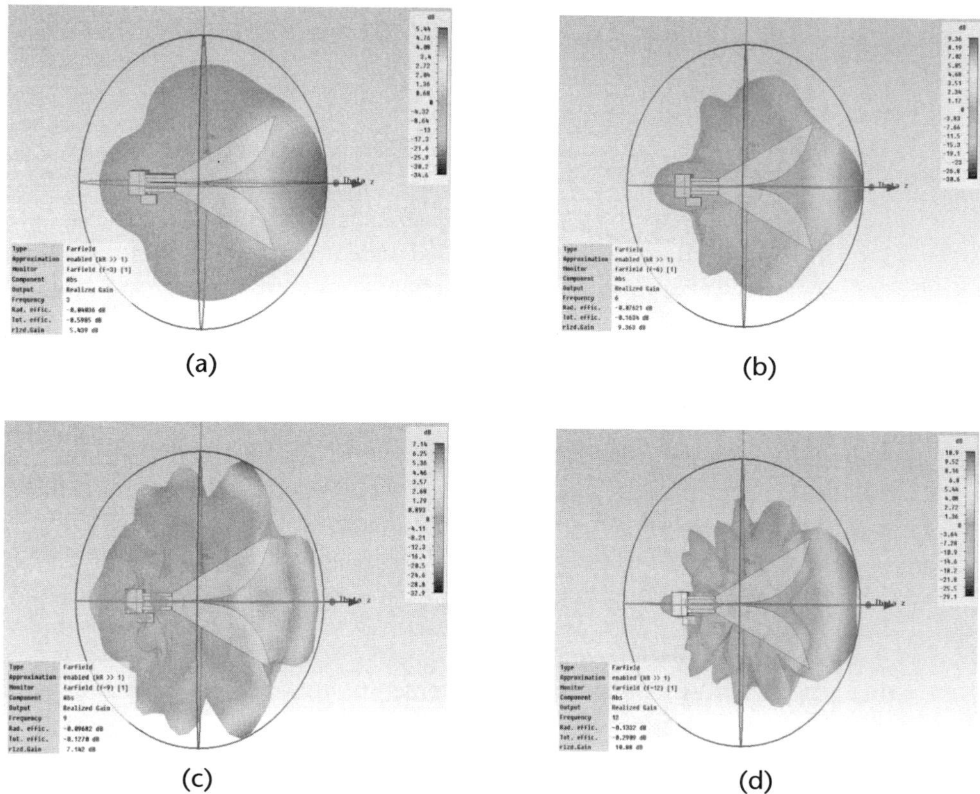

(a)

(b)

(c)

(d)

**Figure 7.50**   3D gain patterns of exponentially tapered ridged horn antenna. (Frequencies are indicated in the plots.)

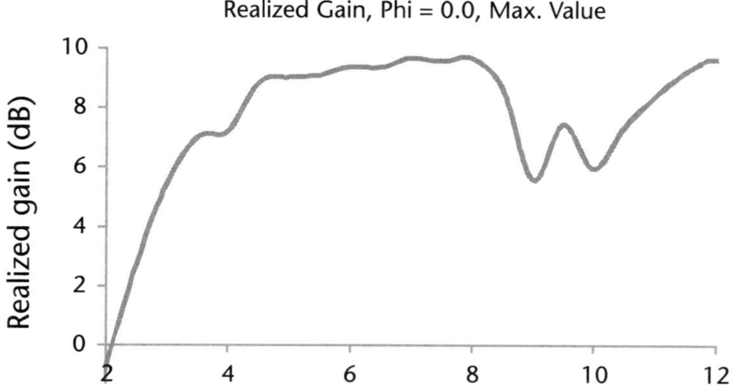



**Figure 7.51**   Realized gain versus frequency plot of exponentially tapered horn antenna.

chipless tags. A few examples of the use of horn antennas as a reader's antenna in laboratory settings can be found in [45, 50].

So far five different varieties of UWB antennas in both planar and nonplanar configurations have been discussed. The planar antennas are disk-loaded monopoles and LPDAAs and the nonplanar antennas are leaf dipole arrays, cavity backed LPDAs, and exponentially tapered ridged horn antennas. These antennas are used in chipless RFID reader systems to read multibit chipless RFID tags. In the following section applications for the UWB antennas that have been developed are described.

## 7.6  Applications

So far various chipless RFID reader antennas have been discussed. The antennas are used in reading the tags using microwave measuring equipment and readers built in-house. A few examples follow of the reader antennas developed and their uses in reading chipless RFID tags in various forms.

### 7.6.1  Disk-Loaded Monopoles as Reader Antennas

A disk-loaded monopole is the most basic antenna that can yield omnidirectional radiation patterns and a uniform return loss over the UWB band. Figure 7.52 shows the setup for measuring a retransmission based chipless RFID tag using an Agilent E8361A PNA at Monash University. As can be seen, two monopole antennas are connected with the two ports of the PNA. The chipless RFID is placed 1 to 2 cm above the antennas. The display of the PNA shows the frequency signature of the chipless tag. The four frequency resonances of the 4-bit chipless tag are clearly visible on the PNA's display. The monopole antenna can be used in a touch-and-go reader.

### 7.6.2  Elliptical Leaf Dipole Arrays as Reader Antennas

Figure 7.53 shows a similar arrangement for measuring chipless RFID tags using high-gain elliptical dipole arrays. To validate the use of the antenna with chipless

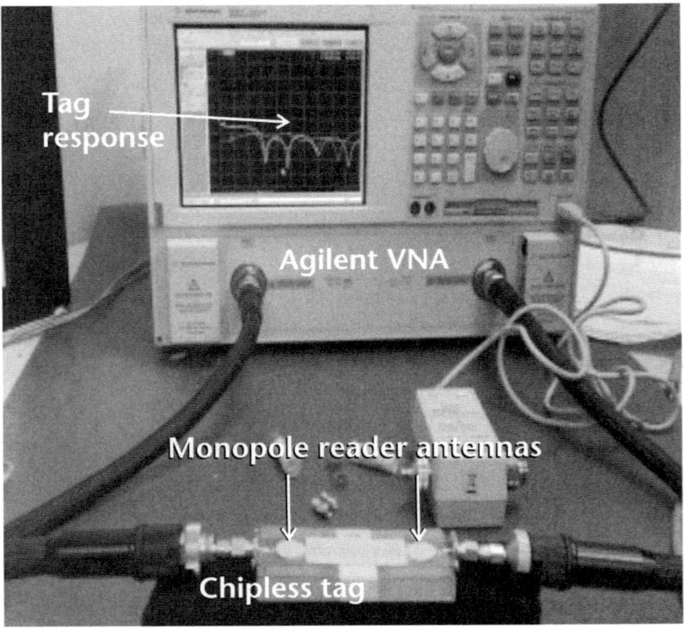


**Figure 7.52**  Disk-loaded monopoles as near-field chipless RFID reader antennas.

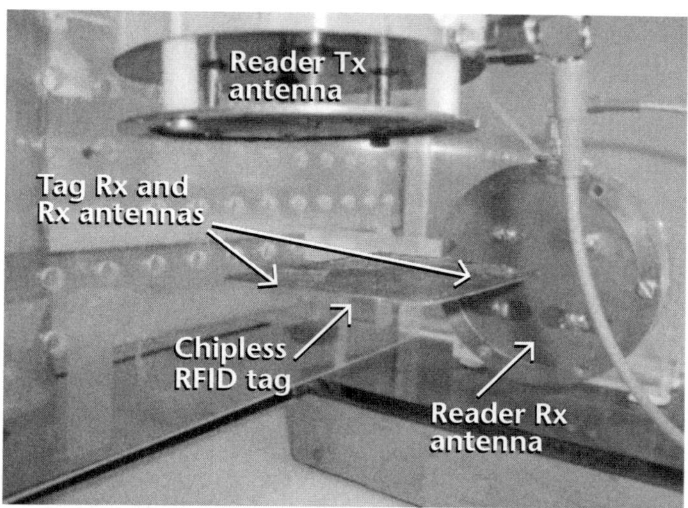


**Figure 7.53**   Experimental setup for reading chipless tags using dipole array antennas.

RFID tags, a chipless RFID tag reader built in-house was used [51]. The antenna was connected with the two antenna ports of the reader (not shown in Figure 7.53). This measurement shows the suitability of the antenna for use as a reader antenna on a chipless RFID tag reader.

### 7.6.3   LPDAAs as Reader Antennas

Figure 7.54 shows the use of a LPDAA to measure a retransmission based chipless RFID tag. Results similar to those of Figure 7.53 were achieved at a reading distance of up to 30 cm with the LPDAA reader antenna.

### 7.6.4   Horns as Reader Antennas

At Monash University the broadband horn antennas are also frequently used to read various chipless RFID tags. The experimental setup used to read retransmission based tags is detailed in [45] and hence not reproduced here.

## 7.7   Conclusion

The chapter presented a comprehensive study of various chipless RFID reader antennas. Classification of the form factors of the antennas based on nonplanar and planar antennas resulted in five varieties of broadband antennas for the chipless RFID reader systems. All of these antennas were developed at the Monash University RFID and Antenna Laboratory for their in-house chipless RFID reader systems [51]. The developed antennas cover the UWB microwave and millimeter-wave frequency bands. Before, introducing the actual design of these reader antennas, basic antenna parameters were defined. These parameters not only helped in developing fundamental understandings of antenna characteristics but also helped in deriving the design specifications for the antennas. In this regard, antenna gain

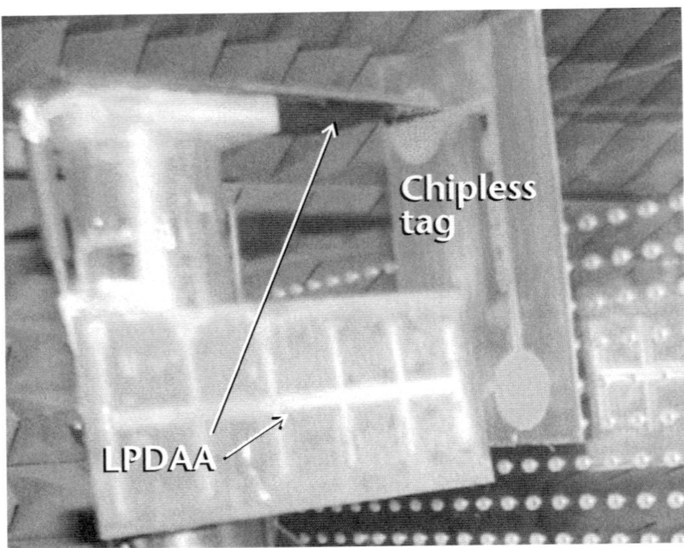


**Figure 7.54**   Experimental setup used for LPDAAs.

and directivity, radiation patterns, bandwidth, polarization, antenna driving point impedance, and array factor were defined.

The low-profile and planar antennas are high-Q passive devices hence the input impedance bandwidth is much smaller than the radiation characteristic bandwidth. Therefore, the antenna design process was initiated with the target of achieving the required input impedance return loss bandwidth. Once the return loss bandwidth was achieved, the radiation characteristics of the antenna were investigated. The surface current investigation using CAD design tools helped our understanding of the radiation mechanism of these antennas. It is not a trivial task to design a low-profile, compact, PCB based antenna with an octave bandwidth. In this regard, a practical design guideline for broadband antenna development was presented before introducing various developed RFID reader antenna types.

The various planar and nonplanar antenna types discussed were disk- and crescent-loaded monopole antennas, high-gain UWB elliptical-shaped dipole antenna arrays, moderate gain compact LPDAs and LPDAAs, and tapered ridged horn antennas. All designed antennas were developed from scratch. A thorough understanding of the characteristics of these antennas was achieved through parametric studies. These exercises helped in obtaining the optimum design for these antennas. Finally, the applications of these developed antennas to the task of reading various chipless RFID tags were presented.

An understanding of the diversity and appropriateness of antenna types for chipless and even conventional RFID readers will benefit design engineers and enrich the new field.


## Questions

1. Name three unlicensed ISM UWB bands.
2. What is an antenna? How does an antenna work?

3. In how many ways can an antenna be made reconfigurable? Explain.
4. What is a rectenna?
5. What are the broad classifications of antennas developed for chipless RFID readers?
6. Discuss the advantages and disadvantages of a fixed-beam antenna compared to a smart reader antenna. Explain why vendors do not make smart antennas for RFID readers.
7. Define important antenna parameters.
8. What is the solid angle of an antenna? From other sources (such as the Internet) find the relationship between antenna solid angle and directivity of antenna.
9. Explain the advantages of steering nulls for antennas. Devise some method of null steering.
10. Define the directivity and gain of an antenna. Get some idea about measuring gain from open sources.
11. What is an isotropic radiator? Does an isotropic radiator exist in reality? If not, name a few radiators that are similar to an isotropic radiator.
12. Define antenna bandwidth. Explain in how many ways antenna bandwidth can be defined.
13. Define polarization of an antenna. Explain the importance of polarization in wireless communication. How do you calculate the polarization loss factor (PLF)? If two antennas with the same polarization are oriented with an angle of 45°, what is the PLF?
14. Define elliptical, linear, and circular polarization.
15. Define axial ratio (AR) and AR bandwidth.
16. Explain why a designer initially looks for the antenna impedance bandwidth instead of the radiation pattern bandwidth.
17. Draw an equivalent circuit model of an antenna system with its symbols and associated circuits. Label all components in the system. Explain their functionality.
18. Explain the practical design procedure for a UWB antenna.
19. Explain how you emulate a UWB dipole with a disk-loaded monopole.
20. Name a few applications of low- and high-gain antennas.
21. Explain a few desirable features of a chipless RFID tag and\or a reader antenna. Does a disk-loaded monopole antenna meet all of these features?
22. Explain the design and operating principle of a disk-loaded monopole. How do you determine the fundamental mode frequency?
23. Explain the advantages of a crescent-loaded monopole antenna. What is the trade-off compared to a disk-loaded monopole?
24. For the practical design of a CPW plastic antenna, explain what precaution is needed to fabricate the antenna.
25. Explain the advantages of a dipole reflect array compare to a disk-loaded monopole antenna.
26. Explain why LPDAs and similar antennas cannot be used in UWB-IR based readers.

27. Explain how a reflector behind a dipole helps improve the performance of telecommunication systems.

28. Explain Babinet's and Rumsey's principles in achieving the frequency independence behavior of antennas.

## References


[1] S. Cataldo and N. C. Karmakar, "Efficient rectenna design for emerging applications," *Proc. 2008 Asia-Pacific Microwave Conference*, Hong Kong and Macau, December 16–20, 2008.

[2] N.C. Karmakar (Ed.), *Handbook of Smart Antennas for RFID Systems*, Hoboken, NJ: Wiley, , 2010.

[3] http://www.alientechnology.com/products/index.php (accessed November 27, 2012).

[4] http://www.omron.com.au/product_info/index.asp?catlvl=36 (accessed November 27, 2012).

[5] R. V. Koswatta, N. C. Karmakar, "A novel method of reading multi-resonator based chipless RFID tags using an UWB chirp signal," *Proc. 2011 Asia Pacific Microwave Conference*, Melbourne, Australia, December 5–8, 2011.

[6] R. V. Koswatta, N. C. Karmakar, "Time domain response of a UWB dipole array for impulse based chipless RFID reader," *Proc. 2011 Asia Pacific Microwave Conference*, Melbourne, Australia, December 5–8, 2011.

[7] C.A. Balanis, *Antenna Theory: Analysis and Design*, 2nd ed., New York: John Wiley & Sons, 1982.

[8] P. Darwood, P. Fletcher, and G. Hilton, "Mutual coupling compensation in small planar array antennas," *IEEE Proc. Microwaves, Antennas and Propagation*, vol. 145, no. 1, pp. 1–6, 1998.

[9] R. Pokuls, J. Uher, and D. Pozar, "Microstrip antennas for SAR applications," *IEEE Trans. Antennas and Propagation*, vol. 46, no. 9, pp. 1289–1296, 1998.

[10] H. G. Schantz, "A brief history of UWB antennas," *Proc. 2003 IEEE UWBST Conference*, 2003.

[11] D. Valderas, J. I. Sancho, and D. Puente, *Ultrawideband Antennas: Design and Applications*, London, UK: World Scientific, 2010, pp. 209–213.

[12] Z. N. Chen, "UWB antennas: Design and application," *Proc. 6th Int. Conf. Information, Communications & Signal Processing*, 2007, pp. 1–5.

[13] J. Liang et al., "Study of a printed circular disk monopole antenna for UWB systems," *IEEE Trans. on Antennas and Propagation*, vol. 53, no. 11, pp. 3500–3504, November 2005.

[14] J. Liang et al., "Study of CPW-fed circular discmonopole antenna for ultra wideband applications," *IEEE Proc. Microwaves, Antennas and Propagation*, vol. 152, no. 6, pp. 520–526, December 2005.

[15] G. Quintero and A. K. Skrivervik, "Analysis of planar UWB elliptical dipoles fed by a coplanar stripline," *IEEE Int. Conf. on Ultra-Wideband (ICUWB 2008)*, Hannover, Germany, September 2008, vol. 1, pp. 113–116.

[16] H. G. Schantz, "A brief history of UWB antennas," *IEEE Aerospace and Electronic Systems Magazine*, vol. 19, no. 4, pp. 22–26, 2004.

[17] J. Volakis, *Antenna Engineering Handbook*, 4th ed., New York: McGraw-Hill Professional, 2007.

[18] C. A. Balanis, *Antenna Theory: Analysis and Design*, 3rd ed., New York: Wiley-Interscience, 2005.

[19] J. Volakis, *Antenna Engineering Handbook*, 4th ed., New York: McGraw-Hill Professional, 2007.

[20]   Z. N. Chen and X. Qing, "Research and development of planar UWB antennas," in *2005 Asia-Pacific Microwave* Conference *Proceedings,* vol. 1, p. 4.

[21]   C. Harrison and C. Williams, "Transients in wide-angle conical antennas," *IEEE Trans. Antennas and Propagation,* vol. 13, no. 2, pp. 236–246, 1965.

[22]   S. Samaddar and E. Mokole, "Biconical antennas with unequal cone angles," *IEEE Trans. on. Antennas and Propagation,* vol. 46, no. 2, pp. 181–193, 1998.

[23]   Z. N. Chen et al., "Planar antennas," *IEEE Microwave Magazine,* vol. 7, no. 6, pp. 63–73, 2006.

[24]   P. Mayes, "Frequency-independent antennas and broad-band derivatives thereof," *Proc. IEEE,* vol. 80, no. 1, pp. 103–112, 1992.

[25]   T. Hertel and G. Smith, "On the dispersive properties of the conical spiral antenna and its use for pulsed radiation," *IEEE Trans. on. Antennas and Propagation,* vol. 51, no. 7, pp. 1426–1433, 2003.

[26]   H. Nazli et al., "An improved design of planar elliptical dipole antenna for UWB applications," *Antennas and Wireless Propagation Letters,* IEEE, vol. 9, pp. 264–267, 2010.

[27]   N. Agrawall, G. Kumar, and K. Ray, "Wide-band planar monopole antennas," *IEEE Trans. on. Antennas and Propagation,* vol. 46, no. 2, pp. 294–295, 1998.

[28]   B. Duffley et al., "A wide-band printed double-sided dipole array," *IEEE Trans. on. Antennas and Propagation,* vol. 52, no. 2, pp. 628–631, 2004.

[29]   X. H. Wu and Z. N. Chen, "Comparison of planar dipoles in UWB applications," *IEEE Trans. on. Antennas and Propagation,* vol. 53, no. 6, pp. 1973–1983, 2005.

[30]   H. G. Schantz, "Bottom fed planar elliptical UWB antennas," in *Proc. IEEE Conf. Ultra-Wideband Systems and Technologies,* 2003, pp. 219–223.

[31]   H. Schantz, "Planar elliptical element ultra-wideband dipole antennas," in *2002 Antennas and Propagation Society International Symposium,* vol. 3, p. 44, 2002.

[32]   H. G. Schantz, "Introduction to ultra-wideband antennas," in *IEEE Conference on Ultra-Wideband Systems and Technologies,* pp. 1–9, 2003.

[33]   Chun-Chi Lee et al., "Broadband printed-circuit elliptical dipole antenna covering 750 MHz–6.0 GHz," in *2008 International Conference on Microwave and Millimeter Wave Technology* (ICMMT 2008), vol. 3, pp. 1207–1209.

[34]   I. Murata, Y. Miyazaki, and N. Goto, *Design of Semi-Circle Type Bow-tie Antenna with Hole Slots for Broadband Electromagnetic Field Measurement Using FDTD Analysis,* Institute of Electronics, Information and Communication Engineers Technical Report, vol. 102, no. 506, pp. 27–32, 2002.

[35]   P. Cerny and M. Mazanek, "Optimized Ultra Wideband Dipole Antenna," in *18th International Conference on Applied Electromagnetics and Communications (ICECom 2005),* pp. 1–4.

[36]   X. N. Low, Z. N. Chen, and T. See, "A UWB dipole antenna with enhanced impedance and gain performance," *IEEE Trans. on. Antennas and Propagation,* vol. 57, no. 10, pp. 2959–2966, 2009.

[37]   Y. Ito et al., "Unidirectional UWB array antenna using leaf-shaped bowtie elements and flat reflector," *Electronics Letters,* vol. 44, no. 1, pp. 9–11, 2008.

[38]   A. Hees and J. Detlefsen, "Ultra broadband dual polarized dipole array with metallic reflector," in *2009 IEEE International Conference on Ultra-Wideband (ICUWB 2009),* pp. 744–747.

[39]   P. Lindberg et al., "Dual wideband printed dipole antenna with integrated balun," *IEEE Proc. Microwaves, Antennas and Propagation,* vol. 1, no. 3, pp. 707–711, 2007.

[40]   W. Wilkinson, "A class of printed circuit antennas," in *Antennas and Propagation Society International Symposium,* 1974, vol. 12, pp. 270–273.

[41]   G. Adamiuk et al., "Dual-polarized UWB antenna array," in *2009 IEEE International Conference on Ultra-Wideband (ICUWB 2009),* pp. 164–169.

[42]  Y. S. Seo et al., "H-shaped dipole array antenna for broadband operation," in *2010 IEEE International Conference on Wireless Information Technology and Systems (ICWITS 2010)*, pp. 1–4.

[43]  S. He and J. Xie, "Analysis and design of a novel dual-band array antenna with a low profile for 2400/5800-MHz WLAN systems," *IEEE Trans. on. Antennas and Propagation*, vol. 58, no. 2, pp. 391–396, 2010.

[44]  M. Gans, D. Kajfez, and V. Rumsey, "Frequency independent baluns," *Proc. IEEE*, vol. 53, no. 6, pp. 647–648, 1965.

[45]  S. Preradovic and N. C. Karmakar, *Multi-Resonator-Based Chipless RFID*, New York: Springer, 2012.

[46]  A. R. Mallahzadeh and A. Imani, "Double-ridged antenna for wideband applications," *Progress in Electromagnetics Research, PIER 91*, pp. 273–285, 2009.

[47]  A. Teggatz, A. Jöstingmeier, and A. S. Omar, "A new TEM double-ridged horn antenna for ground penetrating radar applications," available online: http://duepublico.uni-duisburg-essen.de/servlets/DerivateServlet/Derivate-14694/Final_papers/GM0023-F.pdf

[48]  M. Ghorbani and A. Khaleghi, "Wideband double ridged horn antenna: Pattern analysis and improvement," in *Proc. 5th European Conference on Antennas and Propagation (EU-CAP)*, April 11–15, 2011, pp. 865–868.

[49]  C. Bruns, P. Leuchtmann, and R. Vahldieck, "Analysis and simulation of a 1–18-GHz broadband double-ridged horn antenna," *IEEE Trans. on Electromagnetic, Compatibility*, vol. 45, no. 1, February 2003, pp. 55–60.

[50]  S. Preradovic et al., "Multiresonator-based chipless RFID system for low-cost item tracking," *IEEE Trans. on Microwave Theory & Techniques*, vol. 57, no. 5, part 2, pp. 1411–1419, 2009.

[51]  R. V. Koswatta and N. C. Karmakar, "A novel reader architecture based on UWB chirp signal interrogation for multiresonator-based chipless RFID tag reading," *IEEE Trans. on Microwave Theory and Techniques*, vol. 60, no. 9, pp. 2925–2933, September 2012.

## CHAPTER 8

# Microwave and Millimeter-Wave Active and Passive Components

## 8.1 Introduction

In previous chapters we have discussed chipless RFID technology, principles of operation, and system-level designs. Chapters 2 and 3 described the state of the art of chipless RFID tags and chipless RFID readers. In Chapter 4, two reader designs were presented that operate in the frequency domain, whereas Chapter 5 presented a concept that operates in the time domain. All of the chipless RFID systems described so far in this text use a range of passive and active components that operate in the microwave or millimeter-wave frequency bands.

This chapter presents information on various passive and active microwave and millimeter-wave components used in chipless RFID systems. The chipless RFID system is a collection of passive and active microwave components that is arranged to perform wireless characterization of a passive tag operating in the microwave or millimeter-wave frequency bands. In this chapter the basic operations of different microwave and millimeter-wave components are introduced and the applications of those components to chipless RFID systems are discussed. Many books provide in-depth knowledge about the various microwave passive and active components, so the purpose of this chapter is not to impart complete and thorough knowledge about these types of components, but to introduce how to use them in chipless RFID systems.

Figure 8.1 shows a classification of different passive and active components used in chipless RFID systems. The classification is done for chipless RFID tags as well as readers. As mentioned in previous sections, the chipless RFID tags used in this research work are fully passive and use no active microwave or millimeter-wave components. A detailed description is given for each component mentioned in the figure.

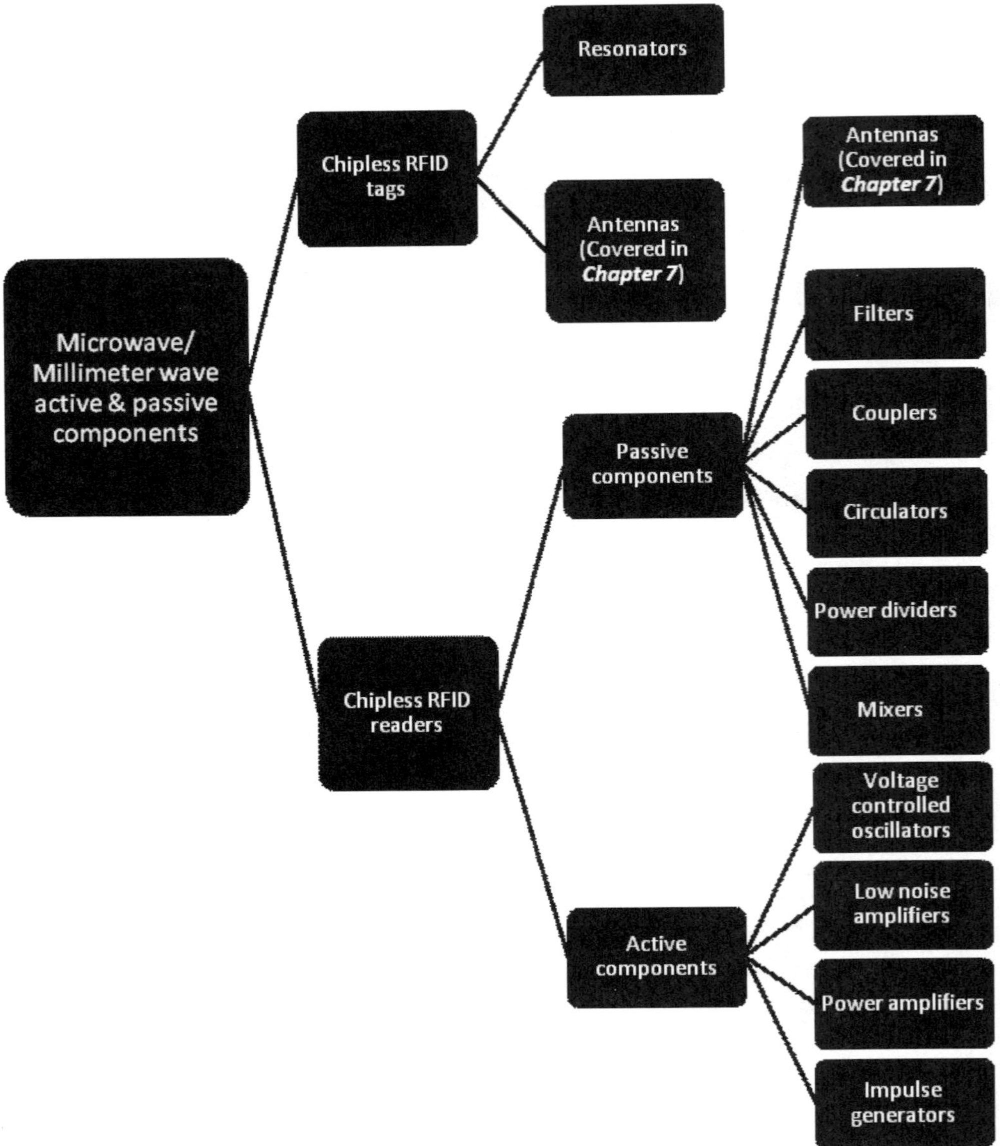


**Figure 8.1**   Classification of microwave/millimeter-wave components used in chipless RFID system.

## 8.2  Passive Components

### 8.2.1  Antennas

Antennas are used in chipless RFID tags as well as readers. Almost all chipless RFID tags or readers described in this book use either one or more antennas. Based on the requirements of the chipless RFID system, the properties of the antenna may change from one antenna to another. However, the function is still the same: either receiving or transmitting electromagnetic waves from or onto the free space. As discussed in Chapter 7, an antenna is a passive component that converts a wave propagating on a transmission line or a cable into a plane wave propagating in free space and receives waves propagating in free space and converts back to propagation in a transmission line.

A wide variety of antennas with varying properties are available. A detailed description of several tag and reader antennas that operate in the microwave and millimeter-wave frequencies and are used in chipless RFID tags and readers was given in Chapter 7. Several designs for antennas that can be used in chipless RFID tags and readers were also presented in Chapter 7.

### 8.2.2   Filters

Microwave filters are two-port networks used to control the frequency response at a certain point in a microwave system. The frequency range that is allowed to travel through the filter is called the passband and the other frequencies, which are attenuated, are called the stopband of the filter. The filters have four typical frequency responses [1]:

- Lowpass
- Highpass
- Bandpass
- Band-reject (band-stop) filters.

One or more of these types of filters are used in almost all microwave systems. Filters are used in both chipless RFID tags and readers. More information on the types of filters used in chipless RFID tags can be found in Chapter 2. This section, therefore, provides information mainly on the filters that are used in chipless RFID readers.

Filter design involves many theoretical approaches: network theory and synthesis, distributed circuit theory, which describes the operation of planar filters, and electromagnetic theory. As well as the theory, both practical and technological aspects must be considered. The limited space of this chapter is sufficient only to introduce the basic operation of different filter categories and the application of filters.

The theoretical foundation of the filter design method is derived from the theory of lumped electrical networks. This approach supplies the methods to identify the physically realizable networks together with the mathematical formulas to synthesize them. The general representation of a filter is shown in Figure 8.2. Here, the filter is considered to be a two-port, loss-free device placed between resistive

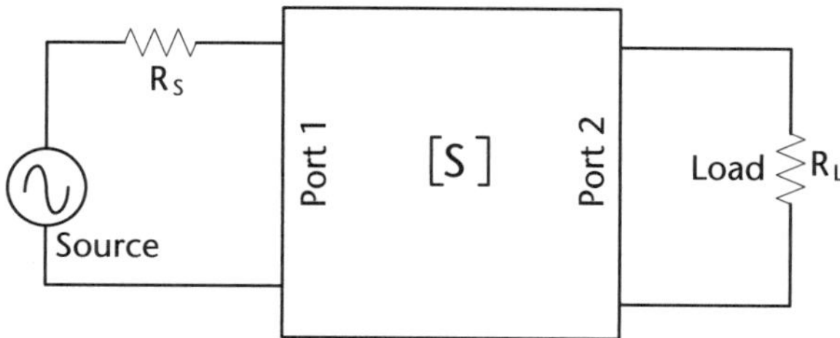


**Figure 8.2**   General block diagram of a filter (two-port network representation).

loads and a source with an internal resistive impedance of $R_L$ and $R_S$, respectively. The filter has a function of frequency-dependent transfer of power. As mentioned earlier, the frequency band that provides high attenuation is called the stopband, and the frequency band that provides low attenuation is called the passband. The filter can be characterized by means of the forward transmission coefficient or scattering parameter $S_{21}$ [2]. The four types of filters classified according to the filter responses are described next.

### 8.2.2.1 Lowpass Filters

The ideal lowpass filter has zero attenuation in the passband and infinite attenuation in the stopband. In other words, in the passband, the filter provides 0-dB power gain and $-\infty$-dB gain in the stopband. Five functions can be used to approximate this ideal filter function mathematically: Butterworth, Chebyshev, Cauer, Bessel, and inverse Chebyshev approximations [2]. Each type of approximation has its advantages and disadvantages, which have been well described [1, 2]. Once a suitable filter function has been selected, the synthesis is done either using lumped elements or distributed inductors and capacitors with pieces of transmission lines.

The lumped element filter design generally works for low frequencies because the lumped elements such as inductors and capacitors are available in a limited range of values (Figure 8.3). Generally, they are difficult to implement in microwave frequencies as well. In addition to that, in microwave frequencies, the distances between filter components are comparable with the wavelengths of the operating frequencies. Richard's transformation [1] is one technique used to convert lumped elements into transmission line sections in a microwave circuit. Kuroda's identities are used to separate filter elements by using transmission line sections [1, 2]. Therefore, the difficulties that arise when attempting filter implementation in the higher frequencies of the microwave frequency band can be avoided using this technique.

Using this technique, lowpass filters are implemented using pieces of transmission lines as shunt stubs or stepped-impedance techniques. Lowpass filter implementation using shunt stubs involves loading a series of stubs in a piece of microstrip transmission line section [2, 3] as shown in Figure 8.4(a). The stepped-impedance method of lowpass filter implementation uses alternating transmission line sections with very high and very low characteristic impedance. The stepped-impedance method of implementation of lowpass filters is more popular than the shunt-stub method because it consumes less space in the RF PCB. Figure 8.4(b) shows an example layout for a stepped-impedance lowpass filter.

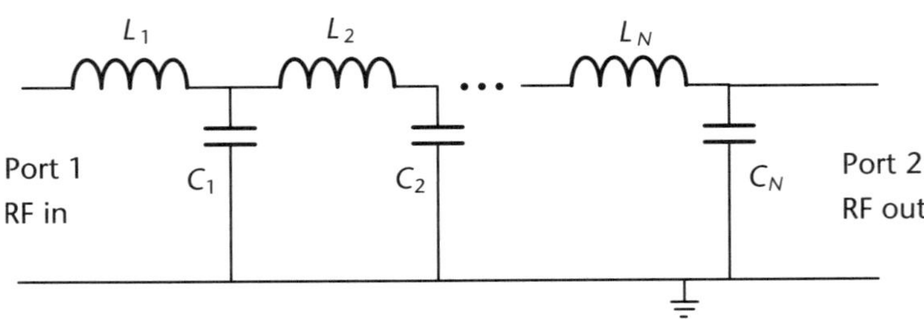


**Figure 8.3**  Lumped element equivalent circuit of lowpass filter.

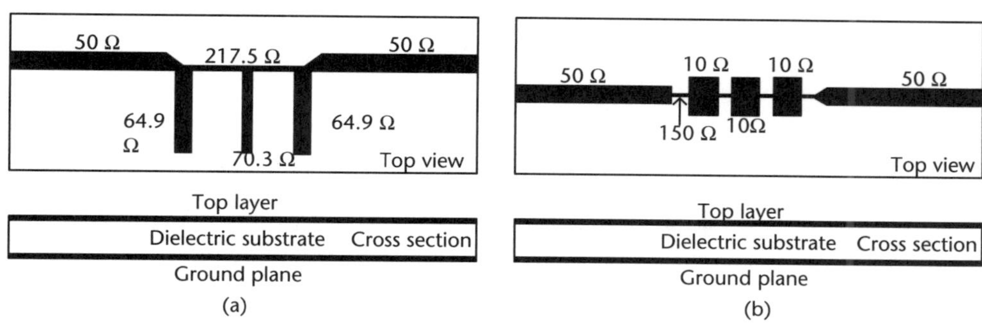

Figure 8.4  (a) Shunt-stub implementation of a lowpass filter and (b) stepped-impedance implementation of a lowpass filter [1].

Figure 8.5 shows a lowpass filter [4] implemented with lumped elements that has a cut-off frequency of 800 MHz. The figure shows a filter chip on a chipless RFID reader board. The measured forward transmission coefficient of the filter ($S_{21}$ parameters) is shown in Figure 8.6.

Other types of filters can be achieved by applying different frequency transformation to the lowpass filters. The application of frequency transformation to lowpass filters has been described extensively [2, 3]. The following sections present brief discussions about other types of filters.

### 8.2.2.2  Highpass Filters

By applying highpass transformation to the lowpass filter design equations, we can achieve highpass filters as described in [2, 3]. The equivalent lumped element circuit of a highpass filter is shown in Figure 8.7. Figure 8.8 shows a lumped element implementation of a commercially available highpass filter that has a lower cut-off frequency than the similar lowpass filter of 730 MHz [5]. The filter is implemented on a package similar to the one shown in Figure 8.5.

Figure 8.5  Mini-Circuits LFCN630+ lumped element lowpass filter on a PCB.

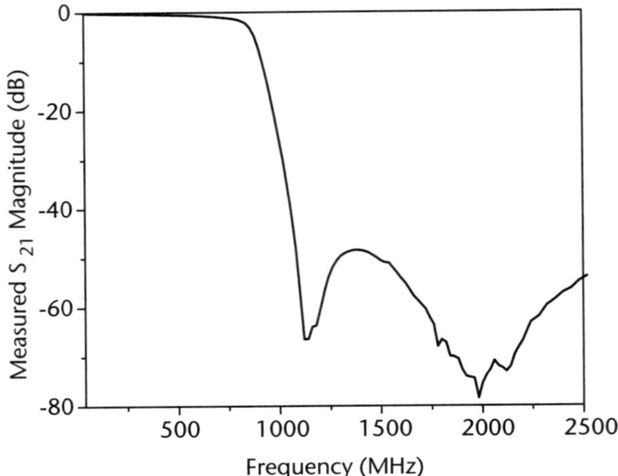

**Figure 8.6** Measured $S_{21}$ parameter magnitude of a Mini-Circuits LFCN630+ lumped element low-pass filter.

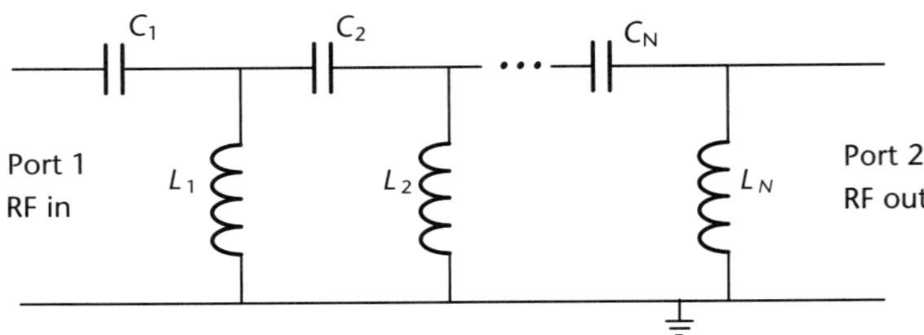



**Figure 8.7** Lumped element equivalent circuit of highpass filter.

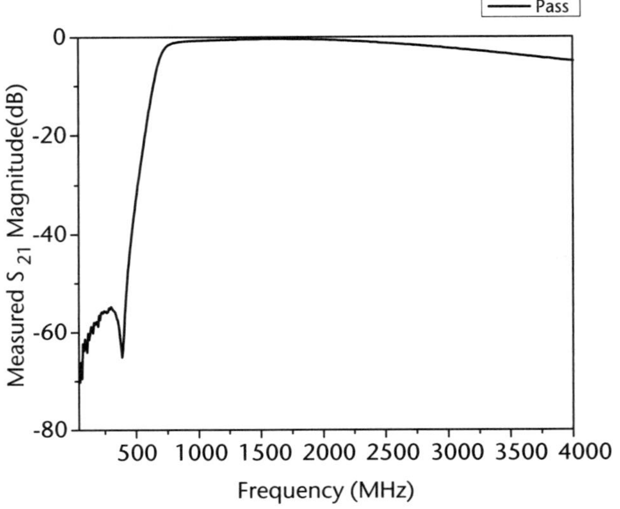



**Figure 8.8** Measured $S_{21}$ parameter magnitude of a Mini-Circuits HFCN630+ lumped element highpass filter.

### 8.2.2.3   Bandpass Filters

Similar to the highpass filter implementation using the frequency transformation method, we can also achieve bandpass filters [2, 3]. The lumped element equivalent circuit of a bandpass filter is presented in Figure 8.9. Similar to the previous two types of filters, for lower frequencies, it is possible to achieve the required frequency bands and performance with lumped inductors and capacitors. When the frequency increases, it is difficult to achieve satisfactory performance of the filters with lumped element implementations.

To address this issue, similar to the lowpass filter implementations with pieces of transmission lines, coupled lines and coupled resonators are used. Narrow bandpass filters can be made using coupled transmission lines. Layouts of two coupled line narrow bandpass filters are shown in Figure 8.10. Coupled line filters are very popular in RF PCBs, since the same PCB substrate can be used to implement the filter without any lumped element. Figure 8.11 shows the measured transmission coefficient of a lumped element implementation of a commercially available bandpass filter [6]. The filter has a passband of 4.2 to 4.7 GHz as shown in Figure 8.11. The filter is packaged in a manner similar to that for the other lumped element lowpass and highpass filters [4, 5].

In microwave receiver sections, sometimes a particular frequency band or a single frequency must be selected from the received signal. Therefore, bandpass filters are widely used in microwave receivers.

### 8.2.2.4   Band-Reject Filters (Band-Stop Filters)

Band-stop filters are used to filter out a particular frequency band from an RF signal. The lumped element equivalent circuit of a band-stop filter is shown in Figure 8.12. Similar to the highpass filters and bandpass filters, it is possible to design band-stop filters by applying a frequency transformation to the lowpass filter design equations. Typically, band-reject (band-stop) filters are used to remove narrowband interferences in RF or microwave systems. Band-stop filters are extensively used in some chipless RFID tag designs. The multiresonator based chipless RFID tags discussed in this book use narrowband band-stop filters to encode data.

As described in previous chapters, the band-stop filter is implemented by loading a microstrip transmission line with a planar spiral. The equivalent circuit for

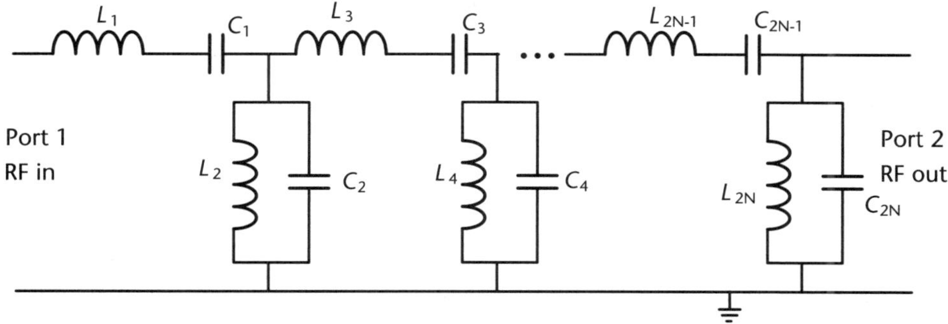

**Figure 8.9**   Lumped element equivalent circuit of bandpass filter.

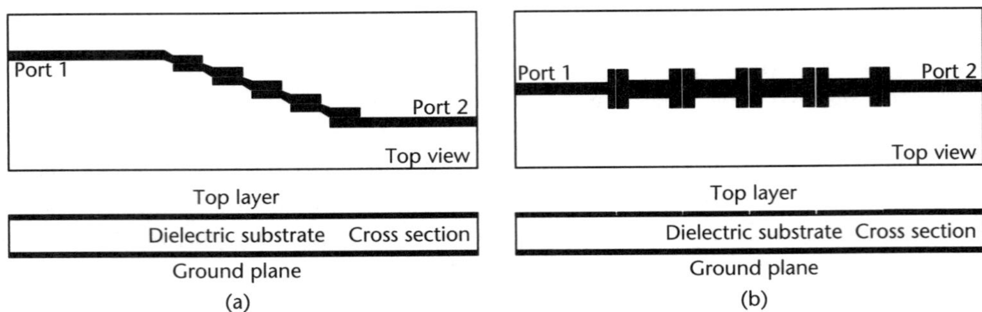



**Figure 8.10**   (a) Coupled line implementation [1] and (b) coupled resonator implementation [2] of two narrow bandpass filter layouts.

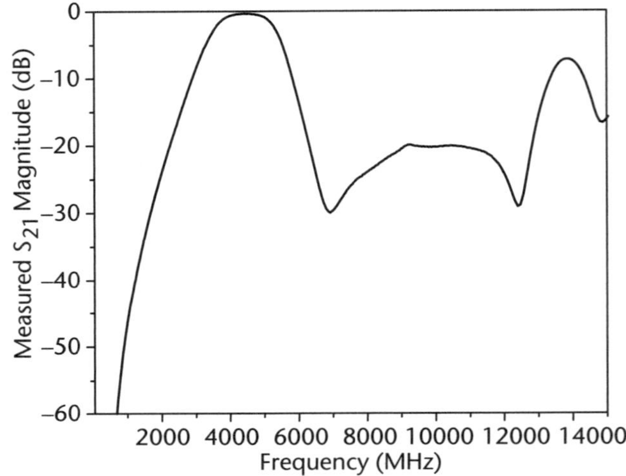



**Figure 8.11**   Measured $S_{21}$ parameter magnitude of a Mini-Circuits BFCN4440+ lumped element bandpass filter.

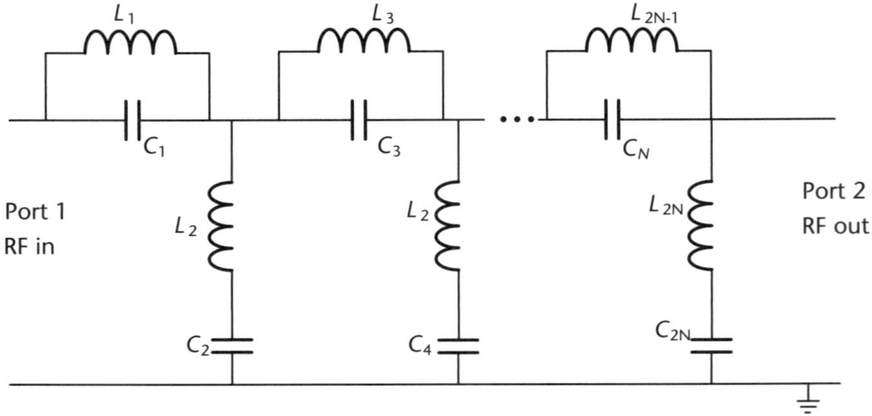



**Figure 8.12**   Lumped element equivalent circuit of band-reject (band-stop) filter.

the spiral loaded microstrip line can be found in [7, 8]. Because the measured transmission coefficient of multiresonator based chipless RFID tags are presented many times in various places of this book, the measured results of a band-stop filter is not presented in this section. Figure 8.13 presents a spiral loaded microstrip band-stop filter section used in a multiresonator based chipless RFID tag.

### 8.2.3   Couplers

Couplers are another passive microwave component that is used for power division. There are three-port and four-port couplers. Directional couplers and hybrid couplers [1] can be made to be three or four port according to the requirements of the application. Directional couplers are widely used in applications where unequal or arbitrary power division is needed. Hybrid couplers have equal power division but the phase of the two outputs differs in 90° or 180°. Waveguide-type couplers are also available and used in a wide variety of applications. However, in chipless RFID systems, waveguide-type devices have very limited or no usability due to the requirement of planar circuits or compact size. Therefore, waveguide-type couplers are not discussed here and only planar couplers are described in this section.

Commonly used circuit symbols and the basic operation of a directional coupler are shown in Figure 8.14. The port notation and the naming of the ports denote the flow of power supplied to the coupler and its basic operation. Power supplied to the port is divided among the "through port" and "coupled port." The coupled port has a portion of its power supplied to port 1 and it is denoted as a coupling factor. The remainder passes through port 2, the "through port." In an ideal directional coupler, there is no power delivered from port 1 to port 4. Therefore, it is referred to as the isolated port. In practice, however, the isolated port has isolation of −40 dB or less. When the directional coupler is used for unequal power division, the isolated port is usually terminated with a matched load. Then only

**Figure 8.13**   Spiral loaded microstrip line band-stop filter section used in a multiresonator based chipless RFID tag.

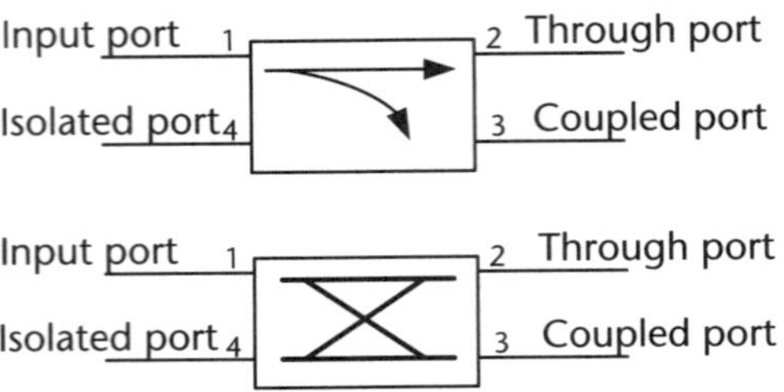


**Figure 8.14**  Widely used circuit symbols for directional couplers.

three terminals are available for use. In most of the commercially available directional couplers, this isolated port is not accessible and it is terminated internally with a matched load as mention before.

These kinds of couplers are reciprocal and are used to isolate the forward and reflected signal paths of RF systems. When the coupler is fed from port 2, port 1 becomes the through port and port 3 becomes the isolated port. Therefore, port 4 becomes the coupled port for the input power from port 2, and the power input from port 2 is not coupled to port 3. Therefore, a directional coupler can be used to identify two signals that travel in opposite directions in a signal path such as a forward traveling signal and a reflected signal. In practice, however, due to the limitations in isolation between ports 1 and 4, it is difficult to separate very weak reflected signals.

A hybrid coupler is a special type of directional coupler that has a coupling factor of 3 dB. However, the coupled port has a 90° phase shift with respect to the through port.

Figure 8.15 shows a photo of a coupler used to extract a portion of a transmitted signal in the chipless RFID reader discussed in Chapter 4. The input port is connected to the output of the VCO. The through port is directly connected to the RF out SMA connector. The isolated port is terminated with a 50Ω resistor as shown in the figure. The coupled port is attached to a mixer to generate the IF to use in the detector section of the chipless RFID reader.

**Figure 8.15**  Coupler used to extract a portion of the transmitted signal in the chipless RFID reader discussed in Chapter 4.

### 8.2.4   Circulators

A circulator is a three-port passive device. The circuit symbol of a circulator is shown in Figure 8.16(a). The signal entering at port 1 is transferred to port 2. The signal entering at port 2 is transferred to port 3. As shown in Figure 8.16(b), in an ideal circulator, port 3 is isolated from port 1 and there is no signal flow in the direction of port 2 to port 1. In a real-world circulator device, however, typical isolation levels are on order of 20 to 30 dB between the ports.

The most common application of a circulator is to use it in RF transceivers to share a single antenna between a transmitter and a receiver section. As shown in Figure 8.17(a), the transmitter and receiver sections are connected to port 1 and port 3, respectively. The transmitted signal travels toward the antenna as shown in the figure. The received signal that travels from the antenna toward the circulator is transferred to port 3. In practice, however, using a circulator to share a single antenna works for systems that have strong transmitted and received signals. Due to the limited isolation levels between ports 1 and 3, the weak received signals that are transferred from port 3 to 2 are suppressed by the signals coupled from port 1. An example application of a circulator that shares a single antenna between the transmitter and receiver of a chipless RFID reader is shown in Chapter 4, Section 4.3.3. Another application of a circulator is shown in Figure 8.17(b). As shown in the figure, when port 3 is terminated with a matched load, the circulator acts as an isolator that blocks the reverse transmission of the signal from port 2 to port 1.

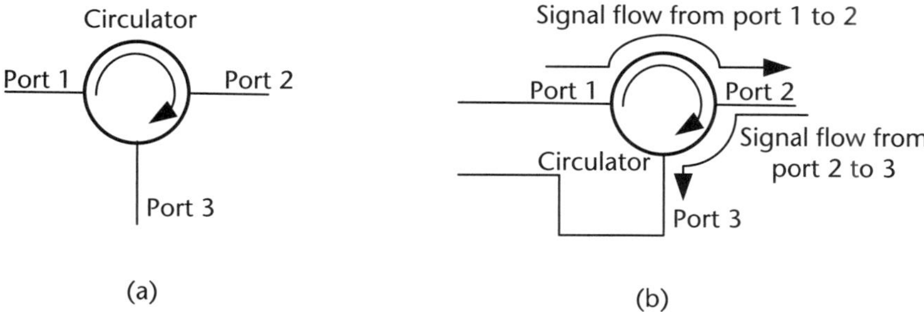


**Figure 8.16**   (a) Circuit symbol of a circulator and (b) direction of signal flow between the ports of the circulator.

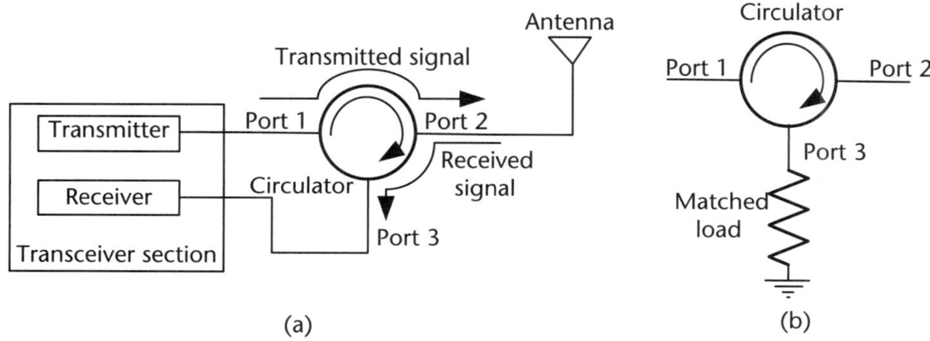


**Figure 8.17**   Applications of circulator: (a) using a circulator to share a single antenna between an RF transmitter and receiver and (b) using a circulator to block the reverse transmission of a signal.

### 8.2.5  Power Divider/Combiner

In RF transceiver systems, performing a linear combination of different signals is a common practice. The subdivision (into equal or unequal parts) of a signal into different components or the combining of signals from different sources into a single one are considered to be linear combinations of signals. The most common power dividers (or splitters) or combiners are the Wilkinson power splitter/combiner and the T-junction power divider/combiner. In this section, the operation of these two types of power dividers and their applications in chipless RFID systems and other systems are discussed.

#### 8.2.5.1  Wilkinson Power Divider

The Wilkinson power divider [9] was invented by E. J Wilkinson. The simplest form is a three-port passive circuit that is capable of dividing the power incident from port 1 into two equal portions between ports 2 and 3 such that all three ports are matched. The most important salient property of the Wilkinson power divider is the high isolation between output ports 2 and 3. However, the T-junction power divider suffers from low isolation at the output ports. Therefore, one output port does not affect the other port due to mismatch or any other phenomenon such as signals traveling in an opposite direction due to other components attached to the ports.

Wilkinson power dividers can be further classified into single-section and multisection dividers. As shown in Figure 8.18(a), the simplest form of the Wilkinson power splitter consists of two quarter-wavelength ($\lambda/4$) sections and one resistor connected between two output ports. The single-stage power splitter has narrowband operation since the design is done based on the quarter-wavelength at a particular frequency. As shown in Figure 8.18(a), just after the input port there is a T-junction splitter that divides the RF power into two equal parts, and the impedance of output lines is kept at $2Z_0$ to obtain good power division and better input matching. Using the quarter-wavelength transformer lines, the $2Z_0$ load impedance is transformed into $Z_0$. All three ports are matched and the $2Z_0$ resistor is applied to provide good isolation between output ports and to minimize reflections [9].

The main problem with the single-stage Wilkinson power divider is its narrow bandwidth performance, which is not suitable for the operating frequency of UWB chipless RFID applications. Several techniques have been used for improving the

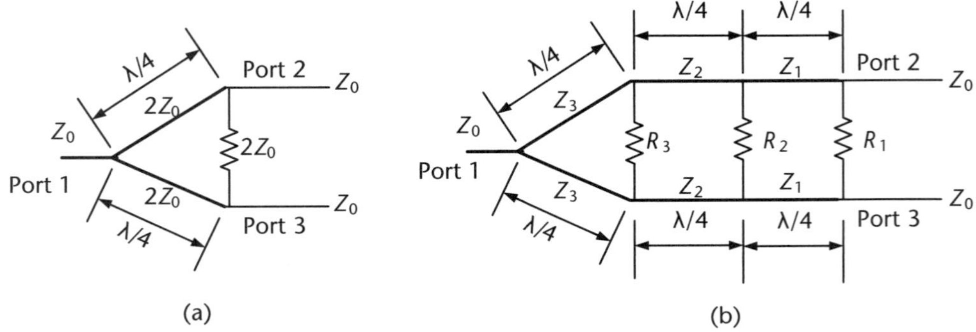

**Figure 8.18**  Schematic diagrams of (a) single-stage and (b) three-stage Wilkinson equal power splitters.

bandwidth of Wilkinson power splitters. Parad and Moynihan [10] suggest using one fixed resistor and two quarter-wavelength sections of strip lines with different impedances. The results of adding stubs to the quarter-wavelength transformer section have been reported [11,12]. Using even-mode and odd-mode analysis, Cohn [13] shows that the use of multiple fixed resistors with multiple sections of quarter-wavelength strip lines can improve the bandwidth of operation to a greater degree. A schematic diagram of his suggestion for the $N$-section Wilkinson power divider is shown in Figure 8.19. Design equations are given up to the two-stage splitter; due to the complexity of the system when the number of stages is greater than two, the method to be followed to calculate design parameters is shown and the calculated values are given in Table 8.1. An extract of the design impedance and resistance values for a three-stage Wilkinson power divider is also given in Table 8.1. In this table, normalized resistance and impedance values are given, and $f_1$ and $f_2$ are the lower and higher frequencies of the operating bandwidth, respectively.

$$R_1 = \frac{2R_2\left(Z_1 + Z_2\right)}{R_2\left(Z_1 + Z_2\right) - 2Z_2} \tag{8.1}$$

$$R_2 = \frac{2Z_1 Z_2}{\sqrt{\left(Z_1 + Z_2\right)\left(Z_2 - Z_1 \cot^2 \varnothing_3\right)}} \tag{8.2}$$

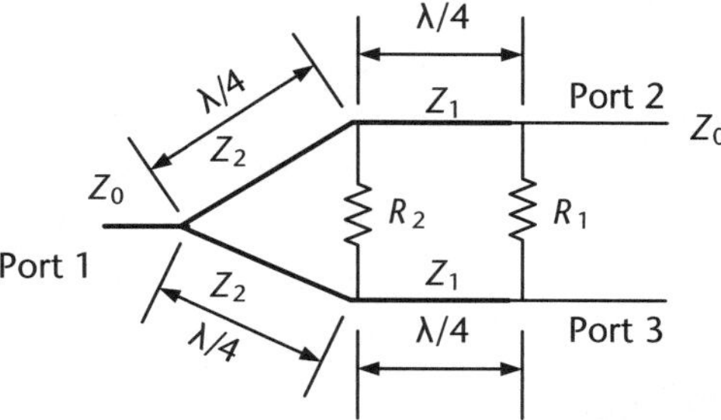


Figure 8.19   Schematic and design equations for a two-stage Wilkinson power divider [5].

Table 8.1   Calculated Design Parameters for Two- and Three-Stage Wilkinson Power Dividers [13]

| $N$ | 2 | 2 | 3 | 3 |
|---|---|---|---|---|
| $f_2/f_1$ | 1.5 | 2.0 | 2.0 | 3.0 |
| $Z_1$ | 1.1998 | 1.2197 | 1.1124 | 1.1497 |
| $Z_2$ | 1.6670 | 1.6398 | 1.4142 | 1.4142 |
| $Z_3$ | — | — | 1.7979 | 1.7396 |
| $R_1$ | 5.3163 | 4.8204 | 10.0000 | 8.0000 |
| $R_2$ | 1.8643 | 1.9602 | 3.7460 | 4.2292 |
| $R_3$ | — | — | 1.9048 | 2.1436 |

$$\varnothing_3 = 90° \left[ 1 - \frac{1}{\sqrt{2}} \left( \frac{f_2/f_1 - 1}{f_2/f_1 + 1} \right) \right] \tag{8.3}$$

where $Z_1$ and $Z_2$ are selected according to the required bandwidth. For $f_1/f_2 = 1.5$, $Z_1 = 1.213Z_0$ and $Z_2 = 1.639Z_0$.

Power dividers are often required for feeding antenna array elements. Figure 8.20 shows a fabricated one- to four-way feed network used to feed a $2 \times 2$ dipole array antenna. As shown in the figure, single-stage Wilkinson power dividers are used to feed the four dipole elements. It can be seen that the equal power division ($-6$ dB in $S_{21}$ parameters) occurs in the frequency band of 5.0 to 5.6 GHz. The fractional bandwidth of proper operation of the power divider network is about 15%, can be considered narrowband operation.

Figure 8.21(a) shows a one- to four-way power divider designed to feed a $2 \times 2$ dipole array antenna with three-stage Wilkinson power dividers. The measured proper bandwidth of operation with $-6$-dB power division at each port is 2.9 to 6.9 GHz, as shown in Figure 8.21(b). The fractional operating bandwidth is ~82% and it shows the wideband behavior of the multisection Wilkinson power dividers. By increasing the number of sections of the Wilkinson power divider, it is possible to achieve further wideband operation. However, wideband operation with multiple sections is achieved at the cost of the increased area occupied by the power divider.

### 8.2.5.2  T-Junction Power Divider

T-junction power dividers generally show broadband operation. However, output ports are not matched with the input port impedance [1]. The matching of the output ports needs an impedance transformation technique. Quarter-wavelength transformers can be used for narrowband applications. Wideband applications, however,

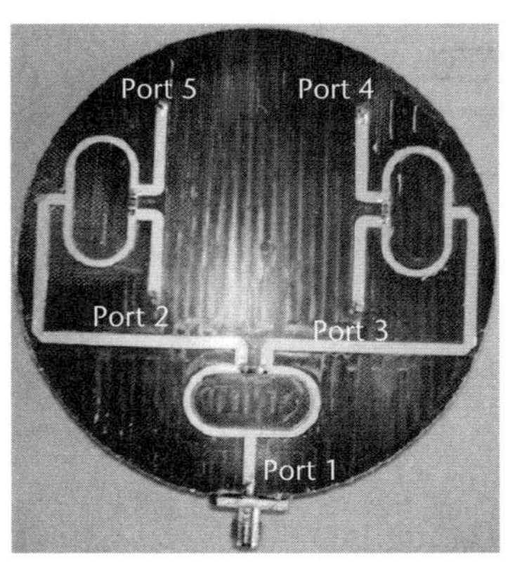


(a)

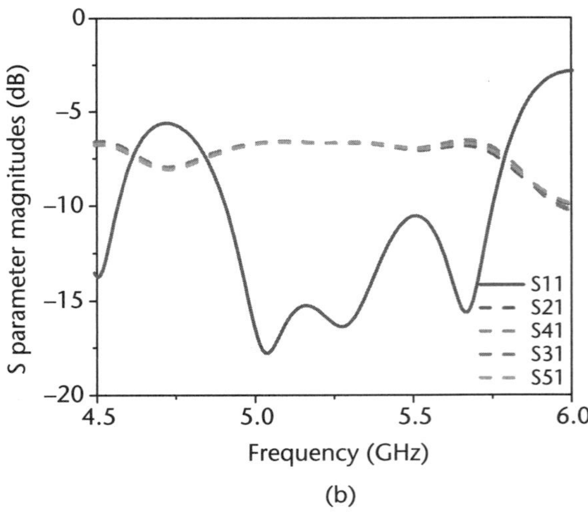


(b)

**Figure 8.20**  (a) Feed network prototype for a $2 \times 2$ dipole reflector array antenna and (b) measured performance of the feed network with three single-stage Wilkinson power dividers in two stages.

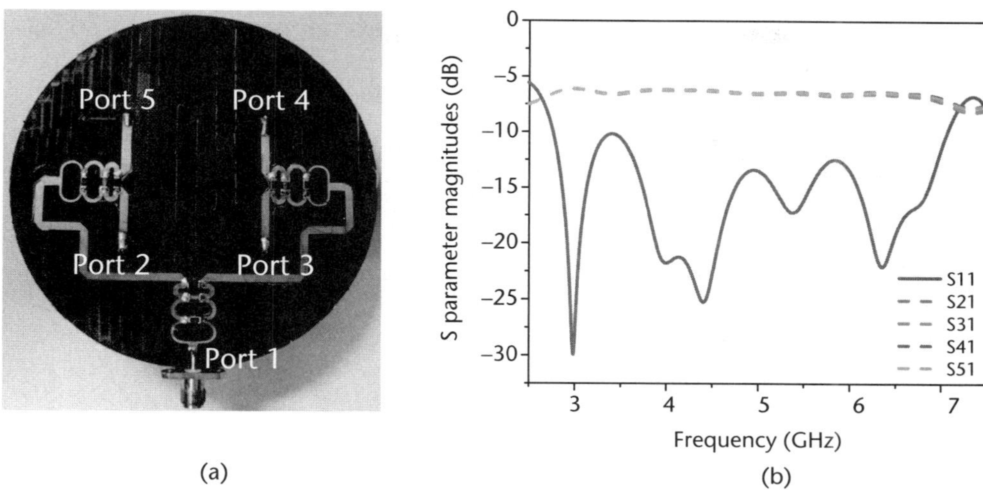


**Figure 8.21**   (a) Fabricated feed network of the antenna array and (b) measured performance of the feed network built with three-stage Wilkinson power splitters.

require different impedance transformation techniques. Using multiple quarter-wave impedance transformers is the wideband impedance matching method used to achieve wideband operation [1]. However, the use of multiple quarter-wavelength transformers at the output ports increases the occupied area of the T-junction. The use of tapered arms at the output ports is the most efficient method for reducing the area occupied by the circuit [14]. In addition to the use of tapered arms, by using an array of vias, through the substrates and with stepped microstrip lines, a very large enhancement of the bandwidth is possible [15]. In [16] a different approach is taken by using a slotted ground plane to improve the bandwidth of operation.

Figure 8.22(a) shows another one- to four-way feed network designed to feed the $2 \times 2$ dipole array antenna. The feed network was designed on Taconic TLX-8 substrate ($\varepsilon_r = 2.4$, tan $\delta = 0.0019$, $h = 0.17$ mm). Tapered arms are used to achieve the proper impedance transformations at each arm. As shown in Figure 8.22(b), the simulated performance of the feed network shows proper operation in the 2- to 14-GHz bandwidth. Since the output port is 75$\Omega$, it was unable to measure the transmission coefficients of each port. However, the broadband matching of the input port could be observed with the $S_{11}$ measurement, which is shown in Figure 8.22(c).

### 8.2.6   Resonators

Resonators of various forms are used in active and passive microwave design. Resonators are used in filter design, matching networks (e.g., a quarter-wave transformer is also a resonant circuit in that sense), oscillator circuits, tuned amplifiers, and frequency meters. A detailed description and analysis of various resonator types can be found in [1]. Figure 8.23 shows a few microstrip resonators used in chipless RFID applications. These resonators are mainly used to characterize the polymer and dielectric substrate materials for chipless RFID tags and many passive components designs. Some of the passive components are presented in this chapter. As can

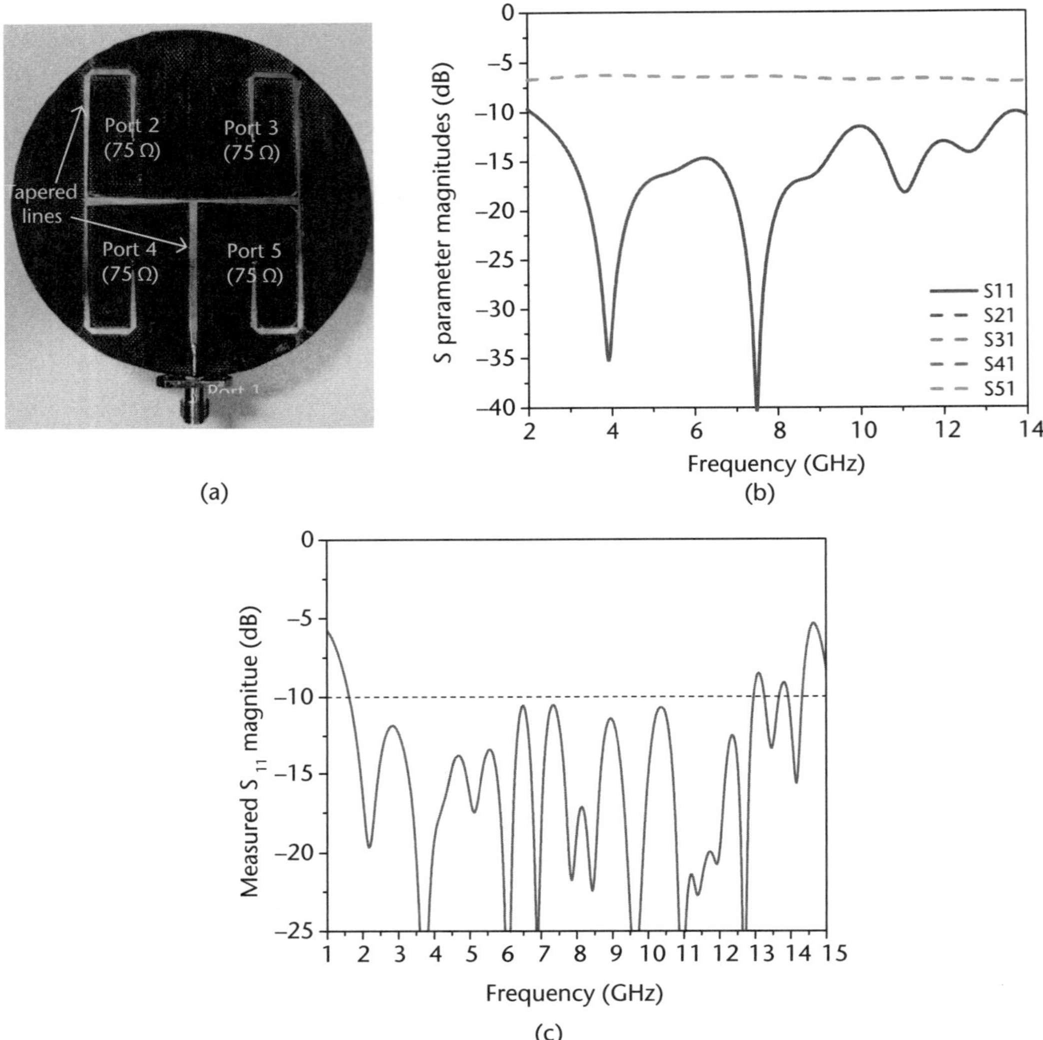



(a)                                    (b)

(c)

**Figure 8.22**   (a) Fabricated feed network with 75Ω output ports used in a dipole reflector antenna array, (b) simulated performance of feed network in CST Microwave Studio 2011, and (c) measured return loss of the fabricated feed network.

be seen in Figure 8.23(a), a line resonator of width $W$ and length $L$ is gap coupled with two transmission lines. Usually, the length of the line is an integer multiple of the guide wavelength. Figure 8.23(b) shows a circular resonator of diameter $D$ gap coupled with two transmission lines. The lines are coupled with the input and output ports of a network analyzer to measure the resonant frequency and quality factor of the resonators. These parameters yield dielectric properties such as relative permittivity (dielectric constant), attenuation, and loss tangent of the substrate materials. Once the material properties are known, they are used in the CAD tools for passive and active circuit design on the materials. Figures 8.23(c) and (d) show a parallel gap coupled line resonator and an annular ring gap coupled resonator, respectively. Dielectric disk resonators are very high $Q$ devices. A dielectric disk-loaded transmission line is shown in Figure 8.23(e). This type of high-Q resonator is used in microwave oscillator design. In microwave high-power devices, rectangular waveguide cavity resonators are commonly used in generators and waveguide

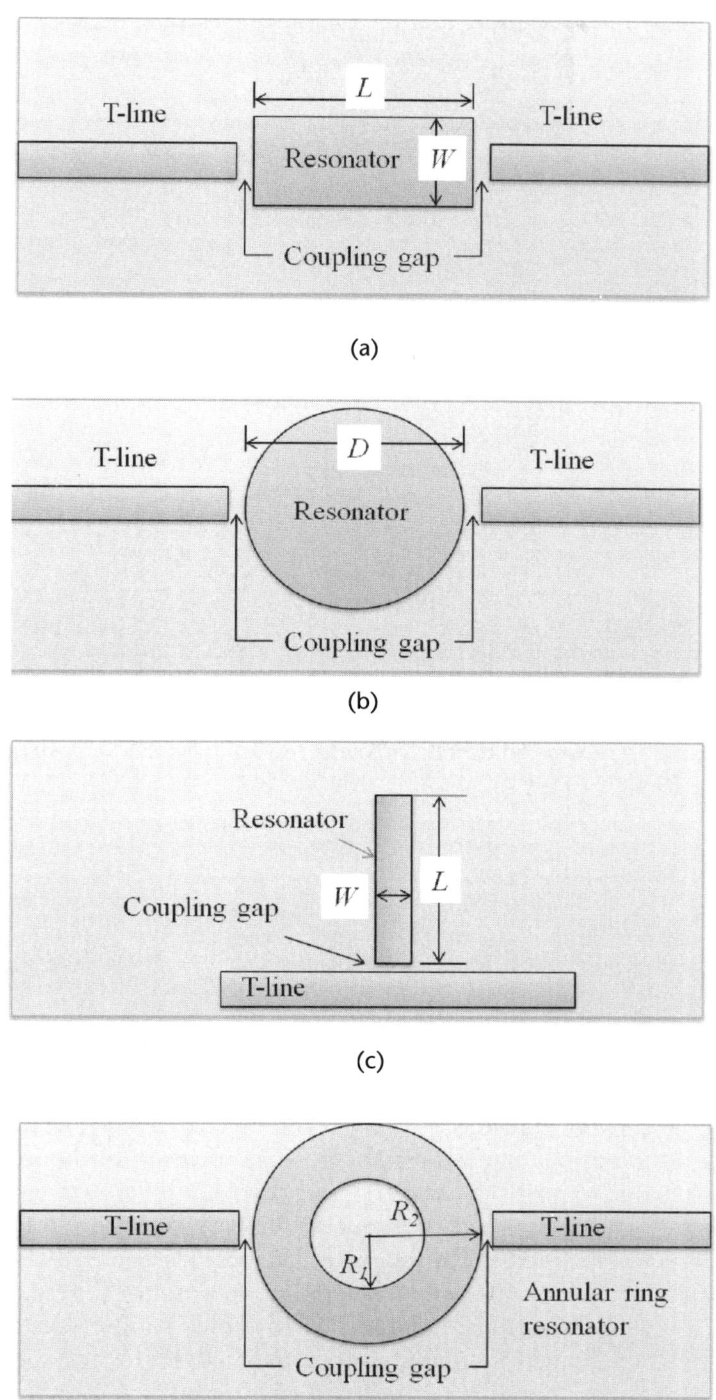



**Figure 8.23**  Various passive resonators: (a) line, (b) circular, (c) parallel line, (d) annular ring, (e) disk, (f) rectangular waveguide, and (g) coaxial line coupled resonators.

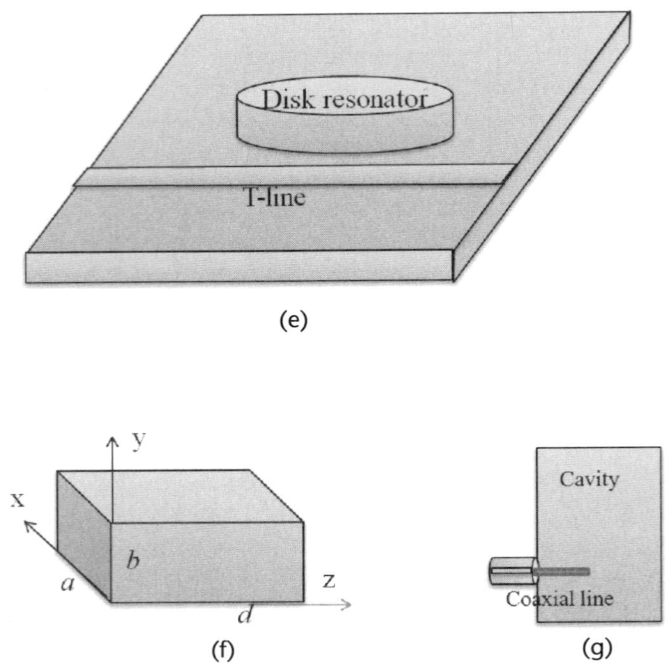



**Figure 8.23**　(continued)

antennas. Figures 8.23(f) and (g) show a rectangular waveguide resonator with dimensions and a coaxial line coupled waveguide launcher, respectively. The tapered and rectangular launchers for use with UWB horn antennas were presented in Chapter 7. Detailed definitions and the analytical procedure for these resonators can be found in [1, 17].

### 8.2.7　Mixers

A mixer is a nonlinear three-port network. The two input ports are the RF and local oscillator (LO) ports; the output is typically called the intermediate frequency (IF) signal. Basically, a mixer produces the product of the two input signals at the output port. Figure 8.24(a) shows a typical schematic symbol for a mixer. The two input ports are fed with signals as shown, and the output port generates the product of the two input signals. Figure 8.24(b) shows the basic operation of a mixer. As shown in the figure, the signal generated at the IF output port contains the frequency components of the sum and the difference of the frequencies of the two input signals. In an RF transmitter and receiver, mixers are used for two purposes: frequency up-conversion and frequency down-conversion, respectively.

　　Based on the purpose of the mixer, the out-of-band frequency components generated at the IF port of the mixers are filtered out. For the frequency down-conversion, a lowpass filter or a bandpass filter is used to filter out high-frequency components. For frequency up-conversion, a highpass filter or bandpass filter operating over a suitable frequency band is used to filter out the high-frequency components of the generated output signal.

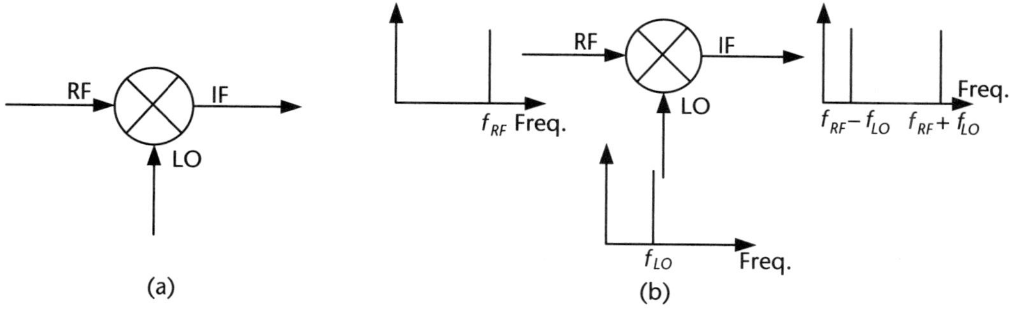



**Figure 8.24**   (a) Schematic symbol of a mixer and (b) basic operation of frequency mixing by a mixer.

Figure 8.25 shows how two mixers are used to down-convert the RF signals into a lower frequency signal in a chipless RFID reader RF section. The complete block diagram of the reader is shown in Figure 4.5. Here only the RF section is shown. The LO signal is divided into two equal portions with a power divider and fed into the mixer LO ports. The received signal is amplified and fed into the Mixer-1 RF port and a portion of a transmitted signal extracted with a coupler is fed into the Mixer-2 RF port. As shown in the figure, two LPFs are used to filter out the high-frequency components from the IF output signal. Figures 8.26(a) and (b) show two RF mixers operating in the 3.7- to 10.0-GHz frequency range with coaxial connector and surface mount packages, respectively. The mixer with the surface-mount package as shown in Figure 8.26(b) [18] is used in chipless RFID readers, which are discussed in Chapter 4.

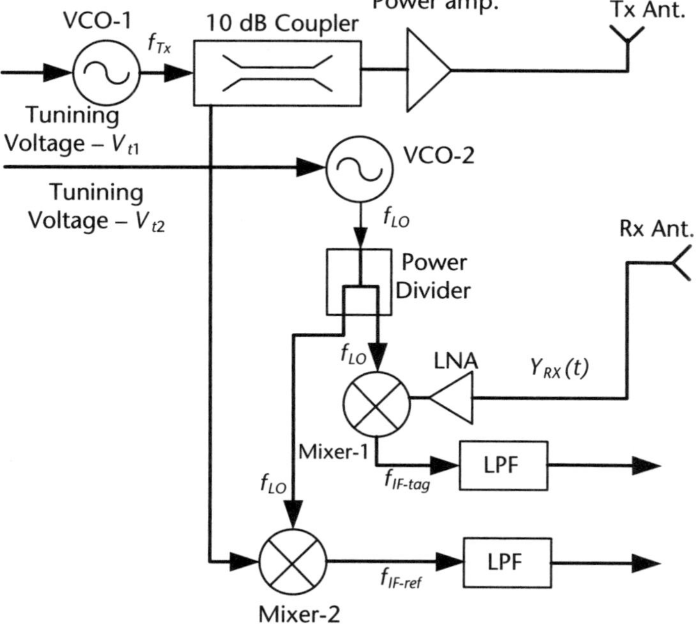



**Figure 8.25**   Using mixers for down-conversion of RF signals in a chipless RFID reader RF section.

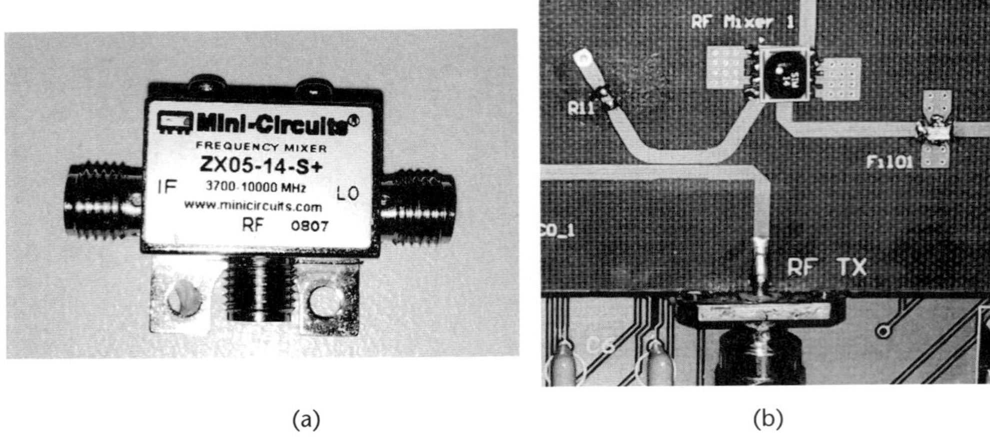

(a)                                                         (b)

**Figure 8.26**    (a) Modular RF mixer with coaxial connectors from Mini-Circuits and (b) a surface-mount mixer with similar performance used in the chipless RFID reader discussed in Chapter 4.

Two kinds of mixers are available: active mixers and passive mixers. Typical schematics for these two types of mixers are shown in Figure 8.27. For active mixers, an external power supply is required and they are transistor based circuits. Passive mixers do not require an external power supply and the operation is based on diodes. The passive mixers have advantages such as low noise figure, wide bandwidth, and low number of components compared to the active counterpart. However, active mixers have the advantages of low cost, low LO power, and higher conversion gain. Therefore, when choosing a mixer for a particular design, all of these advantages and disadvantages should be taken into account to achieve a better design.

There are a few important parameters that require the attention of a designer. LO injection is the amount of power that should be supplied at the LO port of a mixer [2]. As mentioned earlier, active mixers require lower LO injection than passive mixers. Passive mixers have negative conversion gain and, using active mixers, positive conversion gains are possible. The other two important parameters of a mixer are the noise figure (NF) and the operating bandwidth. The part counts and the cost of the mixer are also important factors that need the attention of a designer.

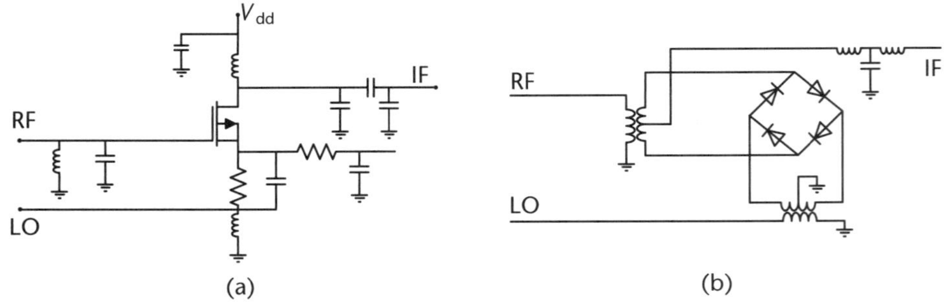

(a)                                                         (b)

**Figure 8.27**    (a) Schematic of a typical active mixer and (a) schematic of a typical passive mixer [19].

## 8.3   Active Components

The components discussed so far have been linear and nonlinear passive components. Because their operation does not require an external power supply they are referred to as passive components. The components that require an external power supply are called active components. Active microwave components can be used for many purposes such as detection, mixing, amplification, multiplication, switching, and as sources. The list is not limited to these tasks; active microwave components and circuits can do much more. Active microwave components are essential in building microwave systems. This section presents the basic concepts and operating principles for commonly used active microwave and millimeter-wave components in chipless RFID systems.

### 8.3.1   Voltage-Controlled Oscillators

An oscillator produces a periodic output. An oscillator that produces high-frequency signals, which are in the RF range, converts DC power into RF power. It produces the high-frequency signals that the other microwave components process. Therefore, oscillators are very important in a microwave or millimeter-wave system. Various types of oscillators are used in various microwave and millimeter-wave systems.

Some of the oscillators are designed to generate a fixed frequency output, whereas others are designed to be tuned according to the requirements. Because most of the chipless RFID systems operate in a very wide frequency band and they encode data in frequency signatures, generation of swept frequency signals is required to interrogate the tags. The generation of swept frequency signals requires an electronically tunable oscillator. A voltage-controlled oscillator (VCO) is a tunable oscillator that is used in many applications requiring generation of swept frequency signals. A VCO produces a variable frequency output depending on a suitable control voltage. The control voltage is sometimes called the *tuning voltage*. In most of the low-cost, compact VCOs, the junction capacitance of a semiconductor diode (or varactor) is used as the tuning device to control the output voltage with the input tuning voltage.

Various oscillator topologies and techniques are used to implement VCOs for different applications. The different types of oscillator topologies and techniques used to implement VCOs are described in [1, 2]. The other, less frequently used techniques are piezoelectric-controlled reactive elements (such as capacitor and inductors) and ferrite resonators (such as YIG resonators) to implement the VCOs. In this section only varactor based and YIG resonator based VCOs are discussed briefly.

Figure 8.28 shows typical schematics for a varactor tuned transistor VCO. The capacitance of the semiconductor junction changes with the applied tuning voltage. Therefore, the series RLC circuit changes and results in a change in the resonance frequency. The main drawbacks of varactor based VCOs are low temperature stability, low phase noise performance, and less accurate output frequencies. However, low-cost, less complicated implantation and operation and a wide tuning bandwidth are advantages of these types of oscillators.

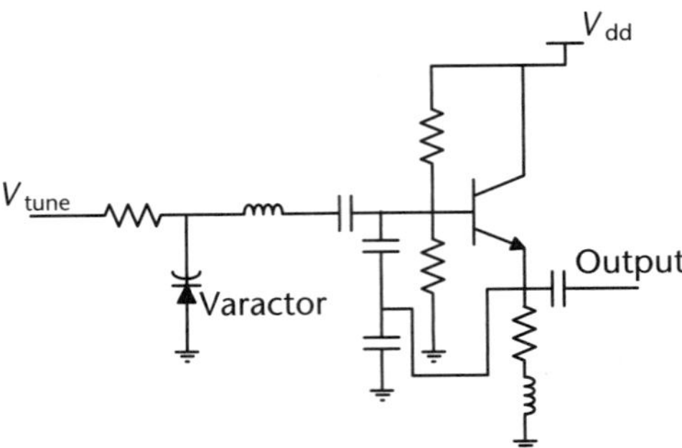


**Figure 8.28**　Typical schematic for a varactor based VCO [2].

Figure 8.29 shows a typical schematic for a yttrium-iron-garnet (YIG) VCO circuit [20]. A YIG sphere is kept inside a magnetic field and the magnetic field is changed electronically with the control or tuning voltage of the oscillator. As the magnetic field varies, the resonance frequency of the YIG sphere changes. The YIG sphere is coupled to transmission lines of the oscillator circuits, and the changes in resonance frequency of the YIG sphere result in a varying output frequency with the control voltage.

YIG tuned VCOs consume higher power than varactor tuned VCOs. In particular, the generation of a varying magnetic field to tune the resonant frequency of the YIG sphere requires special driving circuitry that consumes high current and several voltage levels. In addition to the higher current consumption, YIG oscillators should be powered up with several voltages levels. Therefore, YIG tuned VCOs require complex power supplies. In addition to those disadvantages, YIG oscillators require a certain time after the initial power up to warm up the internal resonant circuitry. However, YIG oscillators have better phase noise performance and better temperature stability of the output frequency signal than do the varactor tuned VCOs. High linearity and very wide tuning capability are the most important properties of the YIG oscillators. However, they are more expensive and bulky than the varactor based VCOs.

The YIG and varactor tuned VCOs have been used in several chipless RFID reader designs [21, 22]. The next section presents some measured results for the YIG and varactor tuned VCOs used in several chipless RFID reader prototypes by the authors.

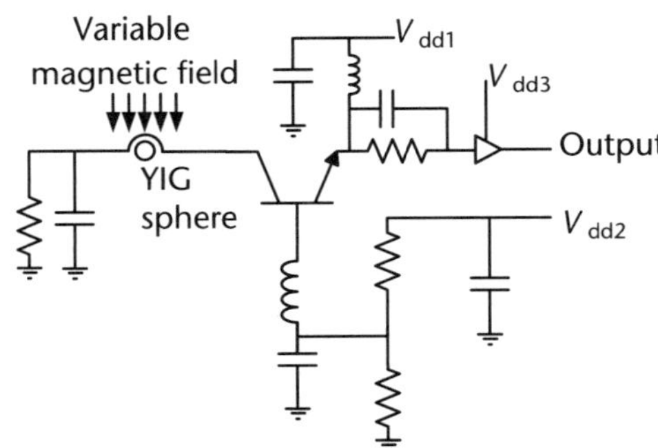


**Figure 8.29**　Typical schematic of a YIG resonator based VCO.

### 8.3.1.1   VCOs Operated at Microwave Frequencies

Figure 8.30(a) shows a Richardson RVCD6000F varactor based wideband VCO used in a chipless RFID reader [23]. As shown in this figure, the VCO is slightly larger than a hermetically sealed transistor. The operating frequency range of the VCO is 4 to 8 GHz. The VCO heats up during operation, so a heat sink should be mounted for the protection of the device. The full assembly of two VCOs with heat sinks is shown in Figure 8.30(b). The variation of frequency with the applied tuning voltage is shown in Figure 8.31. The VCO should be powered up with a 15V DC power supply. The tuning voltage range is from 0 to 20 V for the generation of 4- to 8-GHz frequency output. Based on the experimental results, once the tuning voltage is changed to a different value, VCO requires about a 2-ms time to settle down and to produce the proper frequency output corresponding to the new tuning voltage value.

Figure 8.32(a) shows a Teledyne YIG tuned VCO that is used in the chipless RFID readers discussed in [21, 22]. The bunch of wires in the bottom part of the photo provides the several voltage levels required for the operation of the VCO. Figure 8.32(b) shows the variation of output frequency versus input tuning voltage.

### 8.3.1.2   VCOs Operated at Millimeter-Wave Frequencies

The generation of millimeter-wave signals is primarily done in two ways. Directly using oscillators that generate millimeter-wave frequencies is one method. It is not easy, however, to find oscillators that operate in millimeter-wave frequencies and they are more expensive. Therefore, to generate millimeter-wave frequencies, the technique shown in Figure 8.33 is widely used in millimeter-wave RF systems. The output of a suitable oscillator that operates in microwave frequencies is connected to a frequency doubler. Then the output of the frequency doubler provides millimeter-wave frequencies. A Hittite HMC733LC4B VCO and Hittite HMC576 frequency doubler are attached together in a millimeter-wave VCO as shown in

(a)                                                          (b)

**Figure 8.30**   Richardson RVCD6000F varactor based VCO: (a) mounted on a test board before connecting the heat sink and (b) two VCOs mounted with heat sinks on a chipless RFID reader RF board.

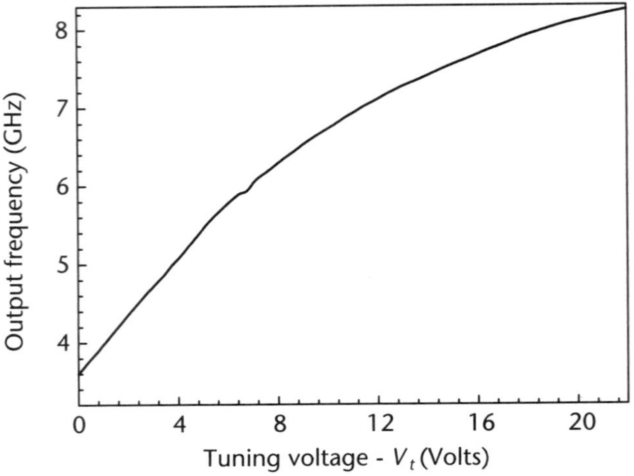

**Figure 8.31**  Variation of the output frequency versus tuning voltage of Richardson RVCD6000F VCO.

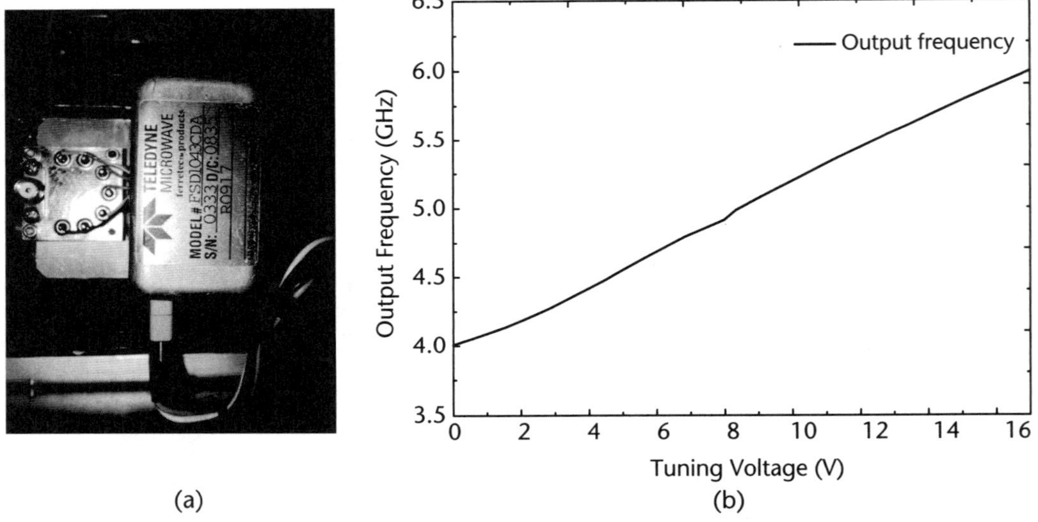

(a)                                                                        (b)

**Figure 8.32**  (a) The Teledyne YIG tuned VCO operates at 2 to 6 GHz and is used in the chipless RFID readers discussed in [21, 22] and (b) variation of output frequency with tuning voltage.

Figure 8.33 to generate frequencies in the 20- to 30-GHz band for a millimeter-wave chipless RFID application. The VCO generates a frequency band of 10 to 20 GHz. With the use of a frequency doubler, the 20- to 30-GHz frequency band is constructed. Figure 8.34 shows the variations in the output frequency with the tuning voltage of the setup. Figure 8.35 shows the variations in output power of the setup with the frequency. Note that this frequency doubler is an active device and it requires a power supply.

Due to the high output power there is no need for a power amplifier to be added at the output of the active doubler. However, when using a passive frequency doubler, the conversion loss must be considered. The conversion loss of a passive

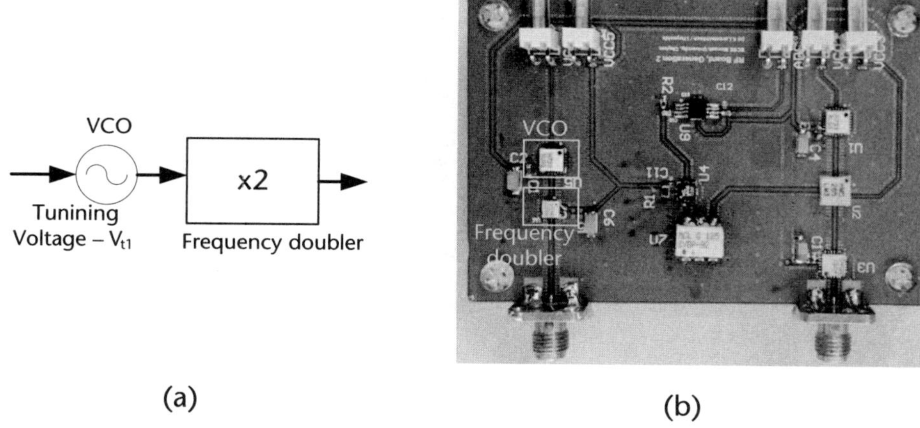


**Figure 8.33**   (a) Block diagram of the setup used to generate millimeter-wave frequencies with a VCO and a frequency doubler and (b) Photo of a mm-wave generation section of a mm-wave chipless RFID reader.

frequency doubler may be as high as 10 dB. In such a situation, a power amplifier is required at the output of the doubler. It addition to the output power, the harmonic levels of the output signal of the frequency doubler are very important. If the harmonic levels are high at the output frequency, a bandpass filter may be required to filter out harmonics. However, in wideband systems, such as the 20- to 30-GHz band, sometimes it may not be practical to use a bandpass filter to cover the whole frequency band. In such cases a frequency doubler with low output levels should be chosen for the design.

Figure 8.33(b) shows the UWB millimeter-wave RF section circuit. The VCO and frequency doubler of the interrogation signal generation section are marked in the figure. The output of the frequency doubler is directly attached to the output RF connector. Figure 8.34 shows the plot of millimeter-wave frequency output versus tuning voltage of the setup shown in Figure 8.33(a). As can be seen, linear swept frequencies from 20 to 29.5 GHz are generated with variations in the tuning voltage (0V to 8V). Figure 8.35 shows the output power spectrum of the millimeter-wave frequency generation section of the RF section. As shown in the figure, it generates quite stable power that averages 13 dBm over the UWB millimeter-wave frequency band of 22 to 26.5 GHz.

### 8.3.1.3   Phase Locked Loop

In many RF systems, accurate generation of frequencies is essential to meet wireless communication regulations and to keep up the performance of the system. In a VCO the output frequency generated is defined by the device parameters of the circuits for a given tuning voltage. Therefore, a VCO driven only by a tuning voltage can be considered to be a free-running oscillator, and the output frequency is defined with internal component parameters. The internal component parameters change due to factors such as temperature, supply voltage changes, and noise induced by the surrounding environment. Using an external control loop, it is possible to compensate for the drifts of the frequencies and achieve accurate frequency

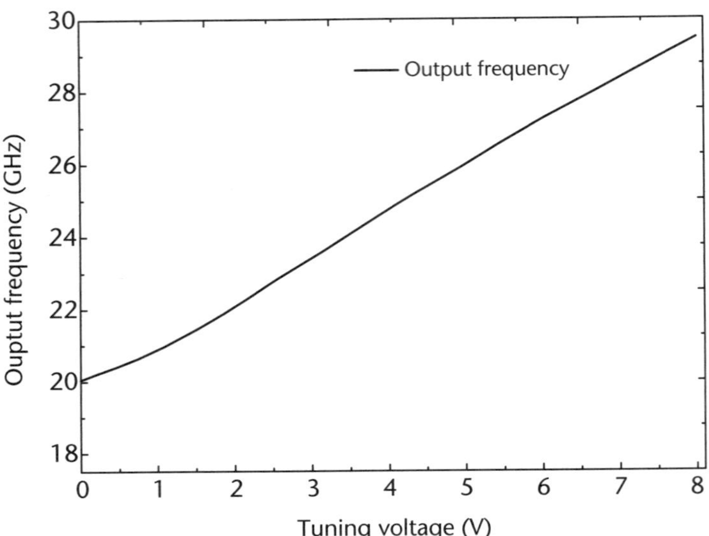

**Figure 8.34**  Variation of the output frequency of the setup with a frequency doubler with tuning voltage.

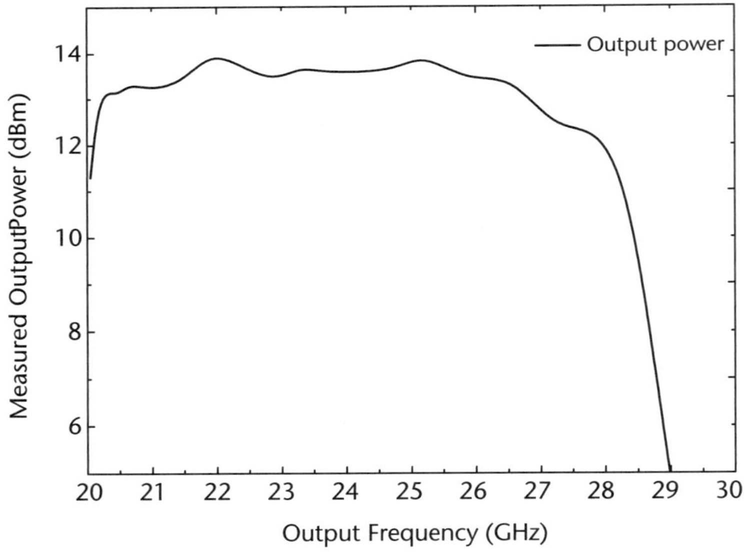

**Figure 8.35**  Variation of the output power with the output frequency of the setup with the frequency doubler.

generation [24]. A phase locked loop (PLL) is such a negative feedback control loop introduced to control the VCO and achieve accurate frequency generation.

A block diagram of a digitally controllable PLL is shown in Figure 8.36. As shown, the output of the VCO is divided with a frequency divider and compares the phase of the signal with a reference signal at the phase frequency detector (PFD). The output of the PFD is sent through a lowpass filter called a loop filter to drive the VCO tuning voltage. Thus, accurate frequency signal generation is

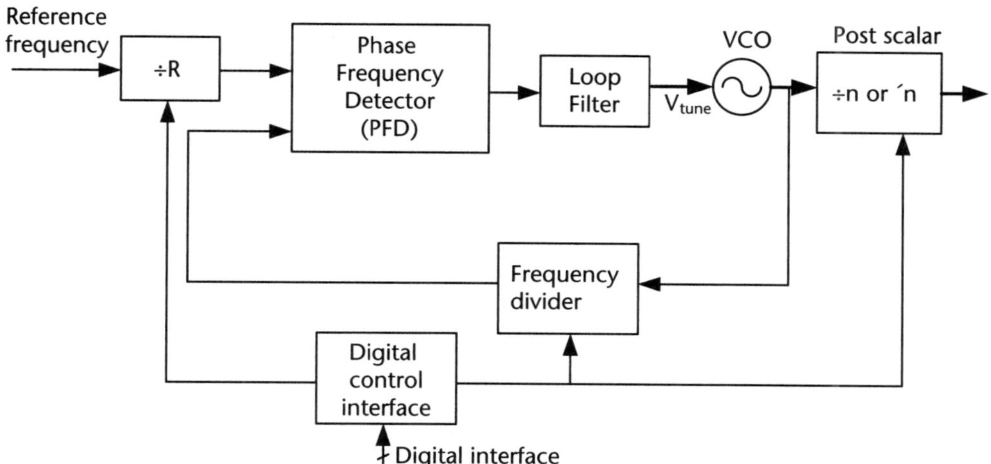


**Figure 8.36**   Digitally controllable PLL with integrated VCO [24].

possible. There are chips available such as Texas instruments TRF3785 and Analog devices ADF4350 that have integrated VCOs with a PLL. In such devices it is possible to configure the feedback loop by setting different divisions of the output frequency of the VCO and the reference signal. Using this digital interface it is possible to change the output frequency in discrete steps with accurate frequency outputs. In most of the VCOs a digital interface is provided to configure the value of the frequency output. Therefore, it is possible to achieve a frequency sweep by incrementing the output frequency setting by configuring the PLL periodically with suitable time intervals. However, it is difficult to achieve very short frequency sweeps due to the time required to lock to the proper frequency output by the PLL and the speed limitations of the digital programming interface.

### 8.3.2   Low-Noise Amplifiers

Because the received signal is the weakest RF signal in a typical RF system, amplification of the signal is required. The performance of an RF system depends very much on the performance the RF front-end. Therefore, the amplification of these weak signals requires more attention if we are to achieve a receiver that provides better performance. A general block diagram of a high-frequency amplifier is shown in Figure 8.37. There is an input impedance matching network before the active device: the transistor. Then before the output another matching network is placed to interface the rest of the circuit with the amplified signal. In addition, there are input and output biases to keep the transistor in the linear region of operation. However, in many real cases matching and biasing circuits share one or more components, making it difficult to separate the impedance matching and bias sections.

In any RF receiver front-end, the low-noise amplifier (LNA) is the first circuit block after the antenna and bandpass filter if there is any. It is well known that amplifiers may add noise to the amplified signal. Since the front-end of the RF system deals with relatively weaker signals than are found in other areas, special attention should be given to the noise performance of the amplifiers. The LNAs are designed

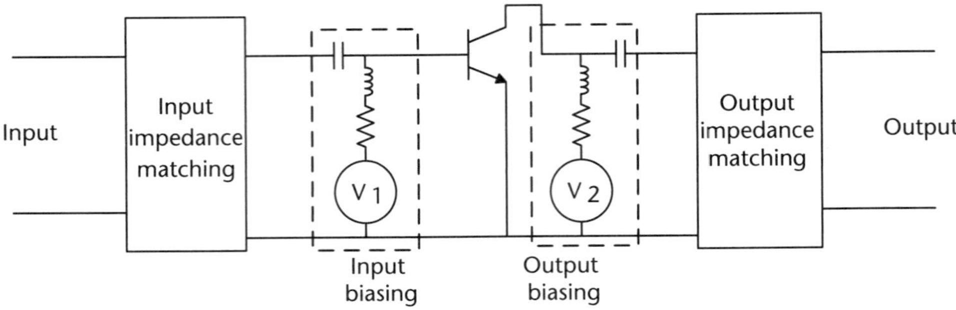


**Figure 8.37**   Block diagram of a general transistor amplifier circuit.

to add noise as low as possible due to the low strength of the input signals mentioned earlier. The amount of noise added by a particular component is given by the value of the *noise figure*. The LNAs should have a low noise figure to perform well in the front-end of the RF section. In addition to the low noise figure, the gain of the amplifier is also important in amplifying the signal. Generally it is not possible to achieve the best noise figure and best possible gain at the same time in an amplifier design [1]. Therefore, in a LNA circuit some sort of compromise is made between gain and noise figure. Most of the chipless RFID systems discussed in this book operate in very wide frequency bands. Therefore, LNAs used in chipless RFID applications must meet the requirement of being wideband as well. Several techniques are used to design wideband LNAs such as compensated matching networks, resistive matching networks, balanced amplifiers, and distributed amplifier topologies.

There are many commercially available amplifier ICs for popular frequency bands in the market. For applications such as chipless RFID readers, it is convenient to use a suitable IC rather than designing an amplifier from scratch. The Gali 2+ amplifier IC was used as the LNA in the chipless RFID readers described in Chapter 4. Figure 8.38(a) shows the biasing circuit required for the amplifier, and Figure 8.38(b) shows an evaluation board and some of the essential components in the biasing circuit. The amplifier works up to 8 GHz and provides a gain of 12.8 dB over the operating band. The noise figure is 4.2 dB.

In summary, when selecting an amplifier for chipless RFID applications, the important parameters that should be considered are operating bandwidth, gain, and noise figure. The cost of the device and the size of the device are also important factors that must be considered in many applications.

## 8.4   Discussion and Conclusions

This chapter has presented definitions and basic operating principles for the important passive and active components used in microwave and millimeter-wave chipless RFID readers. The commonly used microwave and millimeter-wave components in chipless RFID systems are resonators, antennas, filters, couplers, power dividers, mixers, VCOs, and LNAs. Selection of suitable components for a system is very important to achieve the required performance. After deriving the specifications of

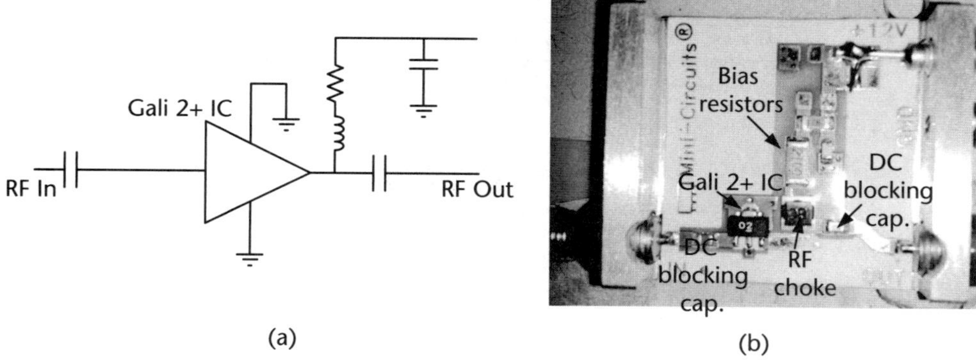



**Figure 8.38**   (a) Schematic of the Gali 2+ RF amplifier used in the chipless RFID reader circuit described in Chapter 4 and (b) evaluation board of Gali 2+ amplifier IC.

the whole system, the next step is selection of the components. When selecting the components, designers must pay attention primarily to the following two things:

1. *Individual performance of the component.* Individual performances of the components are specific to that component. However, sufficient operating bandwidth, insertion, and return losses are common performance indicators for most of the components. In addition to these parameters, other specific parameters of different components such as gain, noise figure, and phase noise should be considered as mentioned in the text.

2. *Integration of components into a system.* In addition to the individual performance of each component, it is essential to consider the interconnections of each component. Some components or devices may have limitations in accepting minimum input power or delivering maximum output power. If the preceding component is not providing the minimum input power or if it exceeds the maximum input power, then the system does not work although the individual performances of the components meet the specifications of the overall system. Therefore, it is important to select the components to meet the requirements of other components while meeting the individual performance criteria.

The summarized component selection process for a chipless RFID system is shown in Figure 8.39. As shown in the figure and explained earlier, the individual performance as well as the interconnectability of the selected components must be considered when building a chipless RFID system. This will reduce the number of times costly revisions of the design will be required. Because a chipless RFID reader comprises RF and digital control sections, the overall performance of the system depends on both modules. If the performance of the RF section is not satisfactory, even with the highest performance digital hardware, the overall performance of the reader system will not be satisfactory. In other words, even with very powerful digital hardware and signal processing algorithms, if the performance of the RF section is poor, the total system performance becomes poor. Most of the time, the performance of the RF section has more limitations than the digital section. Very low cost digital signal processing hardware with decent performance is now available in the

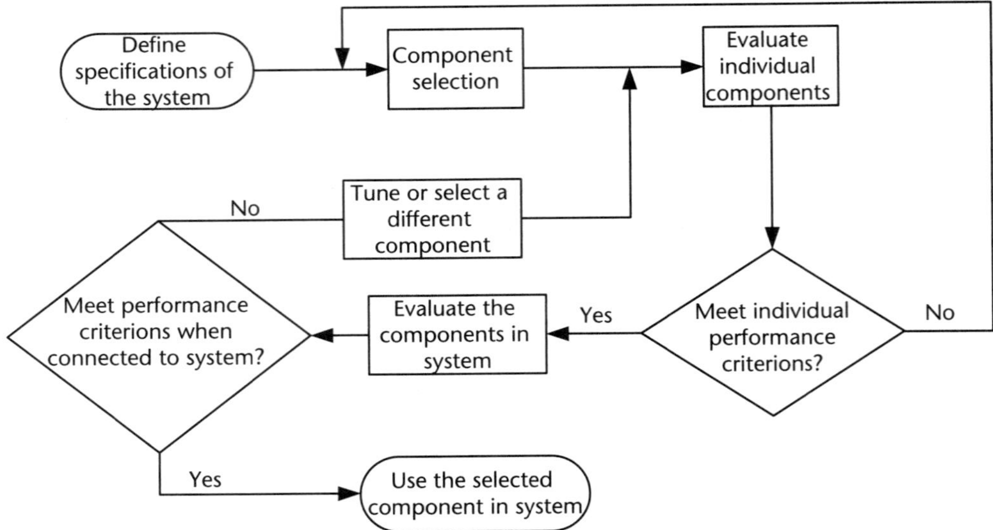



**Figure 8.39** Component selection process for building a chipless RFID system.

market compared to a few years ago. However, the cost of high-performance RF components and devices is still very high compared to the digital hardware.

## Questions

1. Explain the differences between passive and active microwave components.
2. Name six passive microwave components used in chipless RFID systems.
3. Name five active microwave components used in chipless RFID systems.
4. Name three types of passive microwave filters.
5. What are the common problems in lumped element implementations of microwave filters?
6. What are the common problems in microstrip implementations of microwave filters?
7. What types of filters are used in multiresonator based chipless RFID tags?
8. Explain the use of a coupler in a chipless RFID system.
9. What are the typical isolation levels between ports of a directional coupler?
10. Name the special features of a hybrid coupler.
11. Explain why it is difficult to use circulators in chipless RFID readers that operate in the UWB frequency band.
12. Explain the differences between Wilkinson and T-junction power dividers.
13. What are the techniques used to improve the operating bandwidth in Wilkinson and T-junction power dividers?
14. Name three types of resonators.
15. Explain the requirements of the resonators used in chipless RFID systems.
16. Explain the differences between the passive mixers and active mixers used in microwave systems.
17. List the advantages and disadvantages of active and passive mixers.

18. Explain two methods used to control the output frequency in VCOs and name the most suitable method for chipless RFID systems. Justify your answer.

19. Explain a method for generating millimeter-wave frequencies using oscillators operating at microwave frequencies. Explain the limitations of this method.

20. Explain the advantages of using a PLL instead of a free-running VCO.

21. Explain why LNAs are used in the front-end of a receiver circuit.

22. Name three important parameters that should be considered when choosing an amplifier for a chipless RFID application.

23. Explain the process of choosing components for a chipless RFID system.

# References


[1]    D. M. Pozar, *Microwave Engineering*, 2nd ed. New York: Wiley, 1997.

[2]    R. Sorrentino and G. Bianchi, *Microwave and RF Engineering*. New York: John Wiley & Sons, 2010.

[3]    R. J. Cameron, C. M. Kudsia, and R. R. Mansour, *Microwave Filters for Communication Systems: Fundamentals, Design, and Applications*. Hoboken, NJ: Wiley-Interscience, 2007.

[4]    Mini Circuits, "Low pass filter—LFCN-630+," Technical datasheet, October 2012.

[5]    Mini Circuits, "High pass filter—HFCN-740+," Technical datasheet, October 2012.

[6]    Mini Circuits, "Band pass filter—BFCN-4440+," Technical datasheet, October 2012.

[7]    H. Lim et al., "A novel compact microstrip bandstop filter based on spiral resonators," in *2007 Asia-Pacific Microwave Conference (APMC 2007)*, pp. 1–4.

[8]    Y. T. Lee et al., "A compact-size microstrip spiral resonator and its application to microwave oscillator," *Microwave and Wireless Components Letters*, Vol. 12, No. 10, pp. 375–377, 2002.

[9]    E. J. Wilkinson, "An N-way hybrid power divider," *IRE Trans. on Microwave Theory and Techniques*, Vol. 8, No. 1, pp. 116–118, 1960.

[10]   L. I. Parad and R. L. Moynihan, "Split-tee power divider," *IEEE Trans. on Microwave Theory and Techniques*, Vol. 13, No. 1, pp. 91–95, 1965.

[11]   R. Pazoki, M. R. G. Fard, and H. G. Fard, "A modification in the single-stage Wilkinson power divider to obtain wider bandwidth," in *2007 Asia-Pacific Microwave Conference (APMC 2007)*, pp. 1–4.

[12]   O. Ahmed and A. R. Sebak, "A modified Wilkinson power divider/combiner for ultrawideband communications," in *IEEE Antennas and Propagation Society International Symposium (APSURSI'09)*, 2009, pp. 1–4.

[13]   S. B. Cohn, "A class of broadband three-port TEM-mode hybrids," *IEEE Trans. on Microwave Theory and Techniques*, Vol. 19, No. 2, pp. 110–116, 1968.

[14]   G. R. Branner, B. Preetham Kumar, and D. G. Thomas, "Design of microstrip T junction power divider circuits for enhanced performance," in *Proc. 38th Midwest Symposium on Circuits and Systems*, 1995, Vol. 2, pp. 1213–1215.

[15]   K. Chen, B. Yan, and R. Xu, "A novel W-band ultra-wideband substrate integrated waveguide (SIW) T-junction power divider," in *International Symposium on Signals Systems and Electronics (ISSSE)*, 2010, Vol. 1, pp. 1–3.

[16]   A. M. Abbosh, "Planar ultra wideband inphase power divider," *Microwave and Optical Technology Letters*, Vol. 51, No. 5, pp. 1185–1188, 2009.

[17]   R. E. Collin, *Foundations for Microwave Engineering*, 2nd ed. Wiley India, 2007.

[18]    Mini Circuits, "SIM-14LH+ wide band frequency mixer, 3700 to 10000 MHz," Technical datasheet, October 2012.

[19]    R. C. Li, *RF Circuit Design*. Hoboken, NJ: Wiley, 2008.

[20]    F. Losee, *RF Systems, Components, and Circuits Handbook*. Norwood, MA: Artech House, 1997.

[21]    S. Preradovic, N. Karmakar, and M. Zenere, "UWB chipless tag RFID reader design," in *IEEE International Conference on RFID—Technology and Applications (RFID-TA)*, 2010, pp. 257–262.

[22]    S. Preradovic and N. C. Karmakar, "Design of short range chipless RFID reader prototype," in *5th International Conference on Intelligent Sensors, Sensor Networks and Information Processing (ISSNIP)*, 2009, pp. 307–312.

[23]    R. V. Koswatta and N. C. Karmakar, "A novel reader architecture based on UWB chirp signal interrogation for multiresonator-based chipless RFID tag reading," *IEEE Trans. on Microwave Theory and Techniques*, Vol. 60, No. 3, pp. 1, 2012.

[24]    B. Razavi, *RF Microelectronics*, Upper Saddle River, NJ: Prentice Hall, 1997.

## CHAPTER 9

# Digital Module for Chipless RFID Readers

## 9.1  Introduction

In the preceding two chapters, reader antennas and microwave active and passive design considerations were discussed. These two fundamental building blocks form the analog physical layer of the chipless RFID reader. However, these two blocks have no functional capability or intelligence of their own. The third block is the digital control module, which controls the microwave components such as voltage-controlled oscillator (VCO) and provides the biasing voltages to the active components of the RF transceivers such as amplifiers. In this chapter the digital control section of the chipless RFID reader is presented. The next chapter discuss the integration of these fundamental building blocks into a complete chipless RFID reader.

The *digital module* of a reader controls the entire operation of the reader. The operation could be divided into three functions: (1) transmitter control, (2) receiver control, and (3) signal processing and data decoding. The design of the digital control section consists of two parts: hardware design and firmware (also called *middleware*) development for controlling the operation of the reader and for decoding data.

Several different types of chipless tag designs were reported in Chapters 4, 5, and 6. They have a different number of encoded bits varying from 4 to 35 bits. Therefore, the digital control section was designed to adapt the reader for reading different types of tags having a different number of encoded data bits without changing its hardware configuration. Using this design, 4- and 9-bit tags have been successfully read, ad discussed in Chapter 4.

As stated in Chapter 1, the overall chipless RFID system comprises three major components: (1) a chipless tag, (2) an RFID reader, and (3) a back-end database system or enterprise software. Figure 9.1 illustrates the generic chipless RFID tag-reader-enterprise software system [1]. Chipless tags are passive and do not have any power source of their own. The reader sends an interrogation signal to illuminate the chipless tag. The tag receives the signal through its receiving antenna and passes it through its modulation circuit. In a bistatic radar mode, the transmitting antenna of the tag sends the encoded signal back to the reader. The reader receives this tag-encoded signal and processes it to decode the tag information. Finally, the

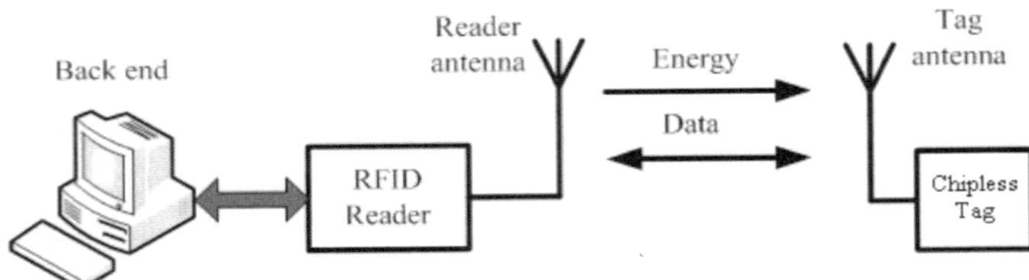


**Figure 9.1**  Generic chipless RFID system comprising tag, reader, and enterprise software.

reader sends the processed tag information to the back-end database system for specific applications.

As stated in Chapters 4, 5, and 6, chipless RFID systems are classified into three types based on their techniques of encoding bits in passive tags. The three types are (1) time domain based tags [2–8], (2) frequency domain based tags [9–19], and (3) their hybrids [20]. Time domain based tags are interrogated by sending a short-duration pulse and observing the received trains of pulses reflected from the chipless tag. Data are encoded in the tag using the time domain reflectometry (TDR) principle. Perturbations in the signal flow path at prescribed delays cause prescribed modifications of the transmitted pulse shapes at different times.

In frequency domain based tags, data bits are mostly encoded based on the absence or presence of resonances originating from the resonating elements located within the tag. The tag responds to a broadband interrogation swept-frequency signal.

In addition to the two main categories of chipless tags—time domain tags and frequency domain tags—another type is the chipless RFID tag in which two or more dimensions such as frequency-phase, frequency-time, and frequency-polarization are used for larger bit encoding in hybrid domain tags. Chapter 6 covered these hybrid domain tags. Different reader types are also used to read these tags. Chapters 4, 5, and 6 covered various reader architectures for these three types of tag reading. All readers use *digital control modules* to interrogate the chipless tags and receive and process the backscatters of the tag into meaningful ID information. This chapter covers the detailed design of the *digital module* of the chipless RFID tag reader.

Figure 9.2 illustrates the detailed functional blocks of a *digital control module* in a chipless RFID reader system [21–23]. As shown in the figure, the *digital module* consists of three parts: (1) *digital control board*, (2) *digital signal processing (DSP) board*, and (3) *middleware*. The middleware is a set of instructions installed in an FPGA or a microcontroller to control the digital board and the DSP board. The functions of the digital module are as follows:

1.  The *digital control section* generates the tuning voltage (also called the VCO control voltage) and feeds into the VCO to generate swept-frequency interrogation signals. The VCO is the heart of the RF transmitter of the reader system. The RF transmit (RF-Tx) module comprises a VCO as the signal generator, followed by a circulator to protect the VCO from unwanted reflection from the reverse path of the transmitter chain, a power

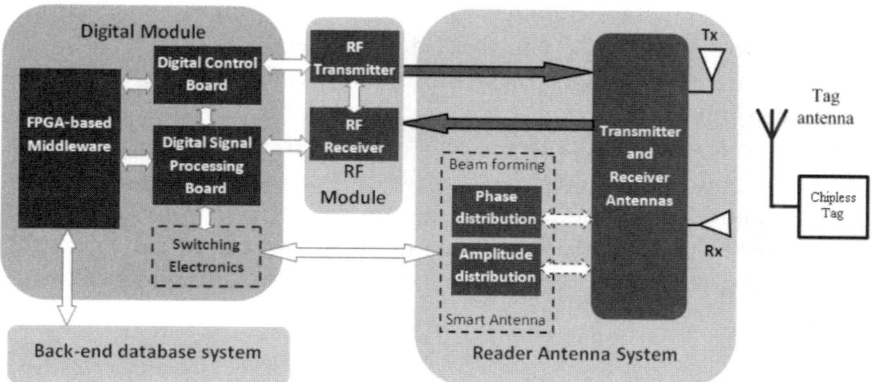

**Figure 9.2**    Functional block diagram of a chipless RFID system. An optional beamforming smart antenna with switching electronics can improve tage visibility, as shown in the dotted boxes.

amplifier, and an antenna. The intended interrogation signal is transmitted through the reader's transmit (Tx) antenna. Therefore, the digital control section provides the intelligence and control of the transmitter. In addition, the bias voltages for the power amplifier in the transmit chain and the low-noise amplifier in the receive chain are also obtained from the digital module's power supply unit. This is discussed in detail in a later section.

2. The *DSP board* performs the signal processing of the encoded data that is carrying signal that comes back from the tag. This backscatter of the tag is received in the receiver (Rx) antenna of the reader. The RF receiver (RF-Rx) module performs some operations in sequence such as low-noise amplification and down-conversion of the RF backscatter to a baseband signal so that the signal can be sampled and processed in the DSP board. The DSP module samples the down-converted signal coming from the RF-Rx and implements a signal processing algorithm for denoising and tag identification on its on-board FPGA or microcontroller.

3. In the case of a mobile chipless tag, the reader antenna needs to track the chipless tag during the tag reading process. In such a scenario the antenna needs to steer the beam electronically toward the tag so that the tag remains in the reading zone all the time. The digital beamforming network of the smart antenna comprises a set of digital phase shifters [23–27]. In this case, the *digital control section* also incorporates the *switching electronics* that digitally control the phase shifters of the antenna array.

4. As shown in Figure 9.2, *middleware* implemented in an FPGA or a microcontroller liaises instructions (control and command) between the back-end database system and the digital control section of the RFID reader. In the uplink, from the reader to the tag communication, the middleware receives the user's instructions from the back-end database system and then generates control and command instructions for the digital control section. In the downlink, from the tag to the reader communication, the middleware receives a tag ID from the DSP module and forwards the ID data of the tag to the backend database system for further processing and use.

These four steps complete the function of a complete chipless RFID tag-reader-back-end system for an enterprise application.

Figure 9.3 shows the organization of this chapter on the digital module. As can be seen, the *control board* that is covered in Section 9.2 is further comprised of four submodules: (1) power supply unit (covered in Section 9.2.1), (2) infrared (IR) sensor module (Section 9.2.2), (3) RF power supply switching electronics (Section 9.2.3), and (4) VCO tuning voltage control unit (Section 9.2.4). Section 9.3 covers the DSP board. Section 9.4 covers middleware, Section 9.5 covers the performance of the digital control section, and Section 9.6 presents conclusions.

## 9.2 Digital Control Board

As seen in Figure 9.2, the digital control board has interfaces with middleware installed in the FPGA and RF module of the reader system. The functional schematic of the digital board is shown in Figure 9.4. It consists primarily of the following three parts:

- Power supply unit;
- IR tag sensor;
- RF power supply control switch;
- VCO control voltage generator.

The *power supply unit* takes a ±12V external power supply and generates all necessary power supplies (±12V, ±5V, 3.3V, and GND) for various components of the digital control board. The IR based *tag sensor* senses the presence of a tag in the reading zone of the antenna and produces digital 0/1 output. The tag sensor's digital output is used by the *power control switch* unit to turn on the power supply of the RF module as well as the VCO control voltage generation part of the

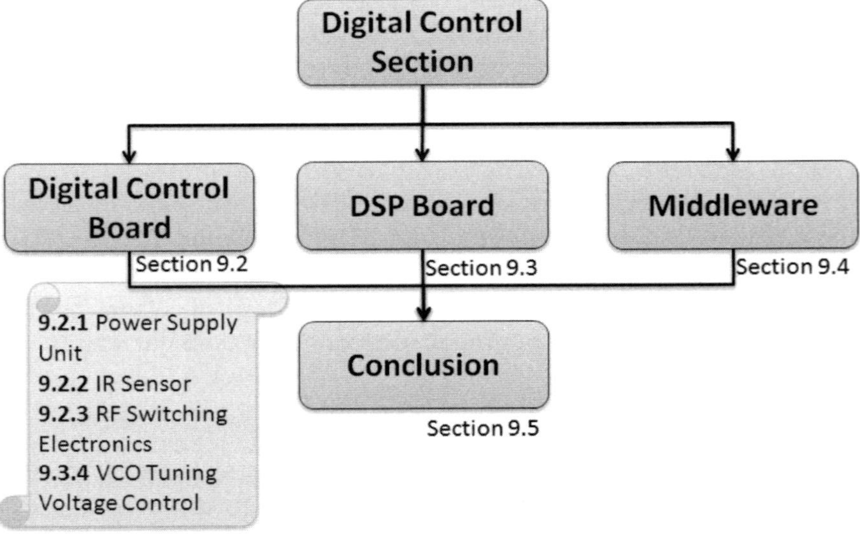


**Figure 9.3**  Organization of this chapter.

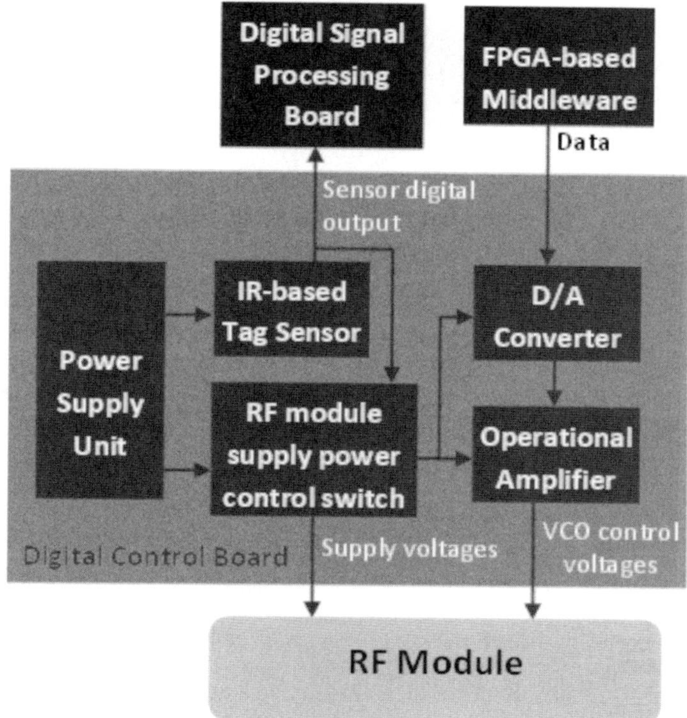



**Figure 9.4**  Components of the digital control board.

digital control board. The VCO control generation part takes digital input from the middleware FPGA and generates one or more VCO control voltages based on the RF board architecture.

### 9.2.1  Power Supply Unit

In the power supply unit, a switching voltage regulator is preferred instead of linear regulators. The operation of linear regulators generates a huge amount of heat, which requires large heat sinks for heat dissipation. Furthermore, the large heat sinks also occupy a large area of the PCB and increase the overall weight of the digital board. The high efficiency of switching regulator removes the need for heat sinks for the reader system and thus reduces the overall size of the digital board. The chosen switching regulator (see Figure 9.5) will be used to generate 5V DC voltage directly from the initial stage 12V DC supply. The 3.3V DC will still be supplied using a linear regulator but no heat sink will be required because the regulator only needs to source a small amount of current (maximum: 300 mA). In addition to the regulators, proper bypass capacitors are included to the power supply of the digital-to-analog converter in order to improve its performance.

### 9.2.2  IR Tag Sensor

The digital control board consists of an IR tag sensor interface to sense the presence of the tag. Figure 9.6(a) shows a Sharp IR sensor. The sensor has a detection

| Features | |
| --- | --- |
| Input range | 7 – 36 VDC |
| Output | 3.3 / 5 VDC (fixed), 1.5 Amps (max) |
| Packaging | Vertical / Horizontal SIP-mount |
| Heat sink | No heat sink required |

**Figure 9.5**   Murata OKI-78SR-5/1.5-W36-C 5V DC switching regulator. (Source: http://www.murata-ps.com/data/power/oki-78sr.pdf, accessed February 22, 2013.)

range of up to 80 cm. It generates an output analog voltage corresponding to the distance of the object that is being detected. Figure 9.6(b) shows the output voltage of the sensor in volts versus the distance of the reflected object in centimeters. As can be seen in the figure, the performance is almost invariant for the two cases of reflective objects white paper with a reflectance ratio of 90% and gray paper with a reflectance ratio of 18%.

For an RFID tag reader, detecting the presence of a chipless tag within the vicinity of the reader is of great interest. Sensing the presence of a tag triggers the reader to initiate the reading process, whereas the absence of a tag keeps the reader in standby mode thus saving power. Figure 9.7 illustrates the schematic of the IR tag sensor circuit. As can be seen in the figure, a comparator along with a potentiometer at the output of the sensor converts the sensor analog output voltage (0V or 3.3V) to digital binary data (0/1). The *RF supply power control switch* uses the control output to turn the reader electronics on and off. The potentiometer adds a degree of flexibility because it allows the operator to modify the measuring distance to the reading range of the reader that is of greatest interest. For example, if the potentiometer is tuned to yield a 1V output to the noninverting side of the comparator, a tag within 28 cm [refer to the graph in Figure 9.6(b)] of the sensor will provide a digital "0" (tag present) at the output of the comparator. Any tag outside this range will output a digital "1" (no tag present). This decision-making capability of the IR sensor provides an additional feature of security to the RFID reader. For example, people carrying chipless RFID tagged banknotes would not like to be tracking them outside of a definitive distance.

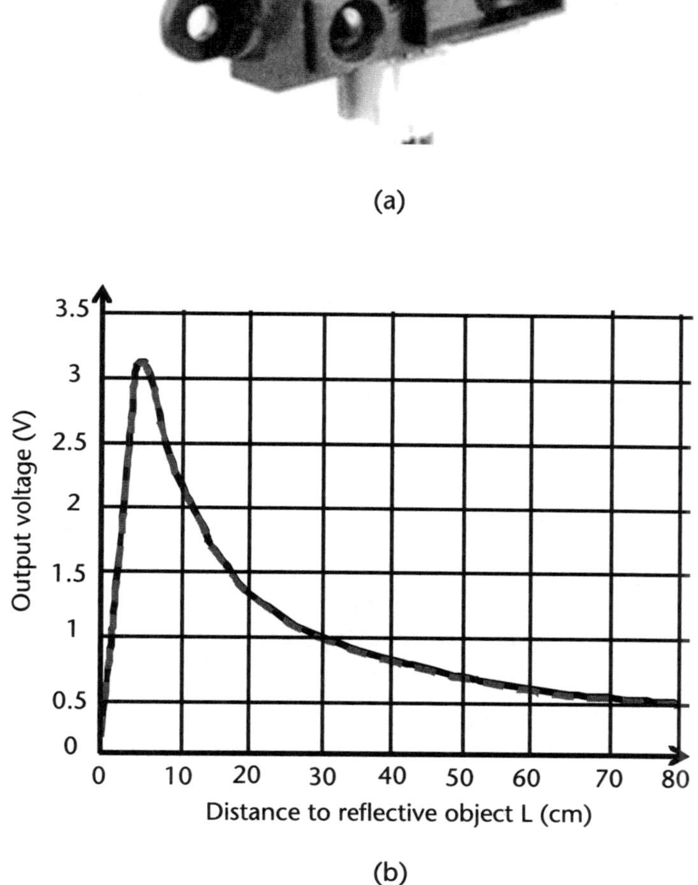

(a)



(b)

**Figure 9.6**   (a) Sharp GP2Y0A21YK0F IR sensor and (b) distance measuring characteristics of the IR sensor. (Source: http://www.sharpsma.com/webfm_send/1489.)

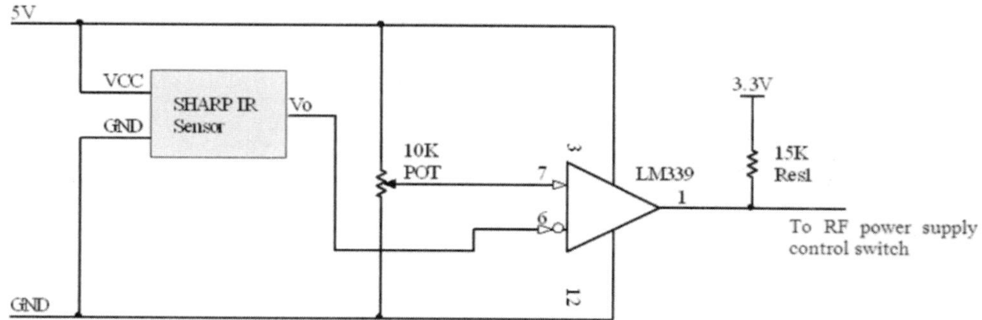



**Figure 9.7**   Schematic diagram for the sensor circuit.

### 9.2.3 RF Power Supply Control Switch

The RF module consumes a significant portion of the overall supplied power to the RFID reader system. Therefore, turning on the power supply of the RF module only at the instance when the tag comes within the vicinity of the reader would minimize the power rating of the overall reader system significantly. The power supply of the RF board would be controlled by using an integrated high-side switch FDC6330L. Its schematic is shown in Figure 9.8. As can be seen in the figure, the input power supply voltage $V_{in}$ will be transferred to output voltage $V_{out}$ if switch control input (On/Off port) is on; otherwise, GND will be connected to $V_{out}$. The FDC6330L switch is capable of sustaining up to 20V of input voltage across its terminals and up to 2.3A

The schematic for the RF power supply control circuit is shown in Figure 9.9. The input power supply VCC5 is transferred to the RF module as RFVCC5 only if switch control input SENSOR_D is on (output from the IR tag sensor); otherwise, GND is connected to the RF VCC5 line and the RF board does not get any supply voltage and, hence, goes into the power-saving standby condition.

### 9.2.4 VCO Control Voltage Generator

The most significant task for the digital control board is to generate the VCO control voltage for the RF module. The generated VCO control voltage, also called VCO tuning voltage, determines the operating frequency of the RF board. The digital control board takes digital input as instructed by middleware that is installed in the FPGA and generates analog output voltage corresponding to the digital input with the digital-to-analog (DAC) converter. The analog outputs from the DAC are passed through an OP-AMP circuit for noise filtering and isolation to protect against any unintended surge propagation between the digital and RF parts of the reader system. In this implementation a serial input DAC is used and interfaced

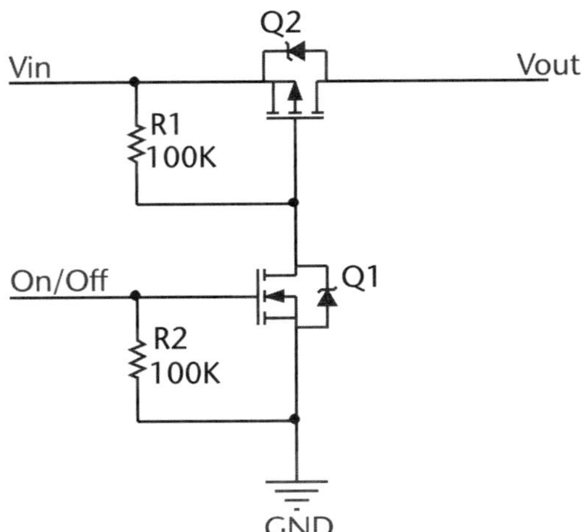


**Figure 9.8** Schematic for FDC6330L high-side switch circuit. (Source: http://www.digikey.com.au/product-detail/en/FDC6330L/FDC6330LTR-ND/979797, accessed February 22, 2013.)load current through its output terminals. The output current handling capability of the switch suffices also as the current rating of the RF transceiver board. The RF board uses approximately 500 mA of current for its successful operation.

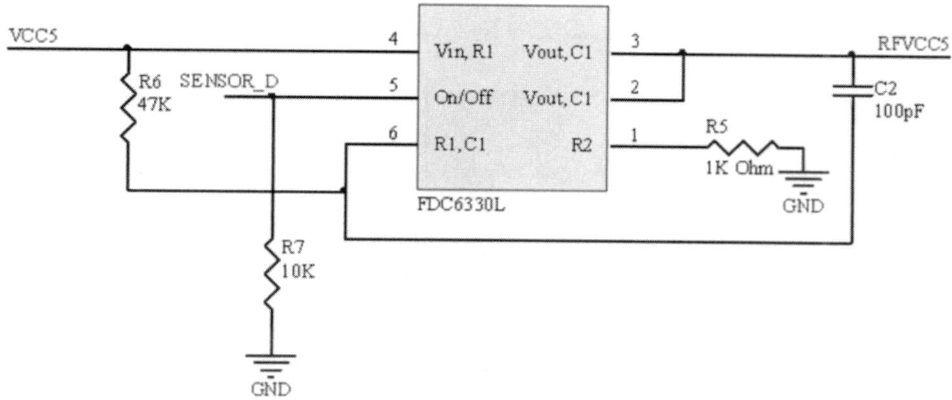

**Figure 9.9**   Schematic of the RF power supply control circuit.

with Altera FPGA general-purpose input/outputs (GPIOs). The schematic of the middleware installed FPGA pins and DAC is shown in Figure 9.10 and the OP-AMP implementation of the DAC output is shown in Figure 9.11.

## 9.3   Digital Signal Processing Board

As shown in Figure 9.2, the DSP board interfaces with the digital control board, RF board, and FPGA-based middleware. The digital control board informs the DSP board about the presence of the tag, so that the ADC present in the DSP board can sample the RF channel coming from the RF board. If the reader uses a beamforming

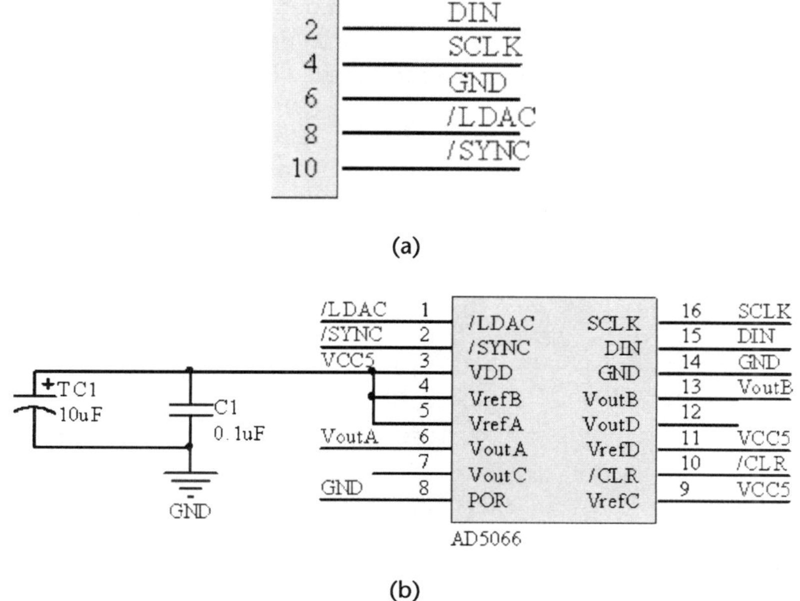

**Figure 9.10**   (a) FPGA GPIO pins from the middleware to the DAC and (b) the DAC circuit schematic.

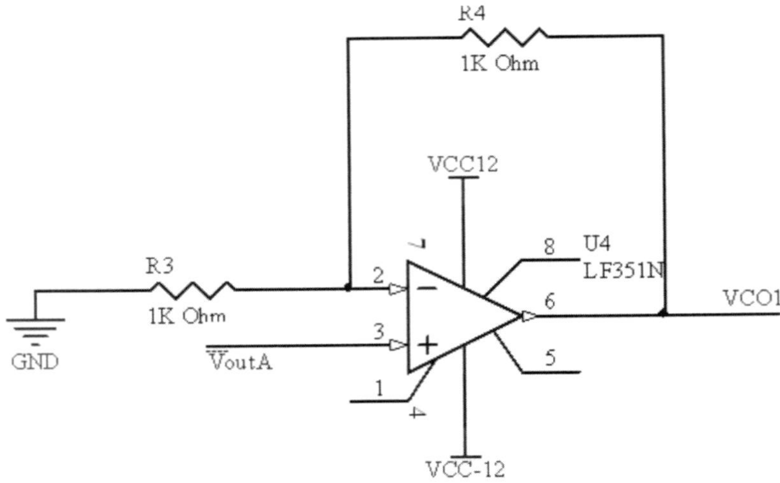

**Figure 9.11** OP-AMP circuit for VCO control voltage generation.

smart antenna, then the digital control board also sends beam direction information to the DSP board, which can be useful for further processing of digital samples from the RF tag signal received from the RF board. However, beam-steering smart antennas are not being used in the current chipless RFID readers. Therefore, this aspect of the digital control module is not covered in this chapter.

The functional block diagram of the DSP board is shown in Figure 9.12. Here a two-channel ADC (ADS2807) capable of sampling up to 50 MSPS in each channel individually is used. The ADC takes only one clock and internally synchronizes the sampling of two channels simultaneously. This is a very important feature that would be very useful for sampling the I and Q components of an IQ modulator based RF system. In an IQ modulation scheme, both the amplitude and phase of the RF signal can be captured. The sampled data are passed through the buffer before the connection to the FPGA GPIO pins to protect the FPGA from external surges. A 50-MHz crystal oscillator is used for the FPGA clock, and PLL is implemented in

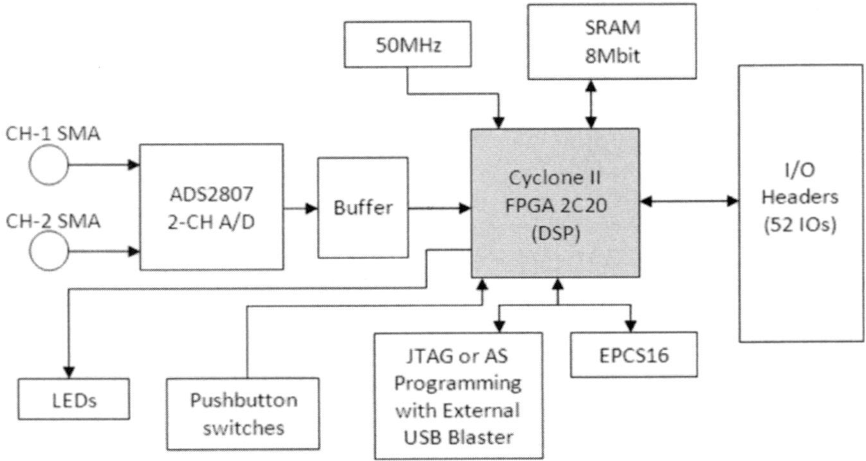

**Figure 9.12** Block diagram for custom DSP board.

the FPGA to generate the clock for the ADC. The DSP board consists of a SRAM to store sampled data from the ADC before processing, in case the internal FPGA flash memory is not sufficient to store a large number of samples. The DSP board also consists of a 52-pin IO header, which would be used for running the switching electronics (options for beam-steering antennas). JTAG, AS, and EPCS16 interfaces are provided for programming the FPGA to implement signal processing and switching electronics control algorithms. The switching electronics will only be used if the reader uses a beam-steering smart antenna.

A custom DSP board is designed and developed as shown in Figure 9.13 and, in summary, incorporates the following components:

- Altera Cyclone II 2C20 FPGA device;
- Altera EPCS16 serial configuration device;
- 8-Mbit SRAM;
- ADS2807 dual-channel ADC, 50 MSPS;
- Two SMA connectors for receiver interface with ADC;
- Two user pushbuttons and one reconfiguration pushbutton;
- 50-MHz oscillator for clock source;
- One 40-pin expansion header and two 10-pin expansion headers for GPIOs
- Powered by 7.5V with 5V, 3.3V, and 1.2V regulators on board;
- Two 10-pin expansion headers for external USB Blaster programming and user API control: one for JTAG and one for Active Serial (AS) programming mode;
- Three indicator LEDs and two user LEDs.

## 9.4   Middleware

As mentioned earlier, middleware is a set of instructions that provides all of the control and command requirements for a chipless RFID system. The operation of the firmware can be divided into two phases: the data acquisition phase and the

**Figure 9.13**   PCB implementation of the DSP board.

data decoding phase. During the data acquisition phase, the output frequency of the VCO is varied from 4.0 to 6.0 GHz in 1024 steps as described in Section 4.3. Then, at each voltage step, the output of the gain/phase detector is sampled using the ADC of the digital control section, and sampled values are stored in the RAM.

This procedure for sampling gain and phase voltage data is illustrated in Figure 9.14. First, a retransmission chipless tag with all shorted spirals is scanned and the gain and phase voltage data are recorded on the RAM. This procedure is called *calibration of the reader*. Then the chipless tag to be read is placed and the data acquisition phase is repeated to store the sampled voltage values into the RAM. During the data decoding phase, a new sample set is generated by taking the difference of the sampled values of the tag data and calibration data. This new data set is then further processed with a simple denoising technique, the moving average filtering technique, to smooth the waveform [28]. This filtering technique enhances the accuracy of the readings as well as the reading range of the reader. Then the firmware starts to search for peaks in the areas according to the number of encoded

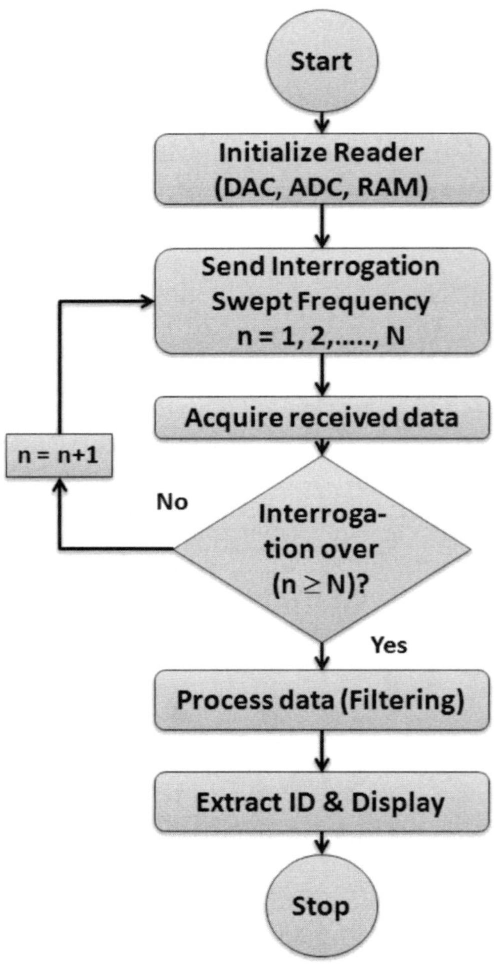


**Figure 9.14** Flowchart of chipless RFID tag detection and control in middleware.

bits and corresponding places to the resonance frequencies of the spirals of the tag. Then the peak value is compared with a threshold value and the threshold value is determined experimentally. If a voltage peak is found, it is identified as a logic "0," and if there is no voltage peak at the searched range, it is identified as a logic "1." Finally, the result is displayed on the LCD screen and sent to the PC connected to the reader. The flowchart for the firmware is shown in Figure 9.14. The firmware was developed using the C language and programmed into the microcontroller.

## 9.5   Performance of Digital Control Section

This section presents some important results on the performance of the digital control section. Figure 9.15 shows the measured voltage ramp generated by the DAC, which is applied to the tuning voltage of VCO or YIG oscillator. Measurements were made using an oscilloscope. Figure 9.16 shows the amplitude spectrum of the swept frequency interrogation signal generated by the VCO. Measurements were taken with an Agilent E4408 spectrum analyzer. As shown in the figure, the solid line represents the measured data and the dotted line is the ideal flat output expected from the VCO. These two results confirm the accurate generation of an interrogation signal by the transmitter. Two retransmission chipless RFID tags having 4 and 9 data bits were successfully read with the digital control section. In the reading process only the number of sampling points (N) was changed in the firmware. Figure 9.17 shows the plot of sampled 10-bit voltage data corresponding to the 4-bit tag-encoded data bits "0000" and "0011." Sampled voltage data for the 9-bit tag obtained from the digital control section are shown in Figure 9.18 for encoded data bits "000000000" and "010101010." These results confirm the multibit reading capability of the reader.

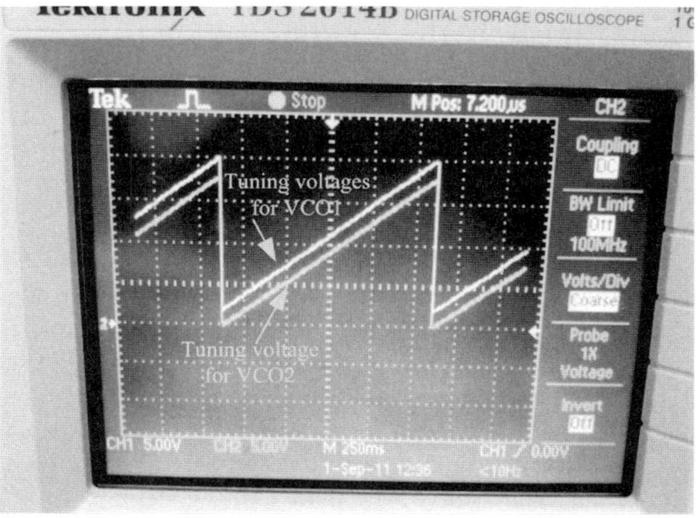


**Figure 9.15**   Screen capture of ramped VCO tuning voltage.

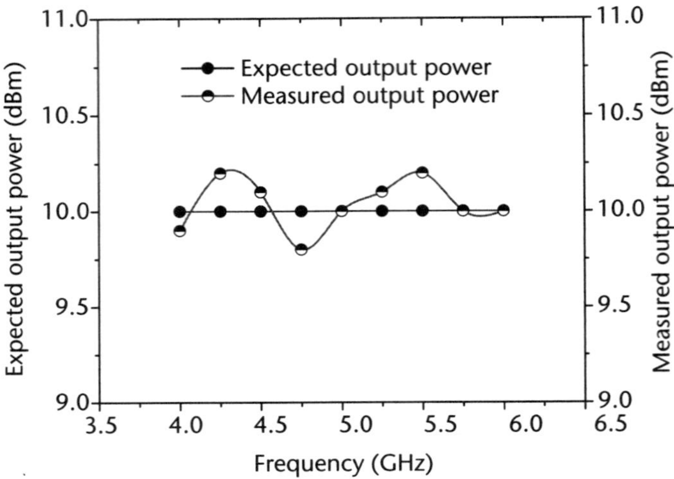

**Figure 9.16**  Amplitude spectrum of VCO output.

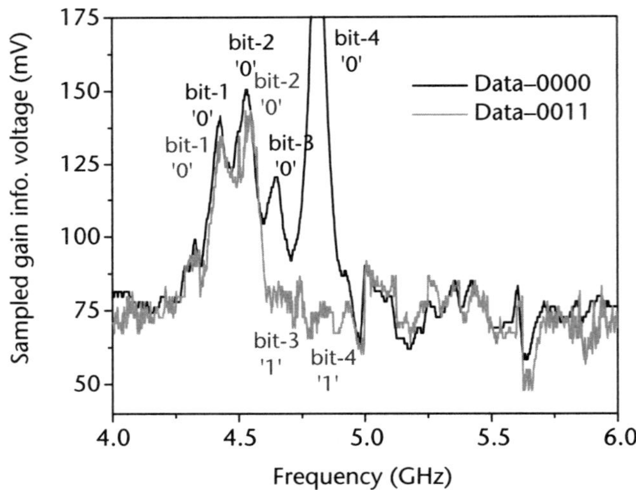

**Figure 9.17**  Stored digitized voltage samples for 9-bit tags—encoded data bits "000000000" and "010101010."

## 9.6  Conclusion

This chapter briefly introduced the overall reader structure of the chipless RFID system and then explained how middleware controls the reader firmware and interfaces the reader with the back-end enterprise software. Detailed descriptions of the functional blocks of the digital module of the chipless RFID reader were provided. The digital module has three main components: (1) the digital control board, (2) the digital signal processing board, and (3) FPGA-based middleware.

The digital control board identifies tag arrival, controls the power supply to the RF transceiver module of the reader, and generates the VCO control voltages required by the RF module to generate tag interrogation swept-frequency signals. In addition, the digital module can have switching electronics in case the reader uses digital beamforming smart antennas for tracking a moving chipless tag. The digital signal processing board receives tag backscatter from the RF receiver module, samples it, and processes it for tag identification. It also implements the switching electronics control FPGA program on its on-board FPGA. The FPGA-based middleware controls both the digital control board and the digital signal processing board. The middleware also interfaces the reader electronics with the back-end user application (enterprise software) that uses tag data.

The modular design approach to the chipless RFID digital architecture presented in this chapter can be adopted to various chipless RFID reader developments. Finally, the design has the potential to handle moving tags or readers or both simultaneously with a digital beamforming smart antenna if switching electronics are implemented.

## Questions

1. Draw the functional block diagram of a chipless RFID reader and briefly discuss the functionality of major components.
2. Describe the function of a digital control board with Figure 9.4.
3. How does the digital control board minimize the overall energy rating of a reader?
4. Discuss the interface of a digital control board with other parts of the reader.
5. Draw the IR sensor circuit and describe how IR sensor sensitivity can be tuned for tag detection.
6. Draw the circuit of the switching regulator and describe its operation for RF module power supply control.
7. Draw the block diagram of the digital signal processing module and describe its operations.
8. What is the middleware in a chipless RFID system? Draw a signal flow graph of the functions of middleware in the digital control section and the digital signal processing board.
9. If you are given a choice to add a smart antenna to a chipless RFID tag reader, explain what extra blocks you would need to perform digital beamforming with the antenna.

## References

[1]   R. Das, "Chip-less RFID—The end game," IDTechEx, February 2006, http://www.idtechex.com/products/en/articles/00000435.asp.

[2]   C. S. Hartmann, "A global SAW ID tag with large data capacity," in *Proc. IEEE Ultrasonics Symposium*, Munich, Germany, October 2002, pp. 65–69.

[3]   S. Preradovic, N. Karmakar, and I. Balbin, "RFID transponders," *IEEE Microwave Magazine*, vol. 9, no. 5, pp. 90–103, October 2008.

[4]   S. Härma, "Surface acoustic wave RFID tags: Ideas, developments, and experiments," PhD thesis, http://lib.tkk.fi/Diss/2009/isbn9789512297436/isbn9789512297436.pdf.

[5]   C. Hartmann et al., "Anti-collision methods for global SAW RFID tag systems," in *Proc. 2004 IEEE Ultrasonics Symposium*, vols. 1–3, pp. 805–808.

[6]   L. M. Reindl and I. M. Shrena, "Wireless measurement of temperature using surface acoustic waves sensors," IEEE Trans. on *Ultrasonics, Ferroelectrics and Frequency Control*, vol. 51, pp. 1457–1463, 2004.

[7]   D. C. Malocha, D. Puccio, and D. Gallagher, "Orthogonal frequency coding for SAW device applications," in *Proc. 2004 IEEE Ultrasonics Symposium*, vol. 2, pp. 1082–1085.

[8]   A. Stelzer et al., "Multi reader/multi-tag SAW RFID systems combining tagging, sensing, and ranging for industrial applications," in *2008 IEEE International Frequency Control Symposium*, vols. 1–2, pp. 263–272.

[9]   D. M. Dobkin, *The RF in RFID: Passive UHF RFID in Practice*, Boston: Newnes, 2007.

[10]  I. Jalaly and I. D. Robertson, "Capacitively-tuned split microstrip resonators for RFID barcodes," in *European Microwave Conference (EuMC 2005)*, Paris, France, 2005, pp. 1–4.

[11]  V. Deepu et al., "New RF identification technology for secure applications," in *2010 IEEE International Conference on RFID—Technology and Applications (RFID-TA)*, pp. 159–163.

[12]  K. Finkenzeller, *RFID Handbook: Fundamentals and Applications in Contactless Smart Cards and Identification*, 2nd ed. Chichester: John Wiley Sons, 2003.

[13]  S. Preradovic et al., "Multiresonator-based chipless RFID system for low-cost item tracking," *IEEE Trans. on Microwave Theory and Techniques*, vol. 57, no. 5, pp. 1411–1419, May 2009.

[14]  S. Preradovic, I. Balbin, and N. Karmakar, "The development and design of a novel chipless RFID system for low-cost item tracking," in *2008 Asia-Pacific Microwave Conference (APMC 2008)*, pp. 1–4.

[15]  S. Preradovic et al., "Multiresonator-Based Chipless RFID System for Low-Cost Item Tracking," *IEEE Trans. on Microwave Theory and Techniques*, vol. 57, pp. 1411–1419, 2009.

[16]  S. Preradovic and N. Karmakar, "Chipless RFID tag with integrated sensor," in *2010 IEEE Sensors*, pp. 1277–1281.

[17]  S. Preradovic, N. Karmakar, and I. Balbin, "RFID transponders," *IEEE Microwave Magazine*, vol. 9, pp. 90–103, 2008.

[18]  S. Preradovic and N. Karmakar, "Fully printable chipless RFID tag," in *Advanced Radio Frequency Identification Design and Applications*, S. Preradovic (Ed.), http://www.intechopen.com/articles/show/title/fully-printable-chipless-rfid-tag.

[19]  T. Singh et al., "A frequency signature based method for the RF identification of letters," in *2011 IEEE International Conference on RFID (RFID 2011)*, pp. 1–5.

[20]  M. S. Bhuiyan, R. Azim, and N. Karmakar, "A novel frequency reused based ID generation circuit for chipless RFID applications," in *2011 Asia-Pacific Microwave Conference (APMC 2011)*, pp. 1470–1473.

[21]  R. V. Koswatta and N. C. Karmakar, "Time domain response of a UWB dipole array for impulse based chipless RFID reader," in *2011 Asia-Pacific Microwave Conference (APMC 2011)*, pp. 1858–1861.

[22]  A. Lazaro et al., "Chipless UWB RFID tag detection using continuous wavelet transform," *IEEE Antennas and Wireless Propagation Letters*, vol. 10, pp. 520–523, May 2011.

[23]  N. C. Karmakar and M. E. Bialkowski, "A beam-forming network for a circular switched-beam phased array antenna," *IEEE Microwave and Guided Wave Letters*, vol. 11, no.1, pp. 1–3, January 2001.

[24]  N. C. Karmakar and M E Bialkowski, "A compact switched-beam array antenna for mobile satellite communications," *Microwave and Optical Technology Letters*, vol. 21, no. 3, pp. 186–191, May 5, 1999.

[25]  N. C. Karmakar and M. E. Bialkowski, "An 8-element switched beam array for mobile satellite communications," in *Proc. of TENCON'98*, Delhi, India, December 17–19, 1998, pp. 241–244.

[26]    N. C. Karmakar and M. E. Bialkowski, "Design and development of low cost components and sub-system for L-band switched beam and phased array antennas," in *Proc. IEEE APCC'98/ICCS'98*, Singapore, November 23–27, 1998, pp. 423–427.

[27]    N. C. Karmakar and M E Bialkowski, "A low cost switched beam array antenna for L-band land mobile satellite communications in Australia," in *Digest of the 1997 IEEE AP-S International Symposium*, Montreal, Canada, July 13–18, 1997, pp. 2226–2229.

[28]    R. Koswatta and N. C. Karmakar, "Moving average filtering technique for signal processing in digital section of UWB chipless RFID reader," in *Proc. 2010 Asia Pacific Microwave Conference*, Yakohama, Japan, December 7–10, 2010 (CD-ROM).

# RFID Reader System Integration and Applications

## 10.1 Introduction

This chapter describes an integrated chipless RFID reader that can be used for commercially viable applications. Up to this point, the book has described various modular components and reading methods for chipless RFID readers. The organization of chapters of this book is illustrated in Figure 10.1. Chapters 1, 2, and 3 provided a general overview of chipless RFID systems, their significance over conventional RFID systems, and the different types of chipless RFID tags and readers that have been proposed so far in the open literature [1–10]. The reading methods of chipless RFID tags can be divided into frequency domain, time domain, and hybrid domain reading techniques. These reading techniques are associated with their reader architectures, which were described in detail in Chapters 4, 5, and 6. The physical layer development of chipless RFID readers involves various UWB antennas [13–16], RF and microwave components, and analog and digital electronics. To augment the reader architecture, Chapter 7 described various types of UWB antennas that can be used in chipless RFID readers. The design, fabrication, and measured results of the antennas were also provided. Chapter 8 described the design specifications and performance evaluation of various active and passive RF components currently available on the market. It also presented some in-house designed and fabricated passive microwave components and performance evaluations. The digital section of the reader was covered in Chapter 9. The digital section of the reader includes a digital control board and the digital signal processing board. Both boards are loaded with in-house developed control algorithms. DSP algorithms will be presented in a companion book titled Chipless RFID Reader Signal Processing. This chapter discusses how to integrate the components and building blocks that were presented in the preceding chapters.

A chipless RFID system comprises two main components: the chipless tags and a reader. Optionally the reader may be connected to a central database system for continuous updating of the information. Because RFID technology is an application-specific technology, the limitations and constraints of system implementation are defined by that application. Even technology wise, the commercially available

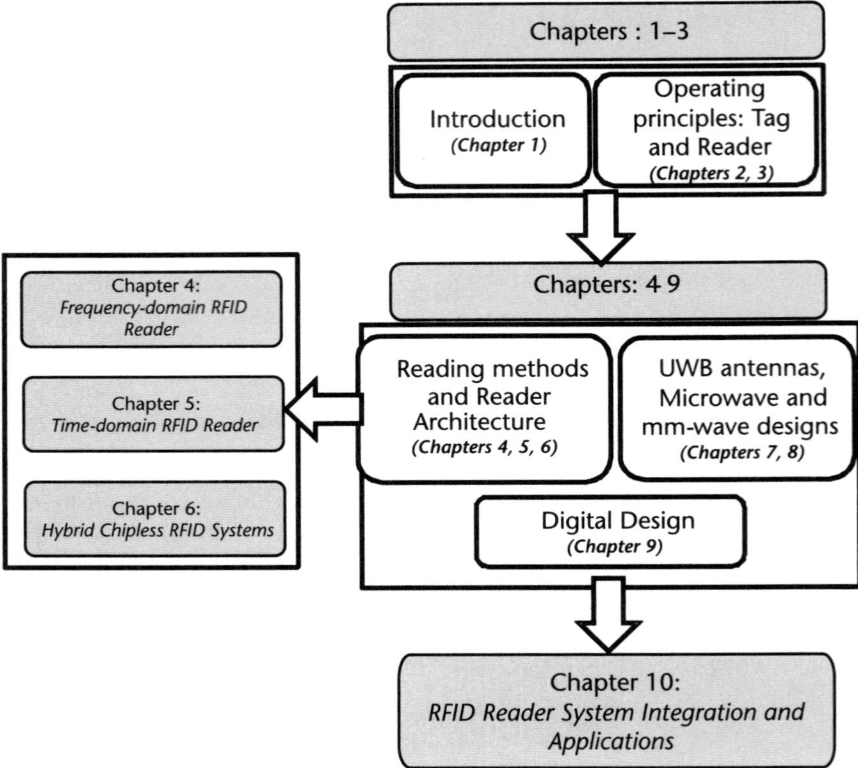



**Figure 10.1** Flowchart of system integration of chipless RFID reader.

chipless tags—RF-SAW, TFTC, SAR-based tags, and radar arrays—are very different from each other. Therefore, the implementation of these tag-reader systems should be addressed carefully during the design process of the system. Further, there are many constraints due to the limitations imposed by technology or by the application itself that should be taken into consideration during the design phase of the system as well. This chapter describes the integration of the total chipless RFID reader system, design considerations in using different technologies, antenna selection, design considerations for chipless tags, and case studies that illustrate the type of applications in which chipless RFID tags can be used.

The chapter is organized as follows. A brief introduction of the components of a chipless RFID system is presented first. The special requirements, limitations, and operating constraints of each component in a RFID system are explained. Then two case studies based on probable commercial application scenarios are presented. In the first case study, a short-range, near-field integrated reader with a chipless tag is described. For this near-field reader, a maximum reading range of 10 cm is targeted. The second case study discusses a long-range RFID reader. A high-gain reader antenna with a high-sensitivity RF board is used for achieving the longer operating range.

## 10.2   The Integrated Chipless RFID Reader

The components of a generic chipless RFID system essentially consist of four parts:

1. Chipless RFID tag
2. Reader antennas
3. Reader device consisting of an RF section and a digital control section
4. Application software or a user interface.

Figure 10.2 illustrates the hierarchy of these main components of a chipless RFID system.

As mentioned in Chapter 8, the individual performance of the selected components as well as their interconnectability must be considered when building a chipless RFID system. This will reduce the number of iterations required for system integration and yield the most efficient system.

## 10.3   Specifications of a Generic Chipless RFID Reader

This section describes the requirements and constraints of different components of a generic chipless RFID system. It also provides some examples of requirements and limitations of the chipless RFID reader prototypes described in Chapters 4 and 5. Table 10.1 shows the specifications for a chipless RFID reader. The reader operates within the UWB frequency band (3.1 to 10.7 GHz) with the FCC power spectral density regulation. Depending on the number of data bits and bandwidth requirement, a portion of the UWB band can be used. Because UWB interrogation with circular polarization over the UWB band is hard to achieve, the system is designed based on linearly polarized (LP) antennas. This makes the reader orientation-dependent like an LP RFID system. However, for reading the retransmission chipless tag, two orthogonal linearly polarized reader antennas are required. The gain of the antennas depends on the reading range of the reader.

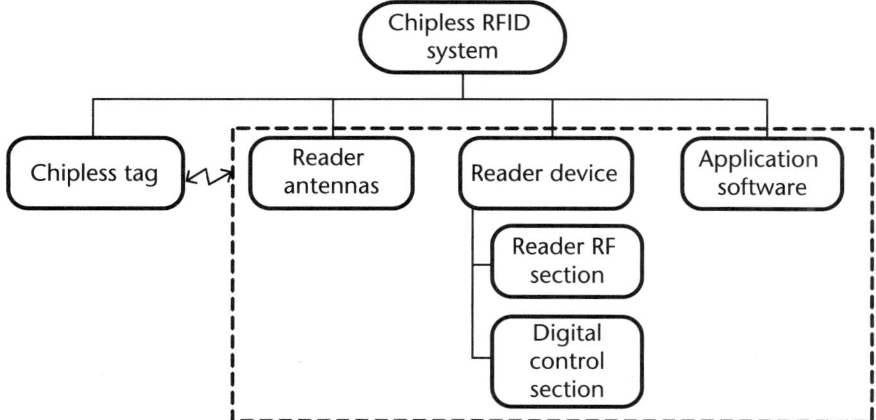


**Figure 10.2**   Main components of a generic chipless RFID system.

**Table 10.1**  Specifications for Chipless RFID Readers

| Frequency of operation | 3.1–10.7 GHz |
|---|---|
| Transmitting power | –41.3 dBm/MHz |
| Number of reader antennas | Two antennas: transmitting and receiving |
| Antenna polarization | Linearly polarized |
| Reader antenna gain | 8 dBi (short range, less than 5 cm)<br>22 dBi (long range) |

### 10.3.1  Chipless RFID Tag

Chapter 2 provided a comprehensive review of various types of chipless RFID tags. Except RF-SAW tag, all tags are in the research phase. Commercial success of these RFID tags will need some time. Monash University–developed retransmission based chipless RFID tags comprise two orthogonally polarized antennas connected to a section of multiresonator as the data encoding device. Although this tag is bulky, it provides supreme performance with a good reading range (60-cm reading range with10-dBm power). It has been used to demonstrate the operation of the integrated chipless RFID tag reader.

### 10.3.2  Reading Techniques

Antennas are used in two components of the chipless RFID system. Tags use one or more antennas to receive and transmit the interrogation signal and data encoded signals. Reader devices require antennas for transmitting (Tx) and receiving (Rx) signals from the tags being interrogated. The most important required feature is the operating bandwidth of the antennas. The operating bandwidth of both tag and reader antennas should cover the entire bandwidth of the data encoding circuits of chipless tags to utilize the full capacity of the tag. The radiation pattern is also important depending on the application. An omnidirectional radiation pattern is preferred for tag antennas so that the placement of reader antennas is flexible. A directional antenna is preferred for a reader antenna to achieve a large reading distance and less interference from the surrounding environment and tags. However, these specifications may vary with the application.

For time domain based RF front-ends, the reader antennas used should be capable of transmitting ultra-short-duration RF pulses with minimal distortion to pulse shapes [13, 14]. A linear phase response and constant gain are two desirable properties of an antenna used for the transmission of ultra-short-duration pulses. Conical antennas, TEM horns, and monopole antennas are some of the commonly used antennas for the transmission and reception of time domain RF impulses. For short-distance applications that require a higher degree of freedom in tag placement, such as tap-and-go credit card applications where the embedded RFID tag can be placed in front of the reader antenna at a random orientation by a user, antennas having a wider 3-dB beamwidth are required. The wider beamwidth is particularly useful in rectifying the ill effects caused by the shorter distance to the object being interrogated; that is, a small linear translation of the object causes a large angular deviation at the reader. For applications demanding large range, directive antennas need to be employed. High-gain antennas such as horns and patch antenna arrays can be used for long-range applications. The high directivity of these antennas, however, reduces the dimensions of the readable zone of the reader, so the reader will be unaware of an RFID tag placed outside of this zone. As a solution to this problem, the Tx and Rx antenna setup, which gives rise to this readable zone, needs to be mechanically or electronically steered. This will move the readable zone, enabling the reader to scan for chipless RFID tags in its vicinity. Aspects of Tx and Rx antenna design were introduced in Chapter 7.

In addition to the operating bandwidth, beamwidth, and gain of the reader antennas, proper alignment is also essential to achieve proper reading of chipless tags. The requirement for proper alignment of tags arises from the type of polarization that the tag and reader antennas have. The basic requirement for proper reading of a tag is that the tag should be placed in a polarization-matched condition in the zone of the 3-dB beamwidth of the Tx and Rx antennas. Furthermore, the proper use of polarization in the Tx and Rx antennas is required for the proper interrogation process. For example, most of the tag designs discussed in this book use linearly polarized antennas. Therefore, the best pair of Tx and Rx antennas is a linearly polarized antenna pair. When using linearly polarized antennas, the polarization of both tag and reader antennas must be known and the antennas aligned accordingly.

Figure 10.3 shows a possible alignment for chipless RFID reader antennas with a tag that has separate Tx and Rx antennas to achieve the best tag reading results. Figure 10.4 shows the reading setup of the tag. Two tea leaf shaped dipole reflector array antennas are placed in parallel with the orthogonally polarized antennas of the tag [13]. Thus, both tag and reader antennas ensure the lowest possible polarization losses between the Tx and Rx chain of them and improved cross-talk between the two.

In the experimental setup for the dual-planar arrangement as shown in Figure 10.4, two dipole reflector array antennas were attached to the two ports of the Agilent E8361A PNA vector network analyzer. The frequency signatures of the chipless tags were recovered successfully as shown in Figure 10.5 for two different tag IDs: "000000000" and "100110101." The maximum read range with this setup was 15 cm in the interference environment.

Figure 10.6 shows a second alignment of a tag and two monopole reader antennas. This arrangement can be termed a *uniplanar* reading alignment. Here

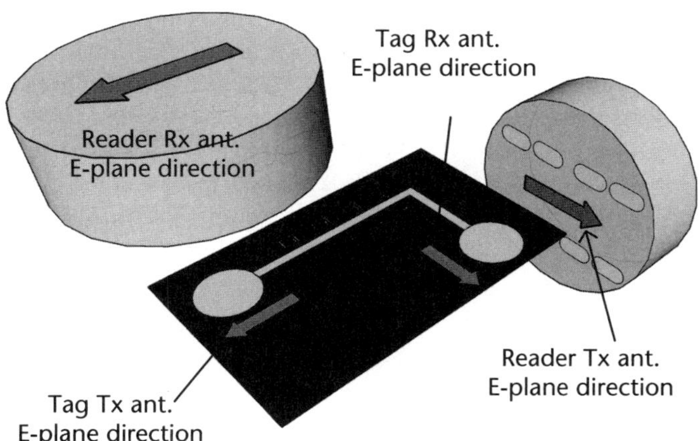


**Figure 10.3**   Dual-planar reader and tag alignment for retransmission chipless tag (type 1).

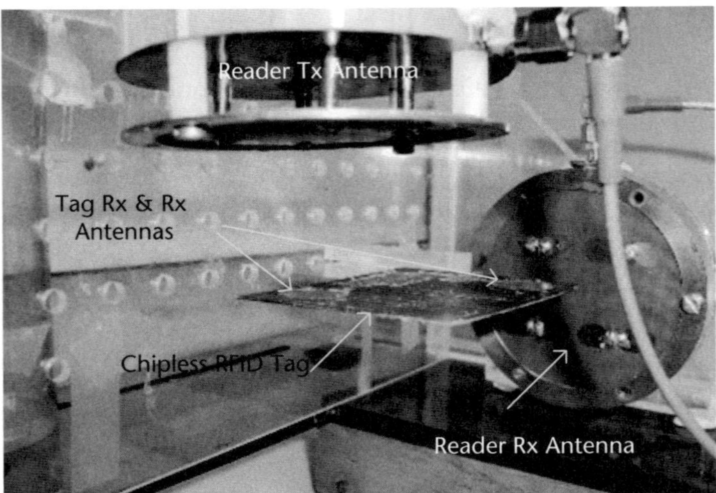



**Figure 10.4**  Reading a 9-bit chipless RFID tag realized on Taconic TLX-0 PCB substrate using tea leaf shaped dipole reflector antenna array (type 1).

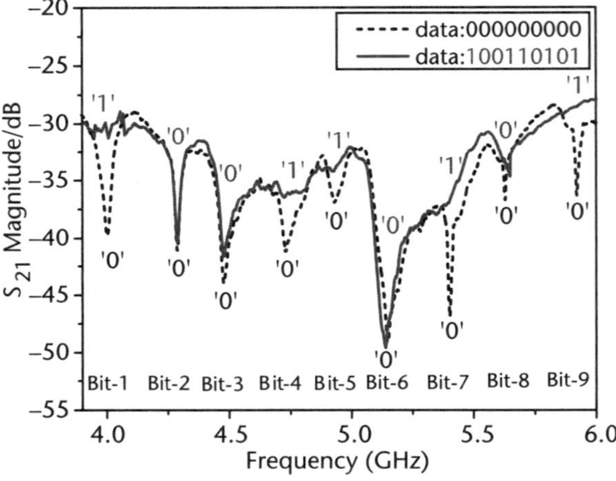



**Figure 10.5**  Measured forward transmission coefficient (S21) magnitude of a 9-bit chipless RFID tag using dual-planar reader setup.

orthogonally polarized reader antennas can be placed on the same plane for compact reader design (i.e., a paralleled reader). Although the disk-loaded monopole antenna yields omnidirectional radiation in parallel with the feed line, the two reader antennas do not directly couple signals between the two antennas since they are kept perpendicular to each other to achieve maximum isolation.

Figure 10.7 shows the reading of chipless RFID tags with monopole antennas. The alignment of tags with reader antennas as shown in Figure 10.6 was used with reader type 2 (uniplanar setup) in this setup. The recovered amplitude and phase responses of the 9-bit chipless RFID tags with the chipless system with two monopole reader antennas are shown in Figures 10.8 and 10.9, respectively.

Another chipless RFID system configured to read printed chipless RFID tags is shown in Figure 10.10. Here two monopole antennas were used as reader antennas

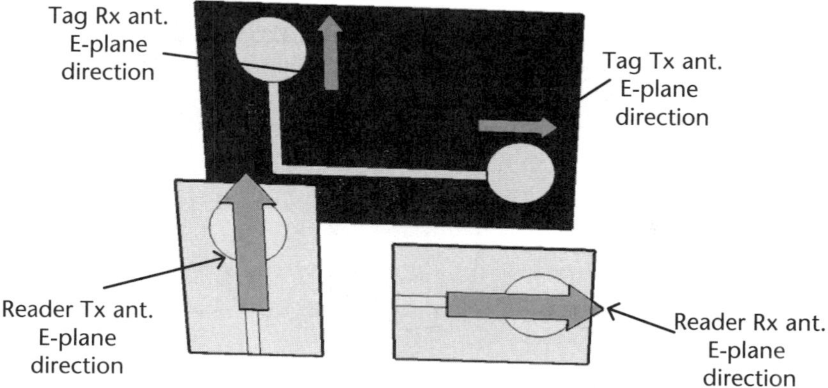


**Figure 10.6**   Uniplanar tag and reader alignment (type 2).

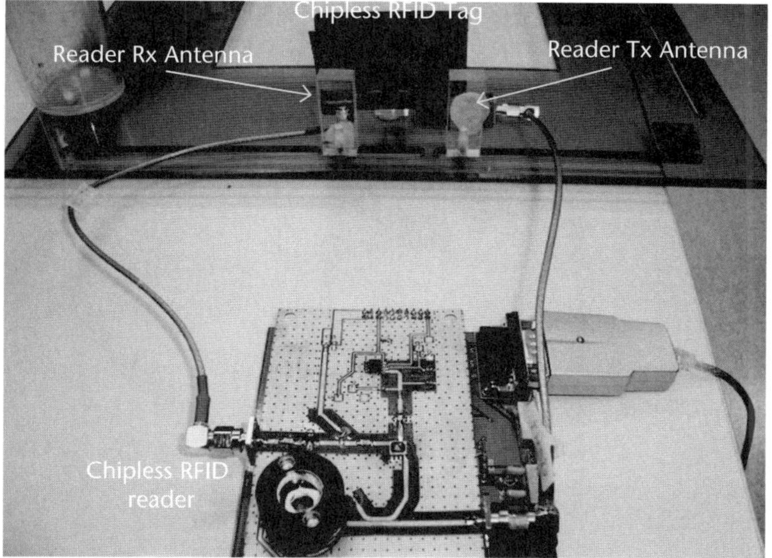


**Figure 10.7**   Reading frequency signatures of chipless RFID tag and reader prototypes realized on Taconic TLX-0 PCB substrate with uni-planar reader set-up (type-2)

and connected to the reader as shown in the figure. The interaction between the two antennas is minimal since the gain is low, and above the circular disk of the monopole antenna, there is a null in the radiation pattern. Because the printed tag antennas have very low gain, it is not possible to read larger read ranges. A read range up to 5 mm with transmitted power of 10 dBm (10 mW) is possible with this setup, and this system is suitable for applications where touch card–type reading is required. Figure 10.11 shows the comparison of amplitude response measurements for a 1-bit printed chipless tag taken from the Agilent E8361A PNA and type 1 reader. The reader is capable of identifying the proper resonant frequency. Due to the lower sampling frequency resolution used in a type 1 reader, the measured result does not provide the proper shape for the amplitude response. This can be

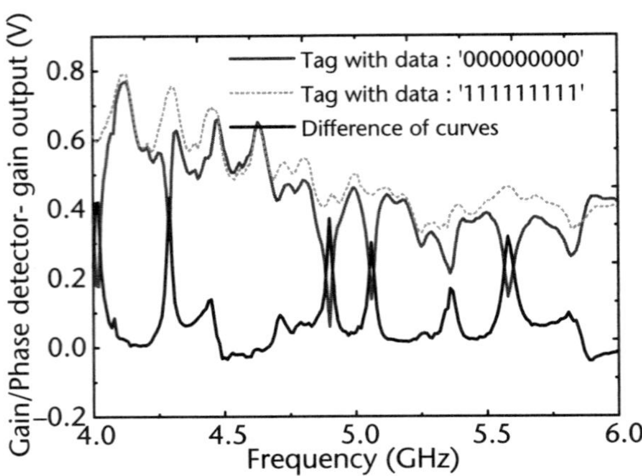

**Figure 10.8**  Measured gain/phase detector (GPD) amplitude information for two tags with encoded data 000000000 and 111111111 using a uniplanar reader system.

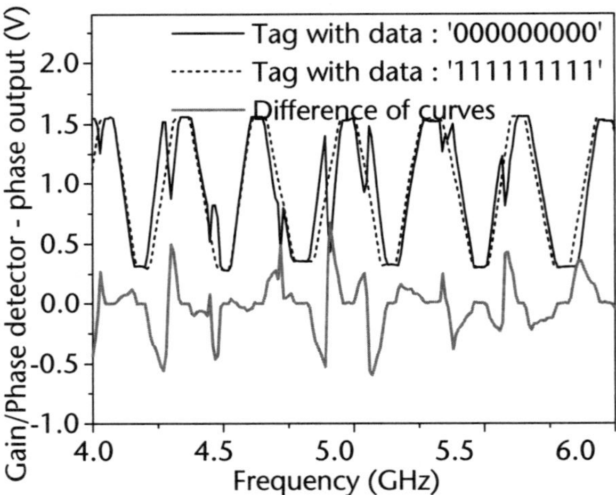

**Figure 10.9**  Measured GPD phase information for two tags with encoded data 000000000 and 111111111 using a uniplanar reader system.

improved by increasing the number of frequency points in the interrogation signal. However, after digitization and display, correct detection of the tag was achieved.

### 10.3.3  Reader Device

As mentioned in previous chapters, reader devices comprise two major sections: an RF section and a digital control section. There should be good interconnectivity between the sections to achieve the desired output. The following description of these two sections provides information on important parameters that should be considered during the integration of the modules into a system.

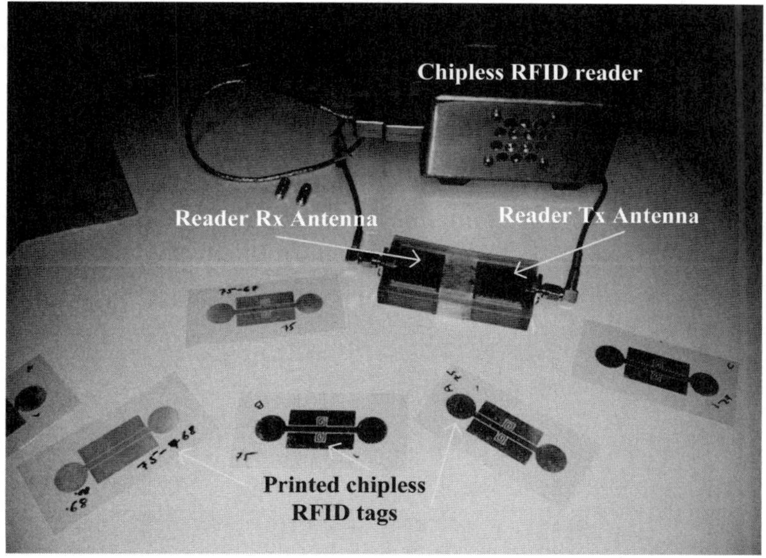



**Figure 10.10**   Reading frequency signatures of chipless RFID tags printed on a polymer with a reader built in-house.

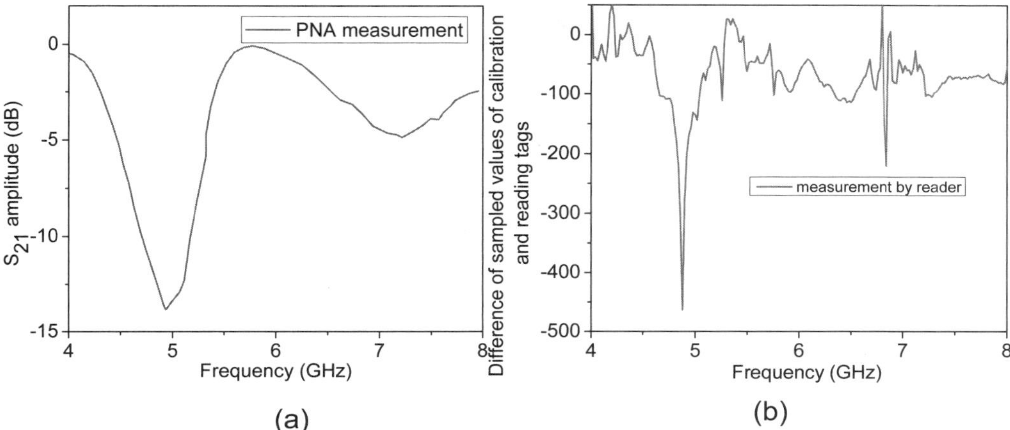



**Figure 10.11**   Reading frequency signatures of chipless RFID tags printed on polymer (a) PNA and (b) dual-planar reader.

### 10.3.3.1   RF Section

The RF front-end should be well separated and shielded from the other sections of the reader in order to minimize the interference and the noise affecting it. The RF front-end consists of colocated Tx and Rx sections that, respectively, transmit an interrogation signal to illuminate the chipless RFID tag and receive backscattered signals from the tag. The theory behind the reader's RF front-end was introduced in detail for frequency domain and time domain RFID readers in Chapters 4 and 5, respectively.

The technology of the RF section plays an important role in the performance of the chipless RFID system. The frequency domain reader design, which is presented in Chapter 4, uses a frequency-swept continuous wave for interrogating the tags. This reader architecture requires a large time for the interrogation process compared to a time domain based system, which transmits an ultra-short-duration burst of RF energy for interrogating the tag. On the other hand, the time domain reader design requires careful synchronization and timing together with complex digital processing algorithms in order to implement high-speed acquisition and digitization of the received time domain backscattering signal. Therefore, the time required for processing the acquired signal may be longer in a time domain reader system than in a frequency domain reader. Generally, even compared with a fast signal processing unit to digitize analog signals in the frequency domain, a time domain reader would still prove to be faster than a frequency domain reader. This is due to the advantage gained through the quick tag interrogation process in the UWB-IR system. In a frequency domain system, the settling time of the voltage-controlled oscillator (VCO) proves to be a bottleneck that slows the interrogation process.

The process can be speeded up by using a VCO with a lower settling time; however, this would increase the cost of the overall reader design. Depending on the application, the time constraints involved in reading a tag change. For example, the time available for reading a chipless tag embedded within a document using a handheld reader is longer than the time available for reading moving items on a conveyer belt. Therefore, system designers need to consider these constraints when choosing the type of technology used for the reader's RF front-end design.

Another fact that should be taken into account is the data encoding techniques used in the tags of the chipless RFID system. If it is essential to obtain both amplitude and phase information for the tag to decode data. The best option for the RF section is frequency domain based interrogation because it provides an accurate and low-cost method to derive both amplitude and phase information. The time domain interrogation technique introduced in Chapter 5 provides only the amplitude information and does not provide phase information. If there is a requirement to consider only the amplitude information of tags, either the frequency domain or the time domain interrogation technique could be considered for the implementation.

### 10.3.3.2  Digital Control Section

The digital control section of a chipless RFID reader is explained in detail in Chapter 9. The digital section operates at very low frequencies compared to the RF front end. The high-frequency signals received at the RF front-end are ultimately demodulated and down-converted to the baseband before they are presented for processing in the digital section. The down-converted signals are first analog to digital converted in the digital section. Once the analog signals have been digitized into binary data bytes, DSP algorithms can be applied to them to extract the frequency content of the backscattered signals as mentioned in Chapter 9.

The fast Fourier transform (FFT) is a commonly used algorithm for this purpose. Other functions for enhancing signal quality such as windowing functions that window and separate signals in the time domain, noise filtering functions, and computation of statistical averages can also be easily performed with DSP algorithms. In addition to these traditional signal processing techniques, special signal processing functions may be required for decoding data as described in previous chapters. The digital section can be implemented using either an industrial micro-controller, microprocessor unit, or a field programmable gate array (FPGA). These devices have limited processing capabilities and processing speeds. Therefore, for applications that require high computational speed and computationally intensive algorithms, the digital section can simply be extended into a digital signal process-ing board, single board computer, or a personal computer. In such a scenario, the digital section residing in the RFID reader system simply does the analog-to-digital conversion of the raw analog signal with some basic noise filtering and forwards the processed data to a larger computer for further processing through an interface such as a wireless adapter, Ethernet, USB connection, and serial cable.

In addition to the operating speed and processing power, the architecture of the digital section and performance of the analog and digital components used in the digital section are important in a system design. The digital section should be able to generate analog and digital control signals for the RF section with sufficient speed as well as within the required range. The digital section also requires suf-ficient dynamic range for the analog front-end for data acquisition and a suitable sampling rate for proper baseband signal acquisition. Therefore, when designing the digital section, certain parameters such as maximum update rates of the DAC, achievable sampling rates of the ADC, and speed of the internal buses such as SPI and I2C need consideration. These parameters are used by the microprocessor or FPGA to send control signals that are also properly matched with the desired per-formance of the overall system. As stated earlier, a chipless RFID reader with time domain based interrogation needs a more powerful digital section than a frequency domain interrogation based system. Therefore, careful selection and a comprehen-sive analysis of the cost and performance are required when integrating a chipless RFID system.

### 10.3.4   Application Software

RFID readers are often connected to a host computer or similar host computing device. The host computer can optionally be connected to a central database system for continuous updating of tag information. To interface the reader with the host computing device, application software is used and sometimes this application soft-ware is called middleware. The application software maintains the software proto-col between the reader and host computing device. Usually, the reader is connected to the host computer using standard input/output ports such as RS232, parallel, or USB ports. However, the software protocol used for data transfer between the application software and the reader is decided by the manufacturer or designer of the RFID system to suit their needs. Depending on the application, the reader can be configured to upload the fully decoded tag ID or the raw data received from the tag. The application software can be used for additional signal processing in cases where a reader is configured to upload raw data.

For the two reader prototypes—dual-planar (type 1) and uniplanar (type-2)—discussed in Chapter 4, two types of application software have been developed and the screen captures of these two are shown in Figures 10.12 and 10.13, respectively. The first type of software (see Figure 10.12) was developed using Microsoft Visual Basic .NET and it receives data from the reader via a USB port. This application software is capable of receiving raw data as well as fully decoded tag IDs from the

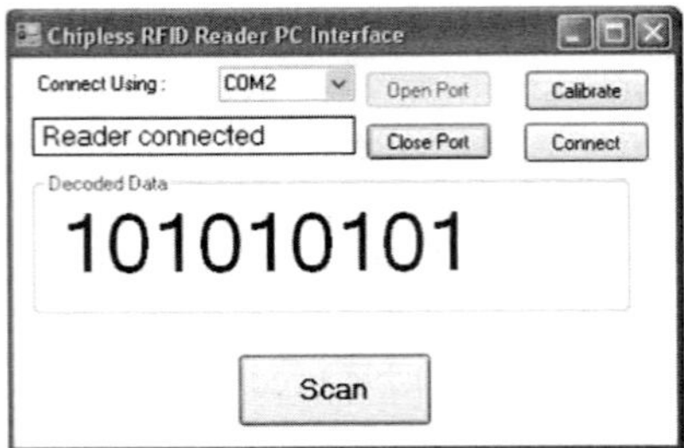

**Figure 10.12**   Screen capture of chipless RFID reader PC interface (for type 1 reader).

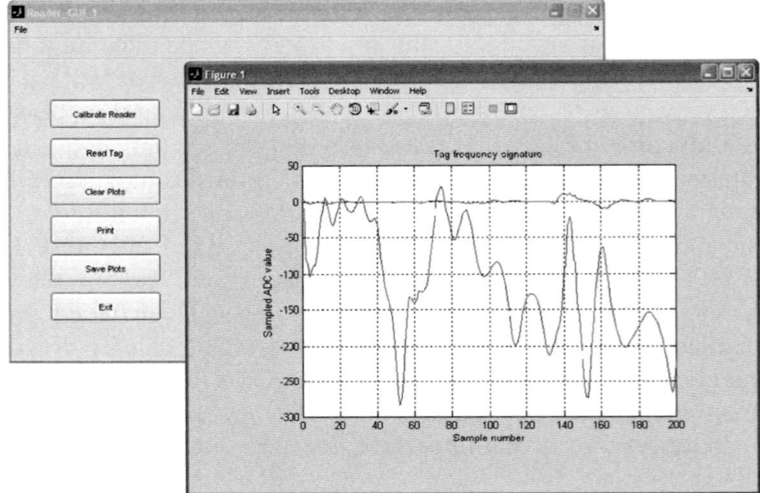

**Figure 10.13**   Screen capture of the application software developed using Matlab (for type 1 and type 2 readers).

reader device and it supports only reader type 1. The software shown in Figure 10.13 was developed using Matlab and it is capable of further processing of raw tag data. This software works with both type 1 and type 2 readers. Because the microcontroller of the digital section is not powerful enough for the signal processing functions required for reader type 2, the software acquires raw data from the reader and the decodes the data. Figure 10.13 also shows an ADC output plot of backscatter.

Figure 10.14 summarizes the design flow for implementing a chipless RFID reader. It is clear that the application requirements and specifications plays vital roles in the selection of key elements and integration involved in the RFID system design process.

## 10.4   Applications

As already discussed, like conventional RFID systems, chipless RFID is application specific as well. Modifications to the RFID infrastructure are needed based on application scenarios. Chipless RFID has the potential to be adopted in different

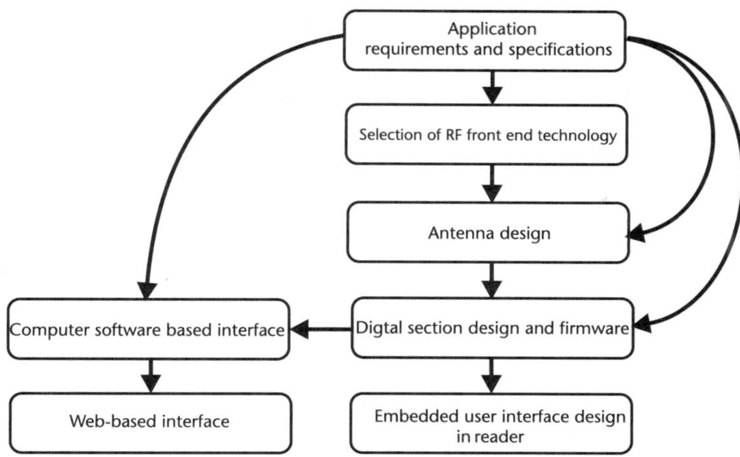



**Figure 10.14** Reader design flow.

application areas such as slot-card readers, touch-and-go type systems, retail, smart library, airport luggage tracking and handling, and vehicle entry systems. However, in this section two types of readers are described: a short-range conveyer belt application and a long-range vehicular tracking application.

### 10.4.1 Conveyer Belt Application

Document authentication is generally considered a near-field reading application in which the tagged item will be within 10 cm of the reader. The time available to perform a tag read, however, depends on each specific application scenario. For example, in a case where the tagged documents are being read over a counter when being handled by human operators, more time is available for a tag read. Consider a scenario, however, in which the documents are in a printing facility and in the process of being printed and tagged. Now consider that the integrity of each tag in each item needs to be checked while the items are still in the production process. Here, the tag reading needs to be performed while the documents are still in their uncut state as sheets of paper or polymer moving along the printing process [10, 11]. In such a situation a very small amount of time is available for a single tag read operation. Therefore, depending on the application the interrogation technology used in the RF front-end of the reader needs to be enhanced. The frequency of operation of the reader is dictated by the chipless RFID tag design, which is in turn determined by the space allowed in a document for the tagging.

Figure 10.15 illustrates an application scenario in which documents embedded with chipless RFID tags are printed on polymer/paper. Multiple pairs of Tx and Rx antennas are used in order to interrogate more than one tag simultaneously. Since the strength of the backscattered signal from the chipless tag printed on the paper/polymer is very low due to the low conductive properties of the ink used and due to the loss introduced by the polymer substrates, high-gain antennas need to be used to enhance the signal quality. The dipole reflector arrays (14-dBi gain) presented in [13, 14] are suitable for this type of application.

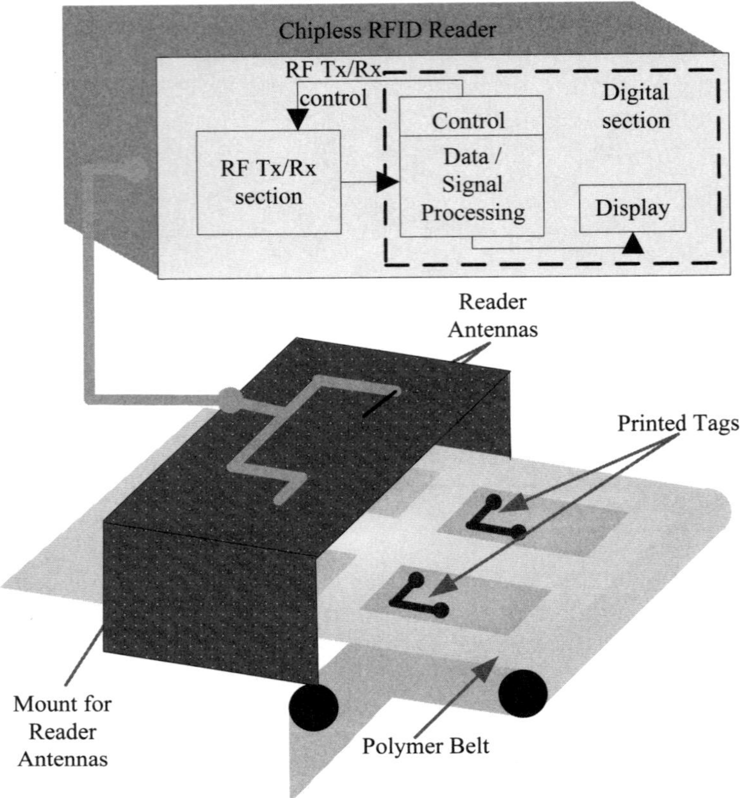



**Figure 10.15** Reader system for interrogating documents/tagged items while the items are in the process of being printed.

## 10.4.2 Vehicle Tracking

This section details the use of chipless RFID for long-range mobile applications such as vehicle tracking. RFID has been widely used in tracking applications ranging from tracking vehicles to wildlife. Long-range applications usually make use of the conventional chipped passive, semipassive, or in some cases active RFID tags for tagging items. Chipless RF-SAW tags are also gaining popularity in toll collection applications in the United States in recent years. These chipped and chipless RFID tags perform very well and operate up to several tens of meters in the case of active tags. We will now look at how a chipless RFID system can be utilized for long-range applications as a cheaper alternative to chipped RFID.

### 10.4.2.1 Long-Range Chipless RFID Tag Design

Most of the chipless RFID tag design considerations discussed in the previous section for authentication and anticounterfeiting applications are generally applicable to many other chipless RFID applications. For longer range applications typically backscatter based chipless RFID tags [15] are chosen as opposed to retransmission based tags. This is because retransmission based chipless RFID tags are more constrained in the degree of freedom available for tag placement with respect to the reader than backscatter based tags. The retransmission based tag consists of two antennas that need to be polarization matched and aligned properly with the respective Tx and Rx antennas at the reader as shown in Figure 10.16(a). Therefore, the tag is constrained both vertically and horizontally, where a small rotation or translation would cause a polarization mismatch and missed alignment, resulting in a degradation of the signal quality received at the reader. With the increase of distance, these effects become more severe. A backscatter based chipless tag is more relaxed in terms of tag positioning constraints since it does not contain a retransmission signal path (via a receiving tag antenna, data encoding filter, and a transmitting tag antenna) and only relies on antenna mode backscatter signal [10] produced by each individual antenna element in the tag. Hence, it provides a greater degree of freedom in the tag reading process. Figures 10.16(b) and (c) show the interrogation of a backscatter based chipless RFID tag.

For long-range chipless RFID tags, maximizing the radar cross section (RCS) of a chipless tag is very important. Techniques for enhancing the RCS of a chipless tag design include use of the following:

1.  Increased reflective area for the chipless tag;
2.  High conductivity of low-loss ink;

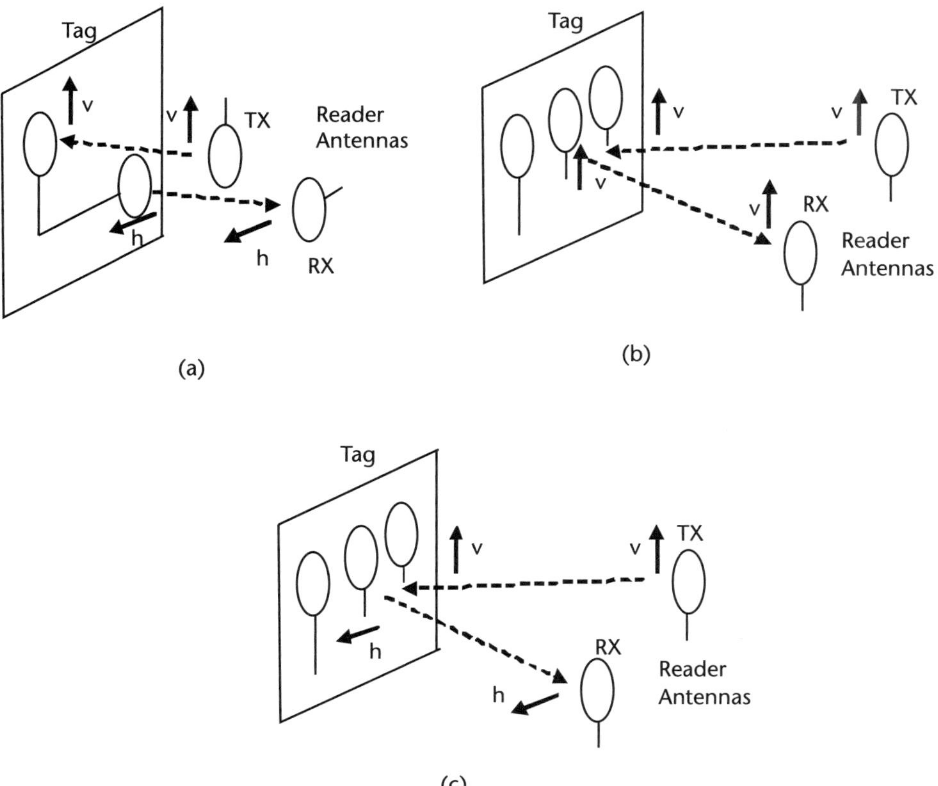


**Figure 10.16**  Interrogation of chipless tags: (a) retransmission based chipless tag, (b) co-polarized backscatter based chipless tag interrogation, and (c) cross-polarized backscatter based tag interrogation. The letters h and v denote, respectively, horizontally and vertically linearly polarized electromagnetic waves.

3.  Low-loss laminates.

Using arrays of reflective patches [15] enhances the RCS many-fold compared to a single reflective patch. As an example, Balbin and Karmakar [15] used a four-element rectangular patch array as a reflective surface to enhance the RCS fourfold compared to a single patch reflector. Use of high-conduction inks [11, 12] also improves the RCS. Choosing a low-loss substrate material is also important in enhancing the backscatter from the tag [1, 2].

Another important consideration for long-range chipless RFID tag design is ensuring an orientation-invariant tag performance. This is because when the object being tracked is several meters away from the reader it is very difficult to control and restrict the orientation of the chipless tag placed on the object. Therefore, the reader needs to be able to read the information contained in the tag regardless of the orientation of the tag in 3D space. To ensure this orientation-invariant tag performance, circularly polarized (CP) resonating elements and antenna elements need to be employed in the chipless tag design. Designing UWB CP backscatterers, however, is a nontrivial task. Some UWB CP antennas can be found in the open literature [16].

### 10.4.2.2  Reader System and Reader Antenna Design

Figure 10.17 illustrates a vehicle tracking application. In such an application the typical distance between the target and the reader is more than 2m. The backscattered energy coming from a tag at such a large distance is very small due to (1) FCC UWB-IR regulation of low-power transmission, (2) limited antenna gain, and (3) the path loss between the tag and the reader. Furthermore, this weak signal is affected by unwanted stray reflections caused by the environmental clutter. To enhance the signal captured by the reader and also to boost the transmitted signal, high-gain Tx and Rx antennas need to be designed. High-gain patch antenna arrays or high-gain horn antennas are suitable reader antennas for these long-range applications. Chapter 7 details such high-gain antennas.

Vehicular applications present numerous challenges in the implementation of a chipless RFID system. The mobile environment causes the RCS to vary with movement. This dynamic RCS of the tag needs to be tracked, acquired, and processed. The metallic body of a vehicle results in a large amount of unwanted stray

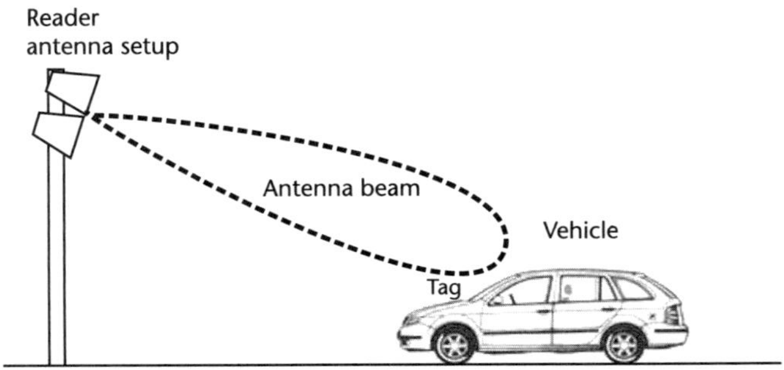


**Figure 10.17**  Vehicle tracking application.

reflections that interfere with the backscatter due to the chipless tag. The reader signal processing needs to cancel these spurious signals and enhance the wanted tag backscatter contained in the total received signal at the reader. Signal processing techniques such as wavelet based processing and noise filtering need to be implemented in the reader digital section firmware for this purpose. Given that the backscattered signal is very weak and is heavily contaminated with interference and noise, many computationally intensive algorithms such as the matrix pencil algorithm [17] will be required to suppress the noise and extract the resonance information contained in the backscattered signal. The processing capabilities of an embedded reader system designed using microcontrollers or FPGAs might not be adequate to perform these complex numerical computations both in terms of speed and memory resource requirements. Therefore, the digital section of the RFID reader would simply forward data that are being processed to some extent via a communication link, such as Wi-Fi or Zigbee, to a centralized processing unit implemented using a computer server for further processing and to apply advance algorithms in order to extract tag information.

## 10.5   Conclusion

The key advantage of chipless RFID technology is the low-cost, printable tags that can be printed on paper or flexible substrates with conventional printing techniques with a fraction of the cost of a chipped tag. Therefore, chipless RFID is creating immense interest among researchers and investors. Chipless RFID has enormous prospects for being widely accepted for item-level tagging, document authentication, and automatic identification. However, for commercial applications an integrated reader that is available at a reasonable cost is required.

In this chapter the design considerations for a complete chipless RFID reader system were discussed. Antenna configurations required for different types of applications were tested and the proper operation was verified with a PNA and two reader prototypes. The selection of an RF front-end technology, digital signal processing requirements, and user interface requirements for an RFID reader system according to application specifications and requirements were discussed.

The application-specific design considerations for chipless tags and chipless RFID readers used for short-range applications, such as document authentication, and long-range applications, such as vehicle tracking, were discussed in detail. These case studies demonstrate the potential of chipless RFID technology in ubiquitous applications. Once fully developed and deployed, chipless RFID technology will penetrate the mass market for low-cost item-level tagging, authentication, security, surveillance, access control, logistics, toll collection, inventory controls, smart logistics, and many more sectors.

## Questions

1.   Name the essential parts of a generic chipless RFID reader. Explain their functions in the reading process.

2. Explain the UWB power regulation, antenna polarization and gain requirements, frequency of operation, and number of bits that can be accommodated in a specific chipless RFID reader system.

3. Why was a retransmission chipless RFID tag selected in the study of chipless RFID systems?

4. Explain the polarization issues and placements of antennas for efficient reading in type 1 and type 2 readers. Suggest your own method of antenna alignment for reading other types of chipless tags.

5. Explain the advantages of a uniplanar (type 2) reader compared to a dualplanar (type 1) reader.

6. Compare the frequency and time domain reading methods. Which one is faster? Which one do you prefer for your desired application?

7. List a few attributes to be considered for selection of the front-end of a chipless RFID reader.

8. List a few attributes of the digital control section of a chipless RFID reader.

9. What is *middleware* in a chipless RFID reader? Explain the function of middleware in the proposed type 1 and type 2 chipless RFID readers.

10. Draw a design flow graph of an integrated chipless RFID reader system for your intended application. Starting from the nomination of your chipless RFID system for your specific applications, formulate the application environment, number of data bits required, reading speed, FCC regulation and power budget issues, selection of RF front-end, reader antenna, digital section, and application software. Expand your design to a web-based interface for ubiquitous access of information.

11. Explain the attributes used in a short-range reader. Develop your own design specifications and layout for a short-range RFID application such as event management, bus ticketing, or luggage handling in an airport.

12. What types of chipless tags do you prefer for long-range (>2m) applications? Beyond RF-SAW, search the Internet for long-range chipless RFID tags and list them.

13. Explain some salient attributes for chipless tag design for long-range applications. Sketch some of your own design.

14. Explain how would you increase the RCS of a chipless tag for long-range applications.

15. Explain how would you design an orientation-insensitive chipless tag for long-range mobile applications. What are the nontrivial design constraints in the design? How would you overcome the limitation?

16. Explain how various signal processing techniques improve the efficiency of a chipless RFID reader for long-range applications.

# References

[1]  S. Preradovic and N. C. Karmakar, "Design of fully printable chipless RFID tag on flexible substrate for secure banknote applications," in *3rd International Conference on Anti-Counterfeiting, Security, and Identification in Communication (ASID 2009)*, pp. 206–210.

[2]   L. Won-Seok et al., "Design of chipless tag with electromagnetic code for paper-based banknote classification," in *2011 Asia-Pacific Microwave Conference Proceedings (APMC 2011)*, pp. 1406–1409.

[3]   M. A. Islam and N. Karmakar, "Design of a 16-bit ultra-low cost fully printable slot-loaded dual-polarized chipless RFID tag," in *2011 Asia-Pacific Microwave Conference Proceedings (APMC 2011)*, pp. 1482–1485.

[4]   M. A. Islam and N. C. Karmakar, "A novel compact printable dual-polarized chipless RFID system," *IEEE Trans. on Microwave Theory and Techniques,* vol. 60, pp. 2142–2151, 2012.

[5]   A. Vena, E. Perret, and S. Tedjini, "Chipless RFID tag using hybrid coding technique," *IEEE Trans. on Microwave Theory and Techniques,* vol. 59, pp. 3356–3364, 2011.

[6]   A. Vena, E. Perret, and S. Tedjini, "A compact chipless RFID tag using polarization diversity for encoding and sensing," in *2012 IEEE International Conference on RFID (RFID 2012)*, pp. 191–197.

[7]   S. Preradovic, I. Balbin, N. C. Karmakar, and G. Swiegers, "A novel chipless RFID system based on planar multiresonators for barcode replacement," presented at the IEEE International Conference on RFID, 2008, Las Vegas, NV, 2008.

[8]   S. Preradovic and N. C. Karmakar, "Design of fully printable planar chipless RFID transponder with 35-bit data capacity," presented at the European Microwave Conference, Rome, 2009.

[9]   S. Preradovic, S. Roy, and N. Karmakar, "Fully printable multi-bit chipless RFID transponder on flexible laminate," in *2009 Asia-Pacific Microwave Conference Proceedings (APMC 2009)*, pp. 2371–2374.

[10]   P. Kalansuriya and N. C. Karmakar, "Time domain analysis of a backscattering frequency signature based chipless RFID tag," in *2012 IEEE International Conference on RFID (RFID 2012)*, Melbourne, Australia.

[11]   *Transparent Conductive Coatings,* http://www.evaporatedcoatings.com/transparent-conductive-coatings (accessed November 27, 2012).

[12]   *Low Ohm Transparent Conductive Coatings,* http://www.jdsu.com/ProductLiterature/lotcc_ds_co_ae.pdf (accessed November 27, 2012).

[13]   R. Koswatta and N. C. Karmakar, "Investigation into antenna performance on read range improvement of chipless RFID tag reader," in *2010 Asia-Pacific Microwave Conference Proceedings (APMC 2010)*,pp. 1300–1303.

[14]   R. V. Koswatta and N. C. Karmakar, "Time domain response of a UWB dipole array for impulse based chipless RFID reader," in *2011 Asia-Pacific Microwave Conference Proceedings (APMC 2011)*, pp. 1858–1861.

[15]   I. Balbin and N. Karmakar, "Multi-antenna Backscattered Chipless RFID Design," in *Handbook of Smart Antennas for RFID Systems*, New York: John Wiley & Sons, 2010, pp. 413–443.

[16]   P. Piksa and M. Mazanek, "A self-complementary 1.2 to 40 GHz spiral antenna with impedance matching," *Radioengineering*, Vol. 15, 2006, pp. 15–19.

[17]   Rezaiesarlak and M. R. Manteghi, "Short-Time Matrix Pencil Method For Chipless RFID Detection Applications," *IEEE Trans. Antennas and Propagation*, Vol. 16, No. 5, May 2013, pp. 2801–2806.

| | |
|---|---|
| **ADC** | analog-to-digital converter |
| **ADS** | advanced design system |
| **Agilent ADS** | Agilent advanced design system (software) |
| **AND** | AND (logic) |
| **ASIC** | application-specific integrated circuit |
| **ASK** | amplitude shift keying |
| **Auto ID** | automatic identification |
| **AWGN** | additive white Gaussian noise |
| **BPF** | band pass filter |
| **BPSK** | binary phase shift keying |
| **CCF** | cross correlation function |
| **CDMA** | code division multiple access |
| **CMOS** | complementary metal-oxide semiconductor |
| **CPW** | coplanar waveguide |
| **CST** | computer simulation technology (simulator software) |
| **CWE** | continuous wave emitters |
| **CWT** | continuous wavelet transform |
| **DAC** | digital-to-analog converter |
| **dB** | decibel |
| **DC** | direct current |
| **DDS** | direct digital synthesis |
| **DDS** | dispersive delay structure |
| **DFT** | discrete Fourier transform |
| **DSO** | digital storage oscilloscope |
| **DSP** | digital signal processor |
| **EAS** | electronic article surveillance |
| **EM** | electromagnetic |
| **EMP** | electromagnetic pulse |
| **ETS** | equivalent of time sampling |

| | |
|---|---|
| **FD** | frequency domain |
| **FDMA** | frequency division multiple access |
| **FFT** | fast Fourier transform |
| **FMCW** | frequency-modulated continuous wave |
| **FPGA** | field-programmable gate array |
| **FrFT** | fractional Fourier transform |
| **FSK** | frequency shift keying |
| **FT** | Fourier transform |
| **GHz** | gigahertz |
| **GPD** | gain/phase detector |
| **GPIO** | general-purpose input/output |
| **HF** | high frequency |
| **HT** | Hilbert transform |
| **ID** | identification |
| **IDT** | interdigital transducer |
| **IF** | intermediate frequency |
| **iFrFT** | inverse fractional Fourier transform |
| **IR** | impulse radio |
| **ISM** | industrial, scientific and medical radio bands |
| **JTAG** | Joint Test Action Group |
| **JTFA** | joint-time-frequency-analysis |
| **kHz** | kilohertz |
| **KSPS** | kilosamples per second |
| **LBC** | linear block code |
| **LF** | low frequency |
| **LFM** | linear frequency modulated |
| **LFMCW** | linear frequency-modulated continues wave |
| **LFSCW** | linear frequency-stepped continues wave |
| **LH** | left-handed |
| **LNA** | low noise amplifiers |
| **LO** | local oscillator |
| **LPDA** | log periodic dipole antenna |
| **LPDAA** | log periodic dipole array antenna |
| **LPF** | lowpass filter |
| **LSB** | least significant bit |
| **LTE** | long-term evolution |
| **MCLK** | master clock |
| **MHz** | megahertz |
| **MMARS** | Monash Microwave, Antennas, RFID and Sensors Laboratories |

| | |
|---|---|
| **MMIC** | monolithic microwave integrated circuit |
| **MRDA** | multiresonant dipole antenna |
| **ms** | millisecond |
| **MSB** | most significant bit |
| **MSPS** | megasamples per second |
| **NF** | noise figure |
| **NLOS** | non-line-of-sight |
| **NRI** | negative refractive index |
| **ns** | nanoseconds |
| **OFDM** | orthogonal frequency division multiplexing |
| **OP-AMP** | operational amplifier |
| **PA** | power amplifiers |
| **PC** | personal computer |
| **PCB** | printed circuit board |
| **PFD** | phase frequency detector |
| **PLL** | phase locked loop |
| **PPM** | pulse position modulation |
| **ps** | picoseconds |
| **PSK** | phase shift keying |
| **PSWF** | prolate spheroidal wave function |
| **QAM** | quadrature amplitude modulation |
| **radar** | radio detection and ranging |
| **RAM** | random-access memory |
| **RCS** | radar cross section |
| **RF** | radio-frequency |
| **RFID** | radio-frequency identification |
| **RH** | right-handed |
| **RLC** | resistive, inductive, and capacitive |
| **RTF** | reader-talks-first |
| **RTOF** | round-trip time of flight |
| **SAR** | synthetic aperture radar |
| **SAW** | surface acoustic waves |
| **SDMA** | space division multiple access |
| **SEM** | singularity expansion method |
| **SNIR** | signal-to-noise plus interference ratio |
| **SNR** | signal-to-noise ratio |
| **SRR** | split-ring resonator |
| **SSI** | selective spectral interrogation |
| **STFT** | short time Fourier transform |

| | |
|---|---|
| **SVD** | singular value decomposition |
| **TD** | time domain |
| **TDMA** | time domain multiple access |
| **TDOA** | time difference of arrival |
| **TDR** | time domain reflectometry |
| **t-f/TF** | time-frequency |
| **TFTC** | thin-film transistor circuit |
| **TLSMP** | total least-squares matrix pencil |
| **TTF** | tag-talks-first |
| **UHF** | ultra-high frequency |
| **UWB** | ultra-wideband |
| **UWB-IR** | ultra-wideband impulse radio |
| **V** | volt |
| **VCO** | voltage-controlled oscillator |
| **VNA** | vector network analyzer |
| **VSAT** | very small terminal |
| **VSWR** | voltage standing wave ratio |
| **WLAN** | wireless local-area networks |
| **WT** | wavelet transform |
| **WVD** | Wigner Ville distribution |
| **XOR** | eXclusive OR (logic) |
| **YIG** | yttrium iron garnet |

# Glossary

**Active components** components that require an external power supply for the operation

**ADC** a device that converts analog signals into a stream of binary numbers

**Advanced design system (ADS)** a software platform that is used to simulate RF active and passive circuits

**Arbitrary waveform generator** a signal generator that can be programmed to generate any arbitrary waveform

**Array factor** a function of the positions of the antennas in the array and the weights used for antenna elements

**Array pattern** the product of a single-element pattern and the array factor

**Axial ratio** ratio between two orthogonal radiating field components of an antenna

**Balun** "balanced/unbalanced"; a network that balances an unbalanced signal

**Bandpass filter** a filter that allows only the intermediate frequency band of a signal to pass through and rejects lower and higher frequency signals

**Band-reject filter** a filter that stops an intermediate frequency band and allows lower and higher frequency bands to pass through

**Band-stop filter** *See* Band-reject filter

**Bandwidth** allowable frequency band over which a device performs satisfactorily

**Beam scanning** technique for steering the beam of an antenna using a mechanical or electrical technique

**Beamwidth** angular spread of the radiation pattern of an antenna

**binary phase-shift keying (BPSK)** a modulation technique based on changing the phase of a continuous signal

**Chipless RFID** An RFID system in which the tag does not contain any type of chip component

**Chipless RFID reader** a device used to detect the tags that does not contain a silicone ASIC

**Circulator** a 3-port nonreciprocal microwave ferrite device that allows a signal to flow in a clockwise or counter-clockwise device. It blocks reflected signals feedback to the input device.

**Coupler**   a device that is used to extract a portion of a signal that travels in a separate signal path; a component used to couple a certain amount of power to another channel

**Digital control**   the section of the digital electronic circuit of an RFID reader that controls the operation of the other subsystems

**Digital interface**   an interface provided to communicate with digital electronic components

**Direct digital synthesis (DDS)**   a method used to synthesize signals using digital electronics based techniques

**Directional coupler**   a coupler that allows separate signals to travel in the forward direction and the backward direction using two separate output ports

**Down-converter**   a device that converts a high-frequency signal into a lower frequency signal

**Elemental encoding structure**   a stub-loaded microstrip patch antenna in which the antenna resonates at a unique resonant frequency and also produces a phase shift due to the loading condition; the phase shift depends on the length of the stub attached to the patch antenna

**EM theory**   a theory based on Maxwell's equations together with the Lorentz force law that forms the foundation of classical electrodynamics, classical optics, and electric circuits

**Equivalent time sampling technique (ETS)**   a sampling technique that is used to sample repetitive signals at much lower sampling rates than the Nyquist sampling rate

**Field programmable gate array (FPGA)**   a chip that has logic gates and a configurable switch matrix, which can be used to implement digital electronic circuits

**Fractal antenna**   an antenna that uses a fractal, self-similar design to maximize the length or increase the perimeter (on inside sections or the outer structure) of material that can receive or transmit electromagnetic radiation within a given total surface area or volume

**Frequency domain readers**   RFID readers that use frequency domain techniques for the detection of tags

**Frequency doubler**   a device that outputs the doubled input frequency

**Frequency signature**   a unique identification that can be obtain using the frequency response of a certain object

**Gain/phase detector (GPB)**   a device that detects the differences of gain and phase between two RF signals

**Gaussian pulse**   a pulse that has the shape of a Gaussian distribution

**Harmonic levels**   the strengths of unwanted multiples of a certain frequency

**Highpass filter**   a filter that allows only the higher frequency band of a signal to pass through and rejects lower frequency signals

**High-Q resonator**   a resonator that has a high quality factor (narrowband resonance)

**Hilbert transform**   In mathematics and in signal processing, the Hilbert transform is a linear operator that takes a function, $u(t)$, and produces a function, $H(u)(t)$, with the same domain

**Horizontally polarized**   polarization angle of 90°

**Hybrid coupler**   a coupler that extracts a signal and outputs it with a 90° phase shift at the coupled port

**Hybrid domain reader**   a reader that uses more than one method of reading for the detection of tags

**Impulse radio**   a receiver that is designed to detect short RF pulses

**Interrogator**   a device that is used to detect objects such as security tags and RFID tags that are used in electronic surveillance

**ISM band**   an free radio-frequency band allocated for industrial, scientific, and medical work

**Isolation**   the level of electrical separation of the signals between two points in a device

**Linear frequency modulated continuous wave (LFMCW)**   type of wave that results when the frequency of a signal is varied linearly over a certain period of time and the continuity is preserved in the signal

**Local oscillator (LO)**   an oscillator used in transceiver circuits to down-convert or up-convert RF signals

**Low-noise amplifier (LNA)**   amplifier used in the front-end of receivers to amplify weak RF signals that are designed to have low-noise figures

**Lowpass filter**   a filter that allows only the lower frequency band of a signal to pass through and rejects higher frequency signals

**Microwave channelized receiver**   a microwave signal receiver that is a combination of a number of narrowband receivers operating in parallel to form a wideband receiver

**Mixer**   a device that is used to up-convert or down-convert signals

**Multiband RF systems**   Radio-frequency systems that are able to operate in multiple frequency bands

**Multisection Wilkinson power divider**   a Wilkinson power divider that has multiple sections designed to achieve broadband operation

**Oscillator**   a device that generates a periodic signal

**Passive components**   components that do not require an external power supply for their operation

**Path loss**   reduction in power density (attenuation) of an electromagnetic wave as it propagates through space; also called *path attenuation*

**Phase jumps**   a sudden variation in terms of a propagation wave's phase

**Phase locked loop**   a control loop that is used to obtain a stable frequency output from a voltage-controlled oscillator

**Phase noise**   the unwanted variations of the phase of an oscillator

**Polarization loss factor**   the squared cosine of the angle between two antennas' orientations

**Power amplifier**    an amplifier that is designed to increase the power of the incoming signal

**Power combiner**    a device that is used to combine multiple incoming signals into a single output signal

**Power divider**    a device that is used to divide power into two or multiple equal or unequal portions

**Pulse generator**    a device that generate short pulses

**Pulse position modulation (PPM)**    a method of modulation used in digital communication in which the position of a pulse is varied in a predefined time window

**Pulse shaping circuit**    an electronic circuit that is used to modify the shape of a pulse in the time domain

**Radar cross section (RCS)**    a measure of how detectable an object is with a radar; a larger RCS indicates that an object is more easily detected; defined as the ratio of the reflected power from an object to the reflected power of a sphere of the same cross section

**Read range**    the distance that can be kept between an RFID reader and tag for proper detection of the tag

**Rectenna**    rectifying antenna; converts a microwave signal into DC power

**Resonance frequency**    the tendency of a system to oscillate with greater amplitude at some frequencies than at others

**Resonator**    a device or system that exhibits resonance or resonant behavior; that is, it naturally oscillates at some frequencies, called its resonant frequencies, with greater amplitude than at others

**RF PCB**    A printed circuit board designed to operate in the radio-frequency bands

**RF receiver**    a device that is used to receive radio-frequency signals.

**RF section**    the section of the electronic circuit of an RFID reader that processes radio-frequency signals

**RFID system**    radio-frequency identification system that uses electromagnetic waves to identify products and goods

**RFID transponder**    another name used for radio-frequency identification tags

**SAW RFID tags**    RFID tags that use surface acoustic wave (SAW) devices for data encoding

**Spatial filtering**    technique of focusing electromagnetic waves (light or RF signals) into a certain area to minimize their interaction with the surrounding areas

**Spectrum analyzer**    equipment that is used to measure the frequency spectrum of an RF signal

**Split-ring resonator (SRR)**    an artificially produced structure common to metamaterials; they produce a desired magnetic response in various types of metamaterials up to 200 THz, creating the necessary strong magnetic coupling or response not otherwise available in conventional materials

**Swept frequency signal**    a signal that is generated by varying the frequency gradually over a finite period of time

**Synthetic aperture radar (SAR)**   a radar imaging technique that uses the relative motion of an antenna and target to generate a large aperture using an antenna that has a small aperture

**Temperature stability**   the stability of the performance of an electronic circuit

**Time domain impulse radio (TD-IR)**   a receiver that detects short RF pulse signals using time domain techniques

**Time domain readers**   RFID readers that use time domain techniques for the detection process

**Time domain reflectometry**   a method of time domain measurement of reflected signals by a certain object or change in the wave propagation medium

**T-junction power divider**   a power divider shaped like the letter "T"

**Tunable filter**   a filter that has a variable frequency response

**Tunable oscillator**   a device that generates a periodic signal; the period of the generated signal can be varied by a mechanical or electrical method

**Tuning voltage**   voltage that is used to vary the output frequency of a VCO

**Up-converter**   a device that converts a low-frequency signal into a higher frequency signal

**Varactor diode**   a semiconductor diode that has a capacitance that can be varied using the bias voltage

**Vertically polarized**   polarization angle of 0°

**Voltage comparator**   a chip that is used to compare two voltages and outputs the result as a voltage level

**Voltage-controlled oscillator (VC)**   an oscillator in which the oscillation frequency can be varied using an external voltage

**Waveguide launcher**   A transitional section that efficiently couples the source to a waveguide horn

**Wilkinson power divider**   a type of power divider that has matched output ports and a high isolation level between two output ports

**YIG tuned oscillator**   a tunable microwave oscillator that use a YIG sphere to determine the oscillation frequency

# About the Authors

**Dr. Nemai Chandra Karmakar** obtained his B.Sc. and M.Sc. in electrical engineering from the Bangladesh University of Engineering and Technology, Dhaka, Bangladesh, in 1987 and 1989, respectively, and his Ph.D. in ITEE from the University of Queensland, Australia, in 1991 and 1999, respectively. He has extensive experience in antennas, microwave active and passive designs, and system integration. He worked as a microwave design engineer at Mitec Ltd., Brisbane, Australia, from 1992 to 1994. In his academic career he taught at Queensland University of Technology and Monash University in Australia and Nanyang Technological University in Singapore. Currently he is an associate professor in the Electrical and Computer Systems Engineering Department, Monash University, Melbourne, Australia.

He is leading a research team in chipless RFID, smart antennas, and microwave biomedical devices. He has edited and wrote 7 referred books, 35 book chapters, 77 journal papers, 180 conference papers, 8 patent applications in chipless RFID, and 5 workshop notes. He co-organized the Workshop on RFID and Smart Sensors at the 2011 Asia Pacific Microwave Conference, Melbourne, Australia, in December 2011. His research interests cover chipless RFID, smart anennas, microwave biomedical devices, and microwave active and passive designs.

**Randika V. Koswatta** obtained his bachelor's degree in electrical and electronics engineering from the University of Peradeniya, Sri Lanka, in 2007. He has completed his Ph.D. studies in electrical engineering at Monash University, Australia. His areas of interests include chipless RFID reader design, UWB antenna designs, UWB transceiver design for chipless RFID applications, digital electronics, and embedded systems design.

**Prasanna Kalansuriya** received a B.Sc. (with first-class honors) from the University of Moratuwa, Moratuwa, Sri Lanka, in 2005, and an M.Sc. degree from the University of Alberta, Edmonton, Canada, in 2009. He is currently working toward a Ph.D. degree in the Electrical and Computer Systems Engineering Department, Monash University, Melbourne, Australia.

From 2005 to 2007, he was an electronic design engineer with Electroteks Global Networks, Pte Ltd., Sri Lanka. He served as a lecturer with the University of Moratuwa, Sri Lanka, in 2007, and as a research assistant with the ICORE Wireless Communication Laboratory, University of Alberta, Edmonton, Canada, from 2008 to 2010. In 2012, he was a visiting researcher with the Auto-ID Laboratory, Massachusetts Institute of Technology, Cambridge, MA. His research interests include wireless communication, signal processing for chipless RFID, and applications of chipless RFID and passive RFID in pervasive sensing.

**Rubayet E-Azim** received her B.S. in electrical and electronic engineering from Bangladesh University of Engineering and Technology, Bangladesh, in 2009 and is currently pursuing a Ph.D. in engineering at Monash University. Her research interests include signal processing, chipless RFID, and antennas.

# Index

**A**

Active components
    defined, 227
    introduction to, 207
    low-noise amplifiers, 233–34
    phase locked loop (PLL), 231–33
    voltage-controlled oscillators (VCOs),
        227–31
    *See also* Microwave/millimeter wave
        components
Active mixers, 226
AD8302 gain phase detector
    functional block diagram, 75
    operation of, 75–76
    output, 76
Agilent ADS2009
    chipless tag modeling, 108
    design of, 105–9
    overview of, 105–6
    receiver section modeling, 109
    simulation model block diagram, 107
    simulation setup, 106–9
    simulation setup illustration, 109
    software package, 107
    system level model, 109
    in TD architecture implementation, 116
    UWB pulse transmitter modeling, 108
Agilent Momentum, 184
Analog-to-digital converters (ADCs)
    low-sampling rate, 103
    quadrature-phase (Q) components sampled
        with, 49
    sampling rates, 267
Antenna arrays
    beamwidth, 156
    dipole, 165–76
    gain, 156
    printed, 156

Antennas
    array factor, 154–55
    axial ratio, 153
    bandwidth, 152
    broad classifications of, 149
    in chipless RFID reader architecture, 44–45,
        147–207
    dipole planar, 156
    directivity, 152
    elliptical-shaped monopole, 156
    fixed-beam, 149
    front-to-back-ratio (F/B), 151
    function of, 44
    gain, 152
    generic chipless RFID reader, 260–61
    input impedance, 153–54
    introduction to, 147–50
    matching, 153–54
    multipath interference and, 150
    panel, 149
    parameters, 151–55
    as passive components, 208–9
    phased array, 150
    polarization, 152–53
    radiation patterns, 151–52
    reconfigurable, 147–48
    rectenna, 148
    sidelobes, 151
    smart, 150
    spatial filtering, 45
    as spatial filters, 147
    in transceiver system, 153
    as transducers, 147
    UWB, 155–202
    vehicle tracking application, 272–73
    in wireless systems, 147
Anticollision, 62

Applications, 268–73
    conclusion, 273
    conveyer belt, 269–70
    low-noise amplifiers (LNAs), 234
    UWB antennas, 200–201
    vehicle tracking, 270–73
Application software, 267–68
Arbitrary waveform generators, 99–100
Array factor, 154
Axial ratio, 153

**B**
Bandpass filters
    coupled line implementation, 214
    coupled resonator implementation, 214
    defined, 213
    frequency spectrum of outputs, 112
    lumped element equivalent circuit, 213
    measured S parameter magnitude, 214
    output, 110
    *See also* Filters
Band-reject (band-stop) filters
    defined, 213
    implementation of, 213–15
    lumped element equivalent circuit, 214
    *See also* Filters
Bandwidth
    defined, 152
    elliptical leaf dipole, 173
    input impedance, 198
    log-periodic dipole antennas, 176
    UWB horn antennas, 193
    *See also* Antennas
Beamwidth, antenna arrays, 156
Biconical antennas, 155–56
Binary phase-shift keying (BPSK), 97
Broadband feed networks, 167–68

**C**
Calibration of the reader, 250
Cavity backing, 192–93
Characteristic frequency, 52
Chemical tags, 20
Chipless RFID reader architecture, 44–47
    antenna, 44–45
    digital section, 46–47

general overview of, 44
    RF section, 45–46
Chipless RFID readers
    antennas for, 147–207
    anticollision, 62
    calibration of, 250
    complete design, 45
    component specifications, 106
    conclusion, 11–12, 63
    cost of, 6, 60–61
    data integrity, 62
    design consideration, 5
    design flow, 269
    developments, 3
    digital section, 239–53
    efficacy of, 148
    error correction, 62
    frequency domain, 50–56, 67–90
    generic, 259–68
    hybrid, 56–57
    integrated, 257–59
    introduction to, 43–44
    limitations and issues with, 60–63
    for multiresonator based chipless tag
        interrogation, 54–55
    read range, 61
    SAR based, 57–60
    tag orientation and, 63
    tag reading speed, 61–62
    tag reading techniques, 47–60
    time domain, 48–50, 93–117
    UWB TD, 93–117
    vehicle tracking application, 272–73
*Chipless RFID Reader Signal Processing*, 62
Chipless RFID systems
    components, 1
    component selection process for, 236
    components of, 44, 239
    generic illustration, 240
    introduction to, 1
    Monash University, xv–xvii
    operating principles, 15–39
    technology prospect, 4
Chipless RFID tags
    backscatter-based, xv
    broad classification, 16
    categories of, 240

chemical, 20
comparison of, 31
conclusion, 38
coplanar strip based, 23
cost, reducing, 15
data encoding methods, 29
dipole array based, 21–22
electromagnetic properties, 15
fractal based, 26–27
frequency reuse based, 142
generic chipless RFID reader, 260
hydrid domain, 29–30
image based, 28–29
ink tattoo based, 29
interrogation of, 271
long-range, 270–72
magnetically coupled LC resonators, 22–23
MRDA-based, 26
multiresonator based, 31–35
operating principles, 15–31
orientation of, 63
radar array, 15
reading, 35–38
reading frequency signatures of, 263
recent development of, xiii–xv
review summary, 30–31
salient features of, 5
SAR based, 28–29
SAW, 3
slot-loaded monopole based, 27
space-filling curves, 20–21
spiral resonator based, 24–26
SRR based, 23, 24
stacked multilayer patch based, 27–28
stub-loaded patch based, 27
TD based, 16–20
TDR based tag, 18–20
TFTC, 3, 15
varieties of, 4
vehicle tracking application, 270–72
Circularly polarized (CP) resonating
    elements, 272
Circulators
application of, 217
circuit symbol, 217
defined, 217
*See also* Passive components

Comparators, 113
Conveyer belt application, 269–70
Coplanar strip based chipless RFID tags, 23
Couplers
circuit symbols, 215, 216
defined, 215
directional, 216
hybrid, 216
illustrated, 216
*See also* Passive components
CPW feed disk-loaded monopole on polymer
conductive epoxy, 163
defined, 162–63
illustrated, 164
return loss versus frequency plot, 164
Crescent-loaded monopole antenna
defined, 162
radiation patterns, 163
C-sections
analytical expression, 126
dispersive delay structure (DDS), 137
in frequency-phase based system, 125–27
hybrid frequency-phase chipless RFID based
    on, 127–28
illustrated, 125
implementation of encoding, 128
resonant behavior of, 126, 127
CST Microwave Studio, 184

**D**

Data integrity, 62
Digital control board
components of, 242–43
defined, 240–41
IR tag sensor, 242, 243–45
power control switch, 242, 246
power supply unit, 242, 243
VCO control generation part, 242–43,
    246–47
*See also* Digital control section
Digital control section, 239–53
amplitude spectrum of VCO output, 252
components of, 239
conclusion, 252–53
defined, 46–47, 239
digital control board, 240–41, 242–47

Digital control section (continued)
  digital signal processing (DSP) board, 241, 247–49
  functional block diagram, 241
  generic chipless RFID reader, 266–67
  introduction to, 239–42
  middleware, 241, 249–51
  performance of, 251–52
  switching electronics, 241
  type-1 frequency domain reader, 82
  type-2 frequency domain reader, 85–86
Digital signal processing (DSP) board
  block diagram, 248
  components of, 249
  defined, 241
  function of, 247–48
  PCB implementation of, 249
Digital-to-analog (DAC) converter, 246
Dipole array based chipless RFID tags, 21–22
Dipole planar, 156
Direct digital synthesis (DDS), 53
Directivity, 152
Discrete wavelet transform (DWT), 143
Dispersive delay structure (DDS)
  C-sections, 137
  frequency-dependent time delay, 136
  group-delay characteristics of, 138
  in obtaining group delays, 95
  pulse position modulation (PPM) encoding technique for, 140

E

Electromagnetic (EM) waves
  antennas and, 44
  reception of, 45
Elliptical leaf dipole
  bandwidth, 173
  dipole arm parameters, 166
  input impedance return loss, 168
  optimized design parameters, 167
  optimum leaf dimensions, 166
  performance, 167
  return loss and gain versus frequency, 173
  simulated results, 173
Elliptical leaf dipole array

defined, 169
  gain versus frequency, 170
  input impedance return loss, 170
  mechanical construction, 170–71
  mutual coupling and grating lobe behaviors, 169
  optimized design parameters, 169
  radiation patterns, 172
  as reader antennas, 200–201
  results for, 171–73
  return loss and gain versus frequency, 172
  surface current distributions, 174–75
Elliptical-shaped monopole, 156
Envelope detectors, 113
Equivalent time sampling (ETS), 62
Error correction, 62
Executive summaries, this book, 7–11
Exponentially tapered horn antenna
  design of, 191–92
  gain patterns, 195–98
  realized gain versus frequency, 199
  3D gain patterns, 199

F

Fast Fourier transform (FFT), 143, 266
Field programmable gate array (FPGA), 47, 267
Filters
  bandpass, 213, 214
  band-reject (band-stop), 213–15
  defined, 209
  design, 209
  general block diagram, 209
  highpass, 211–12
  lowpass, 46, 53, 74, 80, 129, 210–11
  as two-port networks, 209
  types of, 209
  See also Passive components
Fixed-beam antennas, 149
Forward transmission coefficient, 262
Fractal based chipless RFID tags, 26–27
Frequency-dependent time delay, 136
Frequency domain based tags
  chemical, 20
  coplanar strip based, 23

dipole array based, 21–22
features of frequency signature detection
    (type-1 readers), 71–76
fractal based, 26–27
frequency signature recovery (type-2
    readers), 76–81
magnetically coupled LC resonators, 22–23
MRDA based, 26
reading, 36–37
remote measurement of complex impedance,
    28
RFID of letters, 23
slot-loaded monopole based, 27
space-filling curves, 20–21
spiral resonator based, 24–26
SRR based, 23, 24
stacked multilayer patch based, 27–28
stub-loaded patch based, 27
Frequency domain readers
defined, 50–51
design conclusions, 88–90
design of, 81–86
development, 67–90
frequency signature analysis, 70
for multiresonator based chipless tag
    interrogation, 54–55
Nicanti, 55–56
operation of, 69–81
results, 86–88
SAW, 53
Tagsense, 51
time domain readers comparison, 95
type-1, 71–76
type-2, 76–81
See also Chipless RFID readers
Frequency doublers, 231, 232
Frequency-group delay, 138–40
Frequency-phase-based system
based on C-sections, 127–28
chipless RFID reader for, 128–30
constellation diagram, 128
encoding element working principle, 125–27
gain/phase detector (GPD), 128–29
lowpass filters (LPFs), 129
RFID reader illustration, 130
voltage-controlled oscillators (VCOs), 129

See also Hybrid chipless RFID reader
Frequency-polarization-based system
chipless RFID reader for, 134–36
defined, 131
encoding information, 132–34
reader designs, 135
split-ring resonators (SRRs), 131–32
See also Hybrid chipless RFID reader
Frequency-reuse-based chipless RFID tags, 142
Frequency-time-based system
chipless RFID reader for, 142–43
complete chipless RFID used, 141
defined, 136
group delay, 136
nongroup-delay based approaches, 141–42
pulse position modulation (PPM), 140–41
time domain reader for, 143
See also Hybrid chipless RFID reader
Front-to-back-ratio (F/B), 151

G

Gain
antenna arrays, 156
defined, 152
See also Antennas
Gain/phase detector (GPD)
AD8302, 75–76
amplitude information, 264
defined, 46
frequency-phase based system, 128–29
phase information, 264
RF section, 45
type-1 frequency domain reader module, 72
Generic chipless RFID reader
antennas, 260–61
application software, 267–68
chipless RFID tag, 260
defined, 259
digital control section, 266–67
main components of, 259
reader device, 264–67
reading techniques, 260–64
RF section, 265–66
specifications of, 259–68
uniplanar reading alignment, 261–62

Group delay
    characteristics of DDS, 138
    control of, 137
    defined, 136
    frequency, 138–40
    frequency-time-based system, 136–41

**H**

Highpass filters, 211–12
Hybrid chipless RFID reader, 121–44
    4-bit, 122
    classification of, 123
    conclusion, 143–44
    C-section, 123
    defined, 56
    encoding approaches, 122
    encoding element application, 124
    encoding information dimensions, 123, 144
    frequency-phase based system, 125–30
    frequency-polarization based system,
        131–36
    frequency-time based system, 136–43
    introduction to, 121–24
Hybrid domain chipless RFID tags
    defined, 29
    reading, 37
    transmission line length, 30

**I**

Image based tags, 28–29
InkSure chipless RFID tag reader system, 58–59
Ink tattoo-based chipless tags, 29
Input impedance
    bandwidth, 198
    defined, 153
    return loss and gain versus frequency, 168,
        170, 177–78
    *See also* Antennas
Integrated chipless RFID readers
    components of, 259
    defined, 257
    integration flowchart, 258
Interdigital transducer (IDT), 17
Interrogators, 43
Inverse fast Fourier transform (IFFT), 143
IR tag sensor

    defined, 242
    detection range, 243–44
    illustrated, 245
    RF supply power control switch and, 244,
        246
    security and privacy, 244

**L**

Linear frequency modulated continuous wave
        (LFMCW), 53
Linear frequency stepped continuous wave
        (LFSCW), 53
LineCalc, 184
Local oscillators (LOs), 96–97
Log-periodic dipole antennas
    bandwidth, 176
    cavity backed, 178
    circular, 177–80
    configuration of, 177
    defined, 176
    E-plane and H-plane radiation, 182
    as frequency-independent, 176
    gain versus frequency, 183
    radiation patterns, 179–80
    simulated input impedance, 177–78
    simulated return loss, 181
    simulation result, 179
    S-parameter versus frequency, 181
    tight element spacing, 178
    *See also* Antennas; UWB antennas
Log-periodic dipole array antennas
    advantages of, 181–83
    CAD design procedure, 186
    configuration of, 183
    defined, 180–81
    design equation of, 183
    design parameters, 183
    design procedure, 185
    dimensions of, 184
    experimental setup for, 202
    microwave, design of, 186–87
    millimeter-wave, design of, 187–90
    as reader antennas, 201
    spacing factor, 184
Long-range chipless RFID tags, 270–72
Low-noise amplifiers (LNAs)

applications, 234
block diagram, 234
defined, 233
function of, 233–34
noise figure, 234
*See also* Active components
Lowpass filters (LPFs)
in filtering unwanted frequencies, 46
frequency-phase based system, 129
ideal, 210
lumped element equivalent, 210
Mini-Circuits LFCN630+, 211–12
as passive components, 210–11
shunt-stub implementation of, 211
*See also* Filters

**M**

Magnetically coupled LC resonators, 22–23
Matlab application software, 268
Mechanical assembly of antennas, 170–71
Microwave channelized receiver architecture,
103–5
Microwave LPDAA
defined, 186
design parameters, 186–87
illustrated, 187
results for, 187
*See also* Log-periodic dipole array antennas
Microwave/millimeter wave components,
207–36
active components, 227–34
classification of, 208
conclusions, 234–36
individual performance of, 235
integration into system, 235
introduction to, 207
passive components, 208–26
selection of, 235
selection process for building chipless RFID
system, 236
Middleware
defined, 43, 241, 249
flowchart, 250
Millimeter-wave LPDAA
conclusion, 190

CST-generated layout, 188
defined, 187
design of, 187–90
design parameters, 188
experimental setup for gain measurement,
190
forward transmission gain versus frequency,
190
illustrated, 188
input impedance return loss, 189
radiation patterns, 189
results of, 188–90
*See also* Log-periodic dipole array antennas
Mini-Circuits LFCN630+ lumped element
lowpass filter, 211–12
Mixers
active, 226
basic operation of, 225
defined, 224
for down-conversion, 225
modular RF, 226
out-of-band frequency components, 224
parameters, 226
passive, 226
schematic, 226
schematic symbol, 225
*See also* Passive components
Monash Microwave Antennas, RFID and Sensors (MMARS) Laboratories, 148
Monash University chipless RFID system, xv–
xvii
MRDA based chipless RFID tags, 26
Multipath interference, 150
Multiresonant dipole antennas (MRDAs), 26
Multiresonator based chipless RFID tags
chipless RFID reader design for, 45
defined, 32
differences, 31–32
equivalent circuit, 33
measurement of, 34–35
multiresonator circuit operation, 34–35
operating principle for reading, 33–35
schematic diagram, 32
spiral resonators, 32
structure and operation of, 69

**N**

Nicanti swipe reader, 55–56
Noise figure, 234
Non-group-delay based chipless RFID system,
    141–42

**O**

Organization, this book
    chipless RFID readers, 6–7
    executive summaries, 7–11

**P**

Panel, antennas, 149
Passive components
    antennas, 208–9
    circulators, 217
    couplers, 215–16
    filters, 209–15
    introduction to, 207
    mixers, 224–26
    power dividers/combiners, 218–21
    resonators, 221–24
    *See also* Microwave/millimeter wave compo-
        nents
Passive mixers, 226
Phased array antennas, 150
Phase frequency detector (PFD), 233
Phase locked loop (PLL)
    block diagram, 232, 233
    defined, 231–32
    VCO integration with, 233
    *See also* Active components
Polarization, 152
Power dividers/combiners
    T-junction, 220–21
    Wilkinson power divider, 218–20
    *See also* Passive components
Power supply unit, digital control board, 242,
    243
PPM based UWB chipless RFID tag, 50
Pulse generators
    defined, 99
    modeling of, 106
    output, 110
    with pulse shaping circuit, 98–99
    schematic diagrams, 97, 98

system level model, 109
    with upconverter, 96–97
    in UWB short pulse generation, 96–99
Pulse position modulation (PPM), 17
Pulse shaping circuit, 98–99

**R**

Radar array chipless RFID tags, 15
Radiation patterns
    of array, 154
    crescent-loaded monopole antenna, 163
    defined, 151
    elliptical leaf dipole array, 172
    illustrated, 151
    log-periodic dipole antennas, 179–80
    millimeter-wave LPDAA, 189
    UWB antennas, 158
    UWB disk-loaded monopole antennas, 162
    *See also* Antennas
Radio frequency identification (RFID)
    conclusions, 11–12
    defined, xxiii
    infrastructure modifications to, 268
    introduction to, xiii
    technology, 1
    *See also* Chipless RFID readers
Reading
    frequency domain based tags, 36–37
    frequency signatures, 263, 265
    generic chipless RFID reader techniques,
        260–64
    hybrid domain chipless RFID tags, 37
    NiCode, 55–56
    SAR based chipless tags, 37–38
    TD based chipless RFID tags, 36
    uniplanar alignment, 261–62
Read range, 61
Real frequency transfer (RFT), 27
Reception of EM waves, 45
Rectennas, 148
Remote measurement of complex impedance,
    28
Resonators
    defined, 221
    dielectric disk, 222
    high-Q, 222–23

illustrated, 223
line, 222
rectangular waveguide, 223–24
types of, 223
*See also* Passive components
RFID. *See* Radio frequency identification
RFID of letters, 23
RFID readers. *See* Chipless RFID readers
RFID systems. *See* Chipless RFID systems
RFID tags
low-cost, 3
mass deployment of, 2
methods of reading, 35–38
*See also* Chipless RFID tags
RF section
defined, 45–46
generic chipless RFID reader, 265–66
type-1 frequency domain reader, 81–82
type-2 frequency domain reader, 82–84
RF supply power control switch, 244, 246
Richardson RVCD6000F varactor based VCO, 229–30

**S**

SAR based chipless tags
defined, 28–29
digitized images, 28
reading, 37–38
SAR based readers
defined, 57
InkSure, 58–59
Somark, 59–60
Vubiq, 57–58
SAW frequency domain reader, 53
SAW time domain reader
ADC, 49
defined, 48
I and Q components, 48–49
illustrated elements, 48
Sidelobes, 151
Slot-loaded monopole-based chipless RFID tags, 27
Smart antennas, interference problem and, 150
Somark ink tattoo chipless RFID tag reader system, 59–60
Spatial filtering, 45, 147

Spiral resonator-based chipless RFID tags, 24–26
Split-ring resonators (SRRs)
array, 23
behavior of, 132
configured for encoding 2 bits of information, 133
dedicated reference, 133
frequency-polarization hybrid chipless RFID tag design, 134
multiple, resonating at different frequencies, 132–33
outermost, 133
polarization characteristics, 131–32
resonant characteristics, 131–32
resonant modes, 132
split/gap, 132
SRR based tags, 23, 24
Stacked multilayer patch based chipless RFID tags, 27–28
Stub-loaded patch based chipless RFID tags, 27
Surface acoustic wave (SAW) tags, 3, 16–18
advantages of, 17–18
data capacity, increasing, 18
defined, 16
interdigital transducer (IDT), 17
market, 16–17
schematic, 17
Surface current distributions, 173–75
Switching electronics, 241
Synthetic aperture radar (SAR), 28

**T**

Tag reading speed, 61–62
Tagsense chipless RFID readers
defined, 51
detection process, 51
frequency shift detection, 52
photograph, 52
TD-based chipless RFID tags
defined, 16
reading, 36
SAW, 16–18
TDR based tag, 18–20
*See also* Chipless RFID tags
Teledyne YIG tuned VCO, 230

Thin-film transistor circuit (TFTC) tags
   cost of, 15
   HF band, 15
   as tag type, 3
Time division multiplexed approach, 142
Time domain based tags. *See* TD based chipless RFID tags
Time domain readers, 93–117
   Agilent ADS2009, 105–9
   architecture implementation, 116–17
   defined, 48
   frequency domain readers comparison, 95
   for hybrid frequency-time chipless RFID, 143
   introduction to, 93–96
   overview, 93–96
   results, 110–15
   SAW, 48–49
   theory of operation, 96–105
   UWB design, 105–9
   UWB-IR, 49–50
   UWB short pulse based interrogation, 100–105
   UWB short pulse generation, 96–100
Time domain reflectometry (TDR) based tag
   defined, 18
   delay line, 19
   TD reading technique and, 115
   transmission line, 18–19
   *See also* Chipless RFID tags
Time domain waveform, 111
T-junction power divider, 220–21
Transducers, antennas as, 147
Tuning voltage, 227
Type-1 frequency domain reader
   analog-to-digital controller, 82
   application software, 267–68
   block diagram, 73
   conclusion, 88–89
   design of, 81–82
   detection process, 71–74
   digital control section, 82
   fabricated, 81
   feature detection, 71–72
   frequency signature detection, 71–76

   gain/phase detector (GPD) operation, 75–76
   GPD module, 72
   lowpass filters (LPFs), 74
   measured results, 87
   operation of, 72
   results, 86–88
   RF section, 81–82
   signal processing flowchart, 83
   VCO-1 and VCO-2, 74
   wideband interrogation signal generation, 71
Type-2 frequency domain reader
   amplitude response, 80
   application software, 267–68
   block diagram, 77
   conclusion, 89–90
   design of, 82–86
   design parts, 82
   fabricated, 84
   free-space path loss profile, 78
   frequency signature recovery, 76–81
   interrogation chirp signal generator output, 77
   linear RF chirp signal generation, 76
   lowpass filters (LPFs), 80
   measured results, 89
   operation of, 76
   phase response, 80
   results, 88
   RF section, 82–84
   signal processing flowchart, 85
   signal propagation, 78
   summary, 81

**U**

Ultrawideband (UWB)
   channels, 112
   defined, 93
   impulse generator, 94
   impulse response, 112
Ultrawideband impulse radio (UWB-IR), 49
Uniplanar reading alignment, 261–62
UWB antennas
   applications, 200–201
   biconical, 155–56

classification of, 155
conclusion, 201–2
design procedure, 159
design specifications, 158
development, 159–200
dipole array antennas, 165–76
disk-loaded monopole, 159–64
horn, 191–200
log-periodic dipole, 176–80
log-periodic dipole array antennas, 180–90
parameters, 158
practical design procedure for, 157–59
radiation patterns, 158
review of, 155–57
simulation setup, 158
types of, 155
UWB dipole array antennas
    block diagram, 165
    broadband feed networks, 167–68
    defined, 165
    elliptical leaf dipole, 166–67
    elliptical leaf dipole array, 169–70
    mechanical assemblies, 170–71
    quality factor of resonator, 165
    results, 171–76
    surface current distribution, 173–76
UWB disk-loaded monopole antennas
    configuration of, 160
    CPW feed, 162–64
    crescent-loaded, 162
    defined, 159–60
    frequency response of, 161
    input return loss versus frequency response, 161
    operation modes, 161
    radiation patterns, 162
    as reader antennas, 200
    use of, 160
    *See also* Antennas; UWB antennas
UWB horn antennas
    3D gain patterns, 199
    bandwidth, 193
    cavity back, 192–93
    defined, 191
    exponentially tapered ridge, 191–92
    gain patterns, 195–98
    input impedance bandwidth, 198
    performance, 194
    as reader antennas, 201
    realized gain versus frequency, 199
    results, 193–200
    waveguide launcher, 192–93
UWB-IR time domain chipless RFID readers
    block diagram, 50
    defined, 49
    pulse position modulation (PPM) tag, 50
    reading speed, 93
UWB short pulse based interrogation, 100–105
    DSP technique, 101–3
    illustrated, 104
    microwave channelized receiver architecture, 103–5
    signal reconstruction process, 102
    TD signal processing architecture, 102
UWB short pulse generation, 96–100
    with generator and pulse shaping circuit, 98–99
    laboratory equipment used in, 99–100
    modeling of, 106
    overview, 94
    with pulse generator and upconverter, 96–97
UWB TD based chipless RFID reader
    block diagram, 107
    chipless RFID tag, 114
    conclusion, 115–17
    design of, 105–9
    impulse transmitter, 114
    modeling chipless tag, 108
    modeling receiver section, 109
    modeling UWB pulse transmitter, 108
    overview, 105–6
    receiver, 114
    results, 110–15
    simulation setup, 106–9
    system-level considerations, 114
    theory of operation, 96–105

**V**

Varactor based VCOs, 227–28, 229
VCO control voltage generator, 242–43, 246–47

Vehicle tracking application
    antenna design, 272–73
    defined, 270
    illustrated, 272
    long-range chipless RFID tag design, 270–72
    reader system, 272–73
Voltage-controlled oscillators (VCOs)
    defined, 227
    frequency-phase-based system, 129
    implementation of, 227
    operated at microwave frequencies, 229
    operated at millimeter-wave frequencies,
        229–31
    phase locked loop integration, 233
    schematics for, 228
    settling time, 266
    tuning voltage, 227
    varactor tuned transistor, 227–28
    YIG resonator, 228
    *See also* Active components
Voltage standing wave ratio (VSWR), 151
Vubiq chipless RFID tag reader system, 57–58

**W**

Waveform generators, arbitrary, 99–100
Waveguide launcher, 192–93
Wilkinson power divider
    defined, 218
    design equation of, 219
    design parameters, 219
    fabricated feed network, 221
    feed network, 220
    multisection, 218
    schematic diagram, 218
    single-section, 218
    uses of, 220
    *See also* Power dividers/combiners

**Y**

YIG resonator VCO, 228, 230

*Integrated Microwave Front-Ends with Avionics Applications*, Leo G. Maloratsky

*Intermodulation Distortion in Microwave and Wireless Circuits*, José Carlos Pedro and Nuno Borges Carvalho

*Introduction to Modeling HBTs*, Matthias Rudolph

*Introduction to RF Design Using EM Simulators*, Hiroaki Kogure, Yoshie Kogure, and James C. Rautio

*Klystrons, Traveling Wave Tubes, Magnetrons, Crossed-Field Amplifiers, and Gyrotrons*, A. S. Gilmour, Jr.

*Lumped Elements for RF and Microwave Circuits*, Inder Bahl

*Lumped Element Quadrature Hybrids*, David Andrews

*Microstrip Lines and Slotlines, Third Edition*, Ramesh Garg, Inder Bahl, and Maurizio Bozzi

*Microwave Circuit Modeling Using Electromagnetic Field Simulation*, Daniel G. Swanson, Jr. and Wolfgang J. R. Hoefer

*Microwave Component Mechanics*, Harri Eskelinen and Pekka Eskelinen

*Microwave Differential Circuit Design Using Mixed-Mode S-Parameters*, William R. Eisenstadt, Robert Stengel, and Bruce M. Thompson

*Microwave Engineers' Handbook, Two Volumes*, Theodore Saad, editor

*Microwave Filters, Impedance-Matching Networks, and Coupling Structures*, George L. Matthaei, Leo Young, and E. M. T. Jones

*Microwave Materials and Fabrication Techniques, Second Edition*, Thomas S. Laverghetta

*Microwave Materials for Wireless Applications*, David B. Cruickshank

*Microwave Mixer Technology and Applications*, Bert Henderson and Edmar Camargo

*Microwave Mixers, Second Edition*, Stephen A. Maas

*Microwave Network Design Using the Scattering Matrix*, Janusz A. Dobrowolski

*Microwave Radio Transmission Design Guide, Second Edition*, Trevor Manning

*Microwave Transmission Line Circuits*, William T. Joines, W. Devereux Palmer, and Jennifer T. Bernhard

*Microwaves and Wireless Simplified, Third Edition*, Thomas S. Laverghetta

*Modern Microwave Circuits*, Noyan Kinayman and M. I. Aksun

*Modern Microwave Measurements and Techniques, Second Edition*, Thomas S. Laverghetta

*Neural Networks for RF and Microwave Design*, Q. J. Zhang and K. C. Gupta

*Noise in Linear and Nonlinear Circuits*, Stephen A. Maas

*Nonlinear Microwave and RF Circuits, Second Edition*, Stephen A. Maas

*Q Factor Measurements Using MATLAB®*, Darko Kajfez

*QMATCH: Lumped-Element Impedance Matching, Software and User's Guide*, Pieter L. D. Abrie

*Passive RF Component Technology: Materials, Techniques, and Applications*, Guoan Wang and Bo Pan, editors

*Practical Analog and Digital Filter Design*, Les Thede

*Practical Microstrip Design and Applications*, Günter Kompa

*Practical RF Circuit Design for Modern Wireless Systems, Volume I: Passive Circuits and Systems*, Les Besser and Rowan Gilmore

*Practical RF Circuit Design for Modern Wireless Systems, Volume II: Active Circuits and Systems*, Rowan Gilmore and Les Besser

*Production Testing of RF and System-on-a-Chip Devices for Wireless Communications*, Keith B. Schaub and Joe Kelly

*Radio Frequency Integrated Circuit Design, Second Edition*, John W. M. Rogers and Calvin Plett

*RF Bulk Acoustic Wave Filters for Communications*, Ken-ya Hashimoto

*RF Design Guide: Systems, Circuits, and Equations*, Peter Vizmuller

*RF Linear Accelerators for Medical and Industrial Applications*, Samy Hanna

*RF Measurements of Die and Packages*, Scott A. Wartenberg

*The RF and Microwave Circuit Design Handbook*, Stephen A. Maas

*RF and Microwave Coupled-Line Circuits*, Rajesh Mongia, Inder Bahl, and Prakash Bhartia

*RF and Microwave Oscillator Design*, Michal Odyniec, editor

*RF Power Amplifiers for Wireless Communications, Second Edition*, Steve C. Cripps

*RF Systems, Components, and Circuits Handbook*, Ferril A. Losee

*The Six-Port Technique with Microwave and Wireless Applications*, Fadhel M. Ghannouchi and Abbas Mohammadi